STRUCTURE AND METHOD IN ARISTOTLE'S *METEOROLOGICA*

In the first full-length study in any modern language dedicated to the *Meteorologica*, Malcolm Wilson presents a groundbreaking interpretation of Aristotle's natural philosophy. Divided into two parts, the book first addresses general philosophical and scientific issues by placing the treatise in a diachronic frame comprising Aristotle's predecessors and in a synchronic frame comprising his other physical works. It argues that Aristotle thought of meteorological phenomena as intermediary or "dualizing" between the cosmos as a whole and the manifold world of terrestrial animals. Engaging with the best current literature on Aristotle's theories of science and metaphysics, Wilson focuses on issues of etiology, teleology, and the structure and unity of science. The second half of the book illustrates Aristotle's principal concerns in a section-by-section treatment of the meteorological phenomena and provides solutions to many of the problems that have been raised since the time of the ancient commentators.

MALCOLM WILSON is Associate Professor of Classics at the University of Oregon. He is author of *Aristotle's Theory of the Unity of Science* (2000) and numerous articles on ancient philosophy and science.

STRUCTURE AND METHOD IN ARISTOTLE'S *METEOROLOGICA*

A More Disorderly Nature

MALCOLM WILSON

CAMBRIDGE
UNIVERSITY PRESS

CAMBRIDGE
UNIVERSITY PRESS

University Printing House, Cambridge CB2 8BS, United Kingdom

Published in the United States of America by Cambridge University Press, New York

Cambridge University Press is part of the University of Cambridge.

It furthers the University's mission by disseminating knowledge in the pursuit of education, learning and research at the highest international levels of excellence.

www.cambridge.org
Information on this title: www.cambridge.org/9781107042575

First published 2013

Printed in the United Kingdom by Clays, St Ives plc

A catalogue record for this publication is available from the British Library

Library of Congress Cataloguing in Publication data
Wilson, Malcolm, 1961–
Structure and method in Aristotle's Meteorologica : a more disorderly nature / Malcolm Wilson.
pages cm
Includes bibliographical references and index.
ISBN 978-1-107-04257-5 (hardback)
1. Meteorology – Early works to 1800. 2. Aristotle. Meteorologica. I. Title.
QC859.W55 2013
551.5 – dc23 2013023082

ISBN 978-1-107-04257-5 Hardback

For Seth

Contents

List of figures — *page* xi
Acknowledgements — xiii
List of abbreviations — xv

Introduction — 1
The unjustified neglect of the *Meteorologica* — 1
The place of *meteora* in Greek thought — 3
The method and organization of the book — 6
Meteorologica 4 and the date of the work — 8
Overview of the argument and claims of the book — 10

PART I METHODOLOGY AND STRUCTURE

1 The rebirth of meteorology — 19
The Aristotelian renaissance — 25
Appendix: Aristotle's criticism of his predecessors — 31

2 From elements to exhalations — 35
Specification and *sustasis* — 39
Toward a heavenly order — 42
The heat of the sun — 48

3 The exhalations — 51
Recycling the dry exhalation — 60
Meteorological mechanics — 62

4 The biological method — 73
The system of the biology and the *Meteorologica* — 75
Causal interactions and their hierarchical effects — 81

5 Teleology in the *Meteorologica* — 93
The final cause — 93
Of (τινος) and for (τινι) — 98

Non-causal good in the *Meteorologica* 105
Two models 107
Dualizing 113

PART II THE *METEORA*

6 Kapnosphere (1.4–8) 117
The kapnospheric exhalation 118
Meteors (1.4): polemic with Anaxagoras 120
Hierarchy and the rotation of the kapnosphere 126
Chasms, trenches, and bloody colors (1.5) 128
Comets and the Milky Way (1.6–8) 134

7 Condensation and precipitation (1.9–12) 146
Hail 152

8 Fresh waters (1.13–14) 156
The river–wind analogy 158
Theories of river formation 161
Aristotle's geography (1.13) 164
Climate change and land–sea exchange (1.14) 169

9 The sea (2.1–3) 179
Confinement 181
Salinity 191

10 Winds (2.4–6) 196
The vaporous catalyst 197
The efficient cause 200
Disk and sphere 207

11 Earthquakes and stormy phenomena (2.7–3.1) 217
Predecessors 218
Normal winds and earthquake winds 221
The theory of the strait 223
Stormy phenomena 227

12 Reflections (3.2–6) 236
The physical causes of reflection 239
The geometry of haloes and rainbows 246
Appendix: the diagrams and the origin of the interpolation 258

13 Metals and minerals (3.6) 271

A methodological postscript 278

Bibliography 282
Index of principal passages discussed 294
General index 300

Figures

4.1	Aristotle's *meteora*	*page* 74
10.1	Disk and sphere	210
12.1	The halo	249
12.2	Reconstructed diagram (after Merker)	252
12.3	Lee's three-dimensional diagram	253
12.4	Tilting diagram	261
12.5	Lee's locus diagram = Lee's traditional diagram	263
12.6	Lee's three-dimensional diagram	264
12.7	Lee's traditional diagram = Lee's locus diagram	265
12.8	Reconstructed original diagram: aerial view of the horizon	266
12.9	Reconstructed tilting diagram	267
12.10	Laurentianus 87.7 (295v)	269

Acknowledgements

Friends and colleagues have read and commented on various parts of this book in various stages of development. My wife, Mary Jaeger, and my colleague and friend, Scott Pratt, and two anonymous readers for Cambridge University Press read through the entire manuscript and provided invaluable advice. István Baksa, Silvia Berryman, Victor Caston, Alan Code, Paul Connolly, Andrea Falcon, Ron Kaufman, Inna Kupreeva, Henry Mendell, Stephen Menn, Meg Scharle, and Liba Taub have provided comments on various parts. My thanks too to hosts and to audiences in Berlin (István Baksa and Liba Taub), Davis (Rex Stem), and Philadelphia (Jeremy McInerney), and at meetings of the American Philological Association, the American Philosophical Association, the Columbia History of Science Group (especially to James Evans), and the Classical Association of the Pacific Northwest for useful questions and feedback. Dieter Harlfinger, Ingo Steinel, Pantelis Golitsis provided me with access, help, and advice in using the Aristoteles Archiv at the Freie Universität, Berlin. My warm thanks go to Kristine Pellatt and Briony Walker for producing the expert diagrams. I gratefully acknowledge *Apeiron* for permission to incorporate previously published material into Chapter 4. I thank the Oregon Humanities Center and the College of Arts and Sciences at the University of Oregon for financial assistance in producing diagrams and meeting other technical requirements.

It is a great pleasure to be able to acknowledge in this way the debt long owed to my teachers and mentors in Classics and in Aristotle, especially Anthony Littlewood, Ivars Avotins, Alan Code, John Ferrari, Allan Gotthelf, and Jim Lennox; and the boundless help and support of my wife, Mary Jaeger, and of my son and research assistant, Seth Wilson, who was always amused that for Aristotle the sea is a cosmic bladder and that love makes the world go round. αὕτη ἀναξία ἀντίδοσις.

Abbreviations

Works of Aristotle

APo.	*Posterior Analytics*
de An.	*de Anima*
Cael.	*de Caelo*
EE	*Eudemian Ethics*
EN	*Nicomachean Ethics*
GA	*Generation of Animals*
GC	*de Generatione et Corruptione*
HA	*Historia Animalium*
Long.	*On Length and Shortness of Life*
MA	*de Motu Animalium*
Mech.	*Mechanica*
Met.	*Metaphysics*
Mete.	*Meteorologica*
Mir.	*Mirabilia*
Mu.	*de Mundo*
PA	*Parts of Animals*
Phys.	*Physics*
Pol.	*Politics*
Pr.	*Problems*
Sens.	*Sense and Sensibilia*
Somn. Vig.	*de Somno*
Top.	*Topics*

Other works

DBS	C. Gillispie (ed.) (1970–80) *Dictionary of Scientific Biography*. New York.
DG	H. Diels (1879) *Doxographi Graeci*. Berlin.

DK	H. Diels and W. Kranz (1952) *Die Fragmente der Vorsokratiker*. Hildesheim.
KRS	G. S. Kirk, J. E. Raven, and M. Schofield (1983) *The Presocratic Philosophers*, 2nd edn. Cambridge.
LSJ	H. G. Liddell and R. Scott (1996) *A Greek–English Lexicon*. Revised and augmented by H. Jones. Oxford.
RE	A. von Pauly and G. Wissowa (eds.) (1894–1972) *Realencyclopädie der Altertumswissenschaft*. Stuttgart.
ROT	J. Barnes (ed.) (1984) *The Complete Works of Aristotle*. The Revised Oxford Translation (Bollingen Series LXXI.2). Princeton.

Introduction

The unjustified neglect of the *Meteorologica*

Rarely do scholars dismiss the value of their subject matter as frankly as has H. D. P. Lee:

> That the *Meteorologica* is a little-read work is no doubt due to the intrinsic lack of interest of its contents. Aristotle is so far wrong in nearly all his conclusions that they can, it may with justice be said, have little more than a passing antiquarian interest.[1]

Trends in the past three centuries of scholarship are consistent with this assessment. Historians of philosophy and science have never bestowed on the *Meteorologica* the attention they have lavished upon the *Physics* and the *de Caelo*.[2] And even in the last decades, when Aristotle's biological works have been scoured for evidence of his scientific practice, the *Meteorologica* has been largely ignored. With the exception of Hans Strohm's excellent but dated essay, there is no modern monograph in any language devoted to the treatise itself.[3] There has never been an English commentary, and much of the secondary literature is highly specialized and difficult of access.[4] More general treatments of ancient meteorology give invaluable overviews, but by their very nature they do not focus on Aristotle's theory and cannot explore all its relevant features.[5]

[1] Lee 1952: xxv–xxvi. Compare Usener's assessment of Epicurus: "Epicuro ut operam darem, non philosophiae Epicureae me admiratio commovit" (1887: v). Lewis 1996: 1–2 has given a defense of *Mete.* 4.

[2] Typical is Burnyeat's exhortation at the end of his introductory essay on *GC* 1, which entirely ignores the *Meteorologica*: "Read on. Read on to the *De Anima* and beyond" (2004: 24).

[3] Strohm 1936. Solmsen 1960 is first rate. I owe it a great deal both in content and method, but the *Meteorologica* is somewhat buried at the end of the book.

[4] Barthélemy St.-Hilaire's 1863 commentary is mediocre at best. Ideler 1834–6, now over 175 years old, is still useful on many points. Vicomercatus 1565 is not widely available or accessible.

[5] Ideler 1832. Gilbert 1907 is an excellent resource still but suffers from a sort of index-card quality, in which analysis is rather weak. Taub 2003 breaks new ground by setting various forms of meteorology in their appropriate epistemic and social contexts.

The treatise has not always been neglected, of course. Three at least partial Greek commentaries survive from antiquity.[6] Through a "paraphrasing compendium" the *Meteorologica* made its way into the Arabic tradition, from which it was translated into Latin by Gerard of Cremona.[7] Moerbeke produced the *translatio nova* based on the Greek text, and the Western tradition rapidly broadened out and deepened in sophistication.[8] The material here is vast and would require another lifetime to study, but my superficial acquaintance with it indicates that medieval and Renaissance scholars read the *Meteorologica* as part of an active pursuit of truth about the natural world. They rightly used Aristotle's work for their own purposes and did not restrict themselves to interpreting the text in its own right. With the rise of the new science, the treatise was plunged into an obscurity from which it has never emerged. It remains a strange testament to its faded fortunes in modern times that a disproportionate quantity of modern scholarship has been devoted to this rich commentary tradition.[9]

The *Meteorologica* is certainly "little-read," but Lee is also correct that Aristotle is "wrong in nearly all his conclusions." In fact, one can hardly help feeling embarrassed by the obvious mistakes about something so apparently straightforward as the role of rivers in the rain cycle, and it is particularly painful to watch Aristotle rejecting many fundamentally correct theories of his predecessors. With the possible exception of rain and snow, there is not a single theory in the *Meteorologica* that even comes close to the modern truth.

But it does not follow from this, of course, that the subject is intrinsically lacking in interest or merely of antiquarian value. If we took such a view,

[6] Alexander's, often little more than a paraphrase, is available for the whole treatise; Olympiodorus' is complete except for 2.7–9, the beginning of 3.1, and most of 4.10–12; Philoponus' is available for the first book only.

[7] This compendium is very different from the Greek text and has a complicated history of its own (Schoonheim 2000: xiv). The *translatio vetus* consists of *Mete.* 1–3 in Gerard's Latin translation based on Yahya ibn al-Bitriq's Arabic. To this is added Henry Aristippus' Greek–Latin translation of *Mete.* 4, and Alfred of Sareshel's translation of part of Avicenna's *Kitab al-Shifa* (*de Mineralibus* or *de Congelatione et Conglutinatione Lapidum*). See Schoonheim 2000: x–xx; Vuillemin-Diem 2008: 5–16. The other Arabic version, that of Hunayn ibn Ishaq, is even further removed from the Greek (Daiber 1975: 6) and did not influence the Latin translations.

[8] Vuillemin-Diem 2008: 17–41; 256–70 argues that Moerbeke relied mainly on manuscript J (Vindobonensis phil. gr. 100).

[9] Daiber 1975 produced an edition and translation of Hunayn ibn Ishaq's compendium of the *Meteorologica*. Lettinck 1999 contains an edition and translation of Ibn Suwar's *Treatise on Meteorological Phenomena* and Ibn Bajja's *Commentary on the Meteorology*. Schoonheim 2000 contains the Arabic version and Gerard of Cremona's Latin translation from the Arabic. Takahashi 2004 provides an edition of Barhebraeus. See Vuillemin-Diem 2008 for Moerbeke's translation and Rubino 2010 for Henry Aristippus' translation. It was the first Aristotelian treatise translated into Hebrew (Fontaine 2001: 102). Most recently on the Renaissance fortunes, see Martin 2011.

we would ignore perhaps everything in the corpus except Aristotle's logic and his ethics. In fact, a subject becomes antiquarian only if we divorce it from its intellectual or social context and fail to appreciate its broader historical significance. I hope to show that the *Meteorologica* was wrong in profoundly interesting and significant ways, and that it needs to be read carefully by anyone who is concerned with how the Greeks organize their rational thoughts about the physical world.

The place of *meteora* in Greek thought

Today, as in antiquity, our main contact with meteorology is practical: the daily weather, current conditions, predictions for the future. At a professional level too, much of the growth of the contemporary science concerned with climate change is focused on predicting what will happen and what can be done to prevent it. It must strike the reader as strange, then, that Aristotle shows almost no interest in prediction. It is not that these matters excited no interest in ancient times. There was, in fact, a whole literature of weather signs and prognostication. But this was not Aristotle's focus.[10] Instead he concerned himself exclusively with the universal and the theoretical, with the reasons why various phenomena arise, and to the extent that he discusses weather signs, it is always in the service of these causes.

Daily weather, with its observable phenomena and practical effects, always excites interest. Theoretical weather studies are a different matter. Although people talk about the weather every day, meteorology does not figure large in the organization of modern sciences. Few universities have Departments of Meteorology, and when they do, atmospheric studies and oceanography are sometimes joined, sometimes separated. The study tends to be squeezed between physics and chemistry on the one side and biology and environmental science on the other. As we shall see, Aristotle was faced with an analogous squeeze and confronted challenges on the related issues of the scope and integrity of the science. The title *Meteorologica* suggests that the treatise is concerned only with the weather but, in fact, Aristotle's compass is much broader, including some geology and what we take to be parts of astronomy. And broad as Aristotle's conception is, for the earliest Presocratics meteorology was much broader still: the study of things above the earth (*meteora*) was essentially identical with the study of cosmic nature,

10 Taub 2003: 71. Lehoux 2007 for an excellent study and bibliography. Diogenes Laertius 5.26 credits Aristotle with a book on *Signs of Storms* (σημεῖα χειμώνων).

and though it did not include biology (a relatively late study), it sought ultimately to situate man's place in the universe as a whole. However, in the period of Aristotle's more recent predecessors, those who followed in the footsteps of Parmenides, the study of specific meteorological phenomena was losing ground and being reduced to the general laws of material change. One of Aristotle's main goals in writing the treatise was to find a stable compromise between the broad and the reduced version of meteorology, to fix its scope and purview, and provide it with a secure place in the physical world.

For us, who come from a modern perspective, Aristotle's theoretical study holds four major surprises. First, the modern world tends to value complexity and equate it with order. We marvel more at terrestrial life, at the diversity of living species, the sophistication of their behavior, the astounding complexity of their chemistry, and less at the monotony and simplicity of elemental and lifeless matter. For Aristotle, however, the simpler order is the better order. This peculiar feature of Aristotelian thought is alluded to in the subtitle of this book, "A More Disorderly Nature," which derives from Aristotle's comment in the first chapter of our treatise (1.1.338a26–b21): "these things [*meteora*] happen in accordance with a nature (κατὰ φύσιν), but one more disorderly (ἀτακτοτέραν) than that of the first material element [i.e., that of the heavens]."[11] In Aristotle's geocentric system, order decreases as we descend from the heavens into the sublunary world of fire, air, water, and finally earth. And the sublunary world displays a less orderly nature than the heavens in at least three ways: (a) in contrast with the simple rotary motion of the uniform celestial aether, the sublunary realm contains four different elements, which move in two opposite directions and which affect one another in new and complex ways;[12] (b) unnatural or forced motion is disorderly (*Cael.* 3.2.301a4–6), and there is an abundance of unnatural motion among the whirlwinds and other violent phenomena of the *Meteorologica*; (c) temporal cycles, precise and exact in the celestial realm, are much less precise in the sublunary. Though the cycle of evaporation and condensation follows certain regular periods (τισι τεταγμένοις χρόνοις 2.2.355a25–8), water vapor does not necessarily return in the same year or to the same place from which it has evaporated. And the sea's salinity "occurs according to a certain order

11 Here and throughout the translations of Aristotle are my own, though with signficant debts to previous translators. The translators of other authors are indicated.

12 Disorder is the state of the material world when separated from the work of the demiurge (κινούμενον πλημμελῶς καὶ ἀτάκτως -- Pl. *Ti.* 30a3–6). In the *Meteorologica* Aristotle does not discuss the deficiency of the material as a cause of the irregular motions, as he does in *GC* 2.10.336b21–4. Cf. Martin 2011: 42–3.

(τάξιν)," Aristotle says, "to the extent that things here [in the sublunary realm] can have a share in order" (2.3.358a25–6). The complexity of the sublunary world is not a sign of its excellence.

Modern readers may be surprised in a second way by the remarkable degree to which Aristotle's conception of the *meteora* forms an integrated unity, specifically, a unity of cause. Modern scientists, by contrast, invoke a wide variety of causes to explain these phenomena. They claim that the Milky Way is the appearance of our galaxy, a vast number of stars; that the aurora is caused by ionization of atoms in the upper atmosphere; that lightning is an electrical discharge; and that the salt of the sea is caused by the dissolution of minerals in it. Some of these phenomena just happen to take place within the same zone, but they are subject to very different causes, physical, chemical, sometimes even biological. For Aristotle, these phenomena all have a common origin in one of the two fundamental materials, the dry and the wet exhalations. Their shared location and the unifying agency of the sun serve as further assimilative causes. The application here of the principle of economy, usually considered a virtue in scientific explanation, is very powerful and clearly extends beyond what is indicated by the empirical evidence. Indeed, as we examine Aristotle's method more carefully, we find that he is less concerned with furnishing accurate and true causes of each phenomenon individually (something that gives modern meteorology its distinctive interdisciplinary character), and more concerned with how the causes of the *meteora* as a group form an integrated structure. Whereas modern science finds it sufficient to provide whatever happens to be the correct cause, Aristotle found it at least equally important to organize the causes. This is an ancient and medieval style of science very much in keeping with C. S. Lewis' "discarded image."

Third, this structured unity of cause seems to be in conflict with what appears to be a thoroughgoing empiricism. As I argue in my first chapter, when Aristotle reports and refutes the theories of his predecessors (and these reports are reminiscent of modern literature reviews), overwhelmingly his arguments are empirical in nature, that is to say, he characteristically cites observational evidence that conflicts with their theories. In support of his own theories, too, he often provides confirmatory signs drawn from simple observations. This empirical tendency has been regularly noted, but if we suppose that it implies a disciplined, even-handed, and critical scrutiny of meteorological theories, we would be mistaken. Two features of Aristotle's account tell against the empirical character of his method in this modern sense. (i) As just mentioned, Aristotle regularly tailors his evidence to suit both his specific theory and the general principles of the whole discipline. The empirical evidence is present in abundance, but it has already been pre-selected to fit the theory. The structure of the discipline, then, is more important than the individual parts, and rarely do we see Aristotle examining obvious contradictory evidence that might undermine his general principles. (ii) Aristotle composed the *Meteorologica* to stand within a sequence of physical lectures. By using both empirical and philosophical methods of high generality, he had already laid down his groundwork in the *Physics*, *de Caelo*, and *de Generatione et Corruptione* and had drawn conclusions there that he now takes for granted.[13] As a result, when he argues against his predecessors on the basis of empirical evidence rather than theory, it is because he has already put the more general arguments to rest. The upshot is that the *Meteorologica* has an empirical character, only because the basic parameters and foundations of the discipline have already been established by a more general philosophical method elsewhere. Aristotle's empiricism here is not to be mistaken for its modern counterpart.

A final surprise comes from teleology. In view of the integrated place of the *Meteorologica* in the physical works, the modern reader acquainted with ancient teleology will find it strange that the *Meteorologica* is utterly devoid of reference to teleology and the final cause. It is not the purpose of the sun, the rain, the rivers, or the sea to support life on earth or sing the praises of God in heaven. In this respect, Aristotle seems to present a remarkably modern conception of an earth that is an accidental environment, where living things must struggle to make their way if they are to exist at all. But again the modern comparison is deceiving, as I shall argue in Chapter 5; for, while it is true that teleology plays no direct role in the treatise, we shall find that the lack of teleology here merely reflects the *Meteorologica*'s role in the grander cosmic teleological scheme, one that positions it between the celestial teleology of the divine mind (νοῦς) on the one hand and the teleology of terrestrial living organisms on the other. The non-teleological character of the *Meteorologica* cannot be understood apart from the play of teleology around it.

The method and organization of the book

I had long noticed the lack of an English commentary on the *Meteorologica*, and my initial intention when beginning this project was to supply one.

[13] For the argument that even Aristotle's highly general physical works proceed by a largely empirical method, see Bolton 1991 and 2009.

But as I proceeded, Aristotle's hierarchical and integrated causal structure persuaded me that this was a poor strategy: it has been precisely the commentary format that has prevented scholars, from antiquity onward, from understanding the treatise's basic architecture and its significance. It was also clear to me that students of the history of meteorology and geography might be benefited by a guide from outside their discipline, one with broader experience in Aristotle's own habits of thought, one who could provide them with a context in which to understand his peculiar and often flawed arguments.

These considerations and goals have produced a book that falls into two unequal parts (I provide a somewhat more detailed précis below). The first part (Chapters 1 to 5) deals with the context of the treatise and issues of a general nature. I first situate the *Meteorologica* diachronically among the theories of early Greek meteorology (Chapter 1) and synchronically within Aristotle's more general physical works, especially the *de Caelo* (Chapter 2). Then I turn to the material principles, the exhalations (Chapter 3), to the complex causal structure of the whole (Chapter 4), and finally, through a consideration of teleology and the final cause, to the treatise's intermediate position between cosmic and biological nature (Chapter 5). These last two chapters form the core of my argument.

The second part (Chapters 6 to 13) deals with the meteorological phenomena in the order in which Aristotle presents them. The chapters of this section may be read separately, though I do not intend them merely as an independent overview of the several topics of the treatise. I endeavor throughout the second part to stress the major themes of the first, provide amplification and nuance, and show in detail how Aristotle conceives of the meteorological phenomena as a unified whole. In each chapter I focus on one or two problems that have been neglected or unsolved. For these reasons, my treatment of passages is selective and problem-oriented. It is my hope that these discussions will be of aid to historians of science less familiar with the intricacies of Aristotle's thought-patterns and will show students of Aristotle that careful attention to the treatise is worthwhile.

A word about my general approach to the text is necessary, since it might strike some as occasionally more literary than philosophical. As I mentioned above, I have been impressed by the coherence of the grand organizational structures in the *Meteorologica*. I have been much less impressed by the individual arguments, which often seem *ad hoc* and arbitrary. This contrast has led me to conclude that Aristotle's primary interest was the organization of the whole and that the arguments are more or less ancillary to that purpose. For this reason, I have found it profitable to seek Aristotle's motivation for making an odd observation or bad argument in his conception of the whole discipline. I prefer to diagnose rather than to remedy his specific arguments. The reader will quickly detect the pervasive use of the simple techniques of polarity and analogy in my analysis. This is, of course, because they are Aristotle's basic structuring techniques, and their unmistakable similarity to the oppositional structures of Homeric composition confer upon the *Meteorologica* a kind of awesome and epic grandeur, which Aristotle is more than willing to exploit. But they are not for him merely the pattern of folk thought so beloved of structuralists; they are instead highly refined techniques of taxonomy and division joined to a sophisticated apparatus of causation.

My primary interest is the way in which Aristotle's method contributes to the unity of the science of meteorology. When I began this project, I intended to use the *Meteorologica* as a case study in the techniques of "scientific unification" that had been the subject of my previous book. I was especially interested in how Aristotle's construction of the subject genus of meteorology might inform our understanding of his theory of the subject genus in the *Posterior Analytics*. The *Meteorologica* provides ample material to this end, but I found that such a project, whether and how the practice of the *Meteorologica* corresponds or fails to correspond to the theory of the *Posterior Analytics*, became less interesting than the theory of unity implicit in the *Meteorologica* itself. I found that Aristotle's practice here runs far in advance of his theory in the *Analytics*, and for this reason I have not engaged the *Analytics* in any sustained or systematic way. I think it will be clear, however, that the issue of scientific unity and what constitutes a subject genus is always a present concern.

Meteorologica 4 and the date of the work

I do not, of course, discuss the fourth book of the *Meteorologica*, which deals with the transformation of various homeomerous bodies. The textual tradition has always given it a secure place at the end of the *Meteorologica*, but its role in the course of physical lectures has been disputed from antiquity.[14]

[14] Alexander 179.3–11 (cf. Olympiodorus 273.27–9), noting especially its reliance on active and passive powers, sees it as a continuation of the *GC* rather than of *Mete.* 1–3. According to Olympiodorus 272–3, *Mete.* 3.6 treats lifeless homeomeries (metals and minerals), which prompts Aristotle to take up the question of homeomeries in general in *Mete.* 4. *Parts of Animals*, then, treats ensouled homeomeries. Philoponus 3.14–19 likewise sees a continuity between the study of metals (3.6) and Book 4. Lewis 1996: 9–15 argues on a similar basis that *Mete.* 4 follows naturally from the first three books. See Viano 2006: 17–23 for an excellent discussion of the continuities and breaks between 1–3 and 4.

The project of the first three books is left incomplete with Aristotle's hasty and general discussion of metals and minerals, and though he promises an account of their species at the very end of Book 3 (3.6.378b5–6), Book 4 does not provide what we are promised.[15] Whatever continuities may be found, *Meteorologica* 4 is obviously based on different principles and methods from those of the first three books. Not only are the exhalations, so characteristic of the first three books, absent, but the entire structure of cosmic place between the moon and the earth (and below the surface of the earth) has completely disappeared. The agency of the sun and the heavens plays no role. The fourth book carefully straddles the provinces of inorganic and organic matter. In certain respects it continues the work of the *de Generatione et Corruptione* in studying the qualitative changes of matter, and for this reason it has been called the "Chemical Treatise"; in others it prepares the way for the tissues of organic bodies, which require homeomerous materials significantly more complex than the simple elements that are the materials of the cosmos as a whole.[16] Occasionally the fourth book may shed light on problems in the previous three, but for our interests there is little direct interaction. I see no reason to doubt the authorship of either treatise; certainly the authenticity of the first three books has never been seriously doubted.[17]

About the date of composition, I have little to say. The *Meteorologica* was clearly part of a rotating series of lectures (1.1), and that series was undoubtedly delivered more than once in Aristotle's lifetime.[18] We are better supplied for *termini post quem* here than for any other physical treatise. An early datable event is the burning of the temple of Ephesus (3.1.371a31), described as having taken place "now" (νῦν), and dated to 356 BC. A comet is said to have appeared in the archonship of Nicomachus (1.7.345a1) dated to 341/0 BC. The conjunction of Jupiter with one of the stars of Gemini, which "we ourselves have seen" (1.6.343b30) has been dated in recent years to December 337 BC.[19] The second moon-rainbow

15 *Mete.* 4.8 discusses metal formation, but on the basis of different principles (see Chapter 13). A section in Avicenna's *Kitab al-Shifa* is an attempt to fulfill this promise and for this reason becomes attached in translation to the *translatio vetus* of the *Liber Metheorum*.

16 Düring 1944: *Aristotle's Chemical Treatise*; for biological connections, see Furley 1983; Gill 1997.

17 Alexander 179.3–4 affirmed its Aristotelian authorship. Louis 2002: vii–xviii. On the authenticity of 4, see Strohm 1984: 216–18, and at length, Viano 2006: 79–113.

18 Lee 1952: xxiii–xxv; Louis 2002: xviii–xx.

19 Poss 1980 for the occultation of 1 Geminorum. Cohen and Burke 1990, with a somewhat more elaborate study, concur on the star and the date (though ignorant of Poss' contribution). The evidence of Corona being overhead (2.5.362b10) and therefore suggesting an Athenian observer, and therefore Aristotle at Athens, is too ambiguous and imprecise to be useful.

"we have met with in over fifty years" (3.2.372a28–9) could arguably have been observed in the very late 330s.

Stylistically, the treatise has a coherence and uniformity of vision that might suggest a brief period of intense work, but this can hardly be proved. As will become clear, I do think that the treatise in its current form was based on a plan worked out in advance, most likely in a notebook shorter than, though similar to, the *Historia Animalium*. This conjectural *Historia Meteorōn* may have contained a list of problems and predecessors' theories but certainly would have described the phenomena and, most importantly, the list of their differentiae, such as we find in the *Historia Animalium*. This list need not have been the chronological first step toward the composition of the earliest version of the *Meteorologica*, but it was almost certainly drawn up before our *Meteorologica* was completed.[20] At the same time, it is clear that the treatise was not written in isolation from the rest of the physical works. The overall conception of the *Meteorologica* is crucially dependent upon the *Physics*, *de Caelo*, and *de Generatione et Corruptione* on the one side and the *de Anima* and the biological works on the other. The synthesis of the whole set of lectures must have begun early and was certainly the work of a lifetime.

Overview of the argument and claims of the book

Chapter 1 begins by examining the place of the *Meteorologica* in the tradition of early Greek physical investigation. I argue that Aristotle inherited a science in decline and was eager to reinvigorate it. Most controversially, I claim that he attributed meteorology's decline (and since Aristotle largely wrote the history of early Greek philosophy, we have little independent counter-evidence) to the obsession with the general problems of change, which occupied thinkers in the aftermath of Parmenides and which led them to consider the fundamental nature of matter more important than the independent study of meteorology. The neglect of meteorology reached an extreme in Plato's manifest contempt for it even in the physically oriented *Timaeus*. I show that Aristotle aimed his refutations of his predecessors largely at a specific rather than a general level and that he did so because of his desire to avoid engaging with their most general principles and to keep meteorology at the appropriate level of specificity. In this way he prevented meteorology from being reduced to general physics.

20 There may be traces of it incorporated directly into the *Meteorologica* at 3.2. For the practice of ancient writers, see Dorandi 2007, especially Chapter 2.

Chapter 2 demonstrates how Aristotle warded off the threat of reduction within his own comprehensive physical system by insuring that the principles of the *Meteorologica* were based on, but not reducible to, those of general physics laid out in the *Physics* and *de Caelo*. I explore the tension between reduction and autonomy through Aristotle's material principle, especially through the peculiar way in which the four-element theory of the *de Caelo* is transformed into the two-exhalation theory of the *Meteorologica*. I isolate two conceptions of the exhalations in their relationship to the elements, which I call the "*sustasis*" and the "specification" views. In many passages throughout the *Meteorologica* the exhalations behave as concentrated masses (*sustaseis*) of a material different from the four simple elements (earth, water, air, and fire) that surround them. In other passages, especially in the programmatic chapter 1.3, they seem to behave as substitutes for and specified forms of the elements, air and fire, which in consequence are treated as abstract entities with no existence independent of the exhalations. I argue for a reconciliation between these views and provide a new interpretation of chapter 1.3 according to which the heavens play an essential role in the formation of the new material principles and in the regulation of their changes. Under the celestial agency the chaotic and mutually aggressive interaction of the four simple elements is superseded by a new proportion and harmony.

In Chapter 3 I examine the details of these new material principles, the exhalations, and show how they combine the oldest stratum of Ionian speculation with an updated mechanics of the later philosophers, Empedocles, Anaxagoras, and Democritus. I provide new arguments in favor of the controversial view that Heraclitus recognized two exhalations and that Aristotle is indebted to him for them, just as he is indebted to Heraclitus' conception of the meteorological phenomena as homeostatic processes. At the same time, I point out a number of ways in which Aristotle's exhalations differ from those of Heraclitus and I provide reasons for Aristotle's innovations. Specifically, Aristotle's exhalations do not function as effective opposites and are not transformable into one another. I introduce the problem of the dry exhalation, which, in defiance of Aristotle's general laws of material change, does not recycle into other forms of matter (as Heraclitus' does), and argue that Aristotle is beginning here in the *Meteorologica* to introduce the kinds of material, typical of living things, which are restricted in their potentiality for change. The chapter ends with a discussion of Aristotle's mechanics of exhalations, inherited and adapted from the latest part of the Presocratic tradition. Although Aristotle's vocabulary of mechanics is apparently non-technical, my close lexical study, which forms a sort of appendix to the chapter, reveals important patterns in his use.

Chapters 4 and 5 form the core of my argument. Chapter 4 demonstrates that the method of the *Meteorologica* is similar to the method of the biological works. In both fields of study we find a similar system of classification based on "great genera," and the subordinate system of differences of degree or of "more and less." Most importantly, I argue that Aristotle uses three causal factors, material cause, efficient cause, and place, to define the meteorological phenomena and arrange them into a complex unity embracing a number of hierarchies or *scalae naturae* similar to those we find among Aristotle's animals (with man, of course, being the highest animal). Each of the causal factors has internal and hierarchical differentiations. Thus, among the material causes, the dry exhalation is superior to the wet. Among efficient causes, the direct locomotion of the heavens is superior to the indirect motions and changes caused by heating the surface of the earth. Among places, the higher are superior to the lower. The result of applying these principles is more than an episodic set of explanations of the unrelated phenomena; because of hierarchical distinctions among the causes themselves, where one material, one place, one efficient cause is superior to another, Aristotle is able to produce a set of hierarchies among the phenomena themselves, which unifies them as a group and prefigures the *scala naturae* among the living beings of the biological works. The most notable hierarchy among meteorological phenomena is the one that prefigures the biological *scala* of posture. According to Aristotle, man is the most upright of living things and is followed in this respect by mammals, reptiles, the sessile testacea and finally plants, which have, as it were, their heads in the ground. Similarly, I argue, the comets and other fires of the upper region are the most vertically oriented and never descend except by force; they are followed by the cycle of rising vapor and falling rain, by the radially and horizontally oriented rivers and winds, the sessile sea in the hollows of the earth and finally the plant-like metals and minerals. This and other hierarchies corroborate my argument for a correspondence between Aristotle's meteorology and his biological works.

Chapter 5 builds on the observations about this quasi-biological method and asks whether Aristotle saw a role for teleology in the *Meteorologica* corresponding to its important role in biology. I approach the problem through the famous and much-discussed passage in *Physics* 2.8 with its example of rain falling for the sake of the crops. I begin by arguing that the question of the final cause in the *Meteorologica* is best considered through the two recognized senses of the expression "that for the sake of which," the aim and the beneficiary. My examination of the relevant texts indicates that the meteorological phenomena are not end-directed in either of these two senses.

My main purpose in this argument, however, is to show that the absence of the final cause helps Aristotle to position the meteorological phenomena and the sublunary world at the intersection of two different perspectives. In the first, the meteorological phenomena "function" together as if they were the body of the cosmos, where the heavens constitute the soul. (I provide evidence that supports localizing the Aristotelian soul in a restricted location such as the heavens.) On this view the heavens move the sublunary world mechanically, just as the heart moves the limbs of the body (so *de Motu Animalium*). On this perspective the cosmos forms a single unity. According to the second perspective, each meteorological phenomenon, each comet or river, is conceived of as a sort of proto-animal, each being independent of all others. This perspective is already supported by the biological method of the *Meteorologica* revealed in Chapter 4. For additional support, I draw on Aristotle's doctrine of spontaneous generation, in which a common solar heat produces a variety of different kinds of animal according to the specific nature of the putrid matter it acts upon. By locating spontaneous generation within a continuum of being, as a kind of "dualizer" or intermediary between meteorological phenomena and sexually reproduced animals, I am able to provide a new explanation for why Aristotle developed the strange doctrine of spontaneous generation. But more important for the *Meteorologica*, Aristotle, by refusing to assign a role to the final cause among the meteorological phenomena, is better able to stabilize the sublunary world between its role in cosmic unity and its role in prefiguring terrestrial diversity without having it definitively serve the purposes of either.

The second part of the book treats the major phenomena in the order in which Aristotle presents them.

The main purpose of Chapter 6 on the "kapnosphere" (my coinage for the upper smoky region) is to show how Aristotle conceives of these fiery phenomena as quasi-celestial, that is, as intermediary between the celestial and the sublunary realms. This chapter deals extensively with the problems of the non-recycling of the dry exhalation, the composition of the upper region, and Aristotle's friction theory of solar heat. The chapter therefore considers in detail the relationship between general physics and the meteorology. My solution to these problems shows uniformly how Aristotle intended to give the kapnosphere the highest honor in the sublunary realm. My method, here as elsewhere, is to show that Aristotle chose arguments more because they advance his grand structural program than because they are immediately compelling. The chapter ends with a discussion of the Milky Way as the only example in the treatise of a species with a single member. Aristotle's definition of it provides an interesting solution to his problem of the definition of unique eternal things in *Metaphysics* 7.15.

In Chapter 7 I use Aristotle's treatment of precipitation in order to illustrate how he struggled with the limits of the method of "the more and less" as a way of differentiating species within a great genus and was compelled to use instead the analogical relationship between rain and dew, snow and frost, in order to solve the standing problem of hail and its paradoxical nature.

Chapter 8, on fresh waters, takes us to the surface of the earth, where rivers arise not primarily because of the rain, but because of the condensation of vapor under mountains. I argue that Aristotle introduces this new cause, specific to rivers, in order to isolate them from the rain cycle and to make rivers into a great genus in their own right. By making rivers into a great genus corresponding to but lower than the cycle of precipitation, Aristotle builds up the hierarchies of his grand scheme. This chapter also takes up the issue of climate change, where Aristotle's comparison between cosmic and organismic aging looms large and corroborates the intermediate position of meteorological phenomena between the unified cosmos and individual animals. Aristotle's main purpose here is to demonstrate (against his opponents) that even rivers are homeostatic individuals that come to be and pass away and, as such, are of the same ontological kind as the other phenomena of the treatise.

With the sea (Chapter 9) we come to the bottom of the water system. Aristotle makes every effort to isolate the sea from the moving fresh waters above it and to fashion it as a sterile, digestive waste similar to urine. Inasmuch as he likens the sea to the digestive product of a unified organism, while isolating it from other kinds of water, Aristotle is again motivated by the twin conceptions of the sublunary world as a unified organism and a manifold collection of genera.

The winds (Chapter 10) are the dry exhalation's counterpart to the rivers. In an elaborate analogy Aristotle shows that rivers and winds form a low-slung and horizontal pair corresponding to the more vertically oriented kapnosphere and atmosphere. Rivers and winds are both radially oriented around a more or less circular northern landmass with the Mediterranean Sea at its center. I argue that Aristotle accepts this old-fashioned, Ionian, flat-earth view – a view for which he is widely criticized – because of its architectonic advantages and its ability to accommodate the predominantly northern Etesian winds. I show how Aristotle deliberately combined both the flat and the spherical views of the earth but systematically subordinated what he saw as the advantages of the Ionian model to the new and cosmically more accurate sphere. I also propose a new solution to the ancient problem of the winds' efficient cause. According to Aristotle, the winds are driven by the motion of the heavens, with the natural result that all winds would have to blow from east to west. I argue that Aristotle was not thinking of the diurnal motion of the heavens as the driver of the winds, but rather the more appropriate and traditional, seasonal motion of the sun north and south.

In Chapter 11 I argue that earthquakes are the dry exhalation's counterpart to the sea, caused by a chaotic form of wind, which is divorced from the heavenly causal order manifested in the radial system of the winds. I also trace the ways in which Aristotle draws causal connections between earthquakes and the sea in order to accentuate the negative connotations of both.

Reflective phenomena such as rainbows and haloes are the subject of Chapter 12. I explain Aristotle's peculiar placement of reflections near the end of the treatise by means of the theory of mixed or subordinate science. I also examine here how Aristotle's confusing discussion of the properties of the various kinds of reflection can be better understood through his deployment of a system of differentiae. I then support Vitrac and Merker's recent argument that the geometrical locus proposition in 3.5 is a textual interpolation and add overlooked physical corroborations to their mathematical arguments. In an appendix I provide some palaeographical reasons why the locus passage in 3.5 might have been interpolated.

I conclude with Aristotle's brief treatment of metals and minerals (Chapter 13), where I argue that metals are to be understood as debased condensations at the lowest end of a hierarchy that includes rain and riverine condensation.

PART I

Methodology and structure

CHAPTER I

The rebirth of meteorology

By the time that Aristotle wrote the *Meteorologica*, the study of "things above the earth" was already two hundred and fifty years old and in decline.[1] According to one modern account, the early Milesian thinkers, simply by eliminating the traditional gods from meteorological explanation, had made further research unnecessary. Their successors mechanically repeated their old theories with their familiar apparatus then turned to more interesting problems.[2] On this interpretation meteorology was senescent almost as soon as it was born.

It is certainly true that the study had a robust beginning and came quickly into a mature state. As early as Anaximenes, for example, we find theories for rainbows and earthquakes.[3] But meteorology was still advancing, albeit slowly, throughout the fourth century. New problems were being set, and new answers offered: the comets seem first to have been offered natural explanations in the later half of the fifth century.[4] Metrodorus of Chios in the fourth century is the first on record to discuss the colors of the rainbow.[5] Aristotle is the first extant writer to mention the aurora, sundogs, and haloes, though to judge from his descriptions, they had already been discussed before his time.[6] They were probably recent additions.

1 On the name of the science, Capelle 1912a.

2 Kahn 1984: 99: "the main lines of explanation scarcely vary from Anaximander to Roman times. Perhaps no work of Aristotle is so directly dependent upon his Ionian predecessors as is the *Meteorologica*. Whole pages could have been written by Anaxagoras or Democritus." See Strohm 1984: 123 for more judicious comments.

3 DK 13A7.8 (1.92.25–8).

4 Anaxagoras' is possibly the first (1.6.342b27–9). "Some of the Italians among the so-called Pythagoreans," (1.6.342b29–35; cf. Aët. 3.2.1 [*DG* 366.6–12]) claim that a planet is responsible for the appearance of comets. Burkert 1972: 321 and Huffman 1993: 240 point out that this comet-planet is not consistent with Philolaus' ten orbiting bodies. The lack of sophistication of their theory relative to that of Hippocrates of Chios argues for an earlier date.

5 Metrodorus of Chios: DK 70A17 (2.232.29–35) = Aët. 3.5.12 (*DG* 374.6–8) and Schol. Arat. 516 Maass.

6 Aristotle does not claim to have discovered them; they had already been given names in popular discourse, e.g., χάσμα, βόθυνος, αἱματώδη χρώματα 1.5.342a35–6; ῥαβδός, παρήλιος, 3.2.371b19.

In terms of the bulk of material surviving to us, meteorology came to its zenith with the work of Anaxagoras and Democritus. But by the end of the fifth century and through the first half of the fourth – from the time of Diogenes of Apollonia and Metrodorus of Chios to the death of Plato – the study had fallen into a torpor.[7] The doxographical tradition makes little note of the opinions of these last Presocratics, and, where it does, their opinions are little more than recycled versions of earlier theories.[8] Although there remains a continuous tradition of Ionian meteorology through to Epicurus and beyond, our evidence makes it impossible to argue that the discipline was a vibrant focus of innovation. With the exception of Aristotle there is little evidence of significant meteorological research in the fourth century until Epicurus, with whom the character and purpose of the study had markedly changed.[9]

New problems diverted philosophers' energies in other directions and, as time went on, the amount of material concerning *meteora* declines as a proportion of the total research into nature that has survived to us. Early Ionian theories had made the heavens dependent on the earth, but soon enough the heavens gained their independence. By the 430s Philolaus and Hippocrates of Chios began to submit the motions of the heavens to new geometrical methods and establish theories of celestial circles.[10]

7 Capelle 1935: 319 and 339.

8 Diogenes, though contemporary with Democritus, may be put in this later group: Theophrastus (*apud* Simpl. *in Phys.* 25.1–3) says that he "was pretty well the latest of those who studied these matters, and wrote for the most part eclectically in some things following Anaxagoras, in others Leucippus" (trans. Guthrie 1962–81: II.362). As a (not wholly) satisfactory method of measuring meteorological interest, one may consider the number of lines of DK A or "Lehre" material devoted to "meteorological" (i.e., what Aristotle identifies as meteorological) subjects and the percentage they represent of the whole physical "Lehre": Anaximenes (51/137 = 37 percent); Anaxagoras (133/331 = 40 percent); Democritus (175/1367 = 13 percent); Diogenes of Apollonia (42/224 = 19 percent; Simplicius even says that he wrote a *Meteorologia* DK 64A4 [2.52.23]); Metrodorus of Chios (34/88 = 39 percent). Consider too, the number of times each is mentioned by Aëtius on meteorological subjects: Anaximenes: 4; Empedocles: 2; Anaxagoras: 9; Democritus: 6; Diogenes: 2; Metrodorus: 7. Though Metrodorus' proportion is in keeping with the earlier Presocratics' and his representation in Aëtius is robust, his theories are almost entirely reformulations of earlier theories (DK 70A14, 15, 16, 18, 19 [2.232–3], and only his rainbow and earthquakes have something new to offer (A17, 21). Diogenes' theories are likewise unoriginal (DK 64A15, 16 [2.54.2–7]).

9 It is worth pointing out, in light of the argument of Rashed 2005: xxxv–xlviii that in the *GC* Aristotle seems to be arguing with neo-Empedocleans (e.g., ὡς Ἐμπεδοκλῆς καὶ ἕτεροι λέγουσιν 2.1.329b1–2), that in the *Meteorologica* rarely are plurals invoked in the context of specific predecessors, and then only concerning the highest phenomena (comets: οἱ περὶ Ἱπποκράτην τὸν Χῖον καὶ τὸν μαθητὴν αὐτοῦ Αἰσχύλον 1.6.342b36; and the Milky Way: οἱ περὶ Ἀναξαγόραν καὶ Δημόκριτον 1.8.345a25). Unnamed opponents are referred to by ἔνιοι (Empedocles: 2.1.353b11) or τοῖς μέν (Anaxagoras: 1.12.348a14), but where names are given the context is solidly singular. A likely conclusion is that where these predecessors have "schools," their successors' work in meteorology did not merit Aristotle's attention.

10 Huffman 1993: 231–61. Burkert 1972: 313–14 places Hippocrates' *floruit* around 430 BC. He notes (314n77) that Aristotle's mention of the northern comet in the archonship of Euklees (427/6 BC) sets

So powerful did this celestial paradigm become in this period that it was able to coopt closely related phenomena. Thus comets began to be treated as planets.[11] The growing realization of the great distance of the stars from the earth further contributed to the separation of astronomy and the marginalization of the more mundane meteorology.[12] Even if Aristotle did not pursue mathematical methods in his study of the heavens, he too was deeply impressed by them. He describes on several occasions how recent researches had revealed the vastness of space, and he uses these discoveries to argue against the direct material interaction between the celestial and the sublunary worlds.[13] The heavens had grown in importance and the *meteora* were diminished.

New energy was similarly channeled toward psychological and ethical problems. Democritus' atomism required a close investigation of sense perception. Socrates' rejection of natural philosophy and the sophists' interest in a wider range of new questions suggest that the old Ionian research paradigm was losing its vigor. Plato – his *Timaeus* notwithstanding – largely inherited Socrates' contempt for physics, an attitude shared in turn by many in the Academy.

It has been suggested that politics too may have played a role. The trials of Anaxagoras (432 BC) and of Socrates (399 BC), as Capelle has argued, showed how the study of the things above and below the earth, and their associations with sophistry and atheism, could be dangerous and sometimes fatal.[14] Athenian anti-intellectualism may have had a dampening effect on these studies in the late fifth and early fourth centuries.

Most or all of these factors may have contributed to the decline of meteorology in the early fourth century, but closer attention to Aristotle's treatment of his predecessors (his doxographical review) suggests a further explanation and one that more closely touches on our themes. Capelle rightly attributed the flowering of meteorology to Anaxagoras, but Kahn may have been closer to the truth in claiming that the study was fading

Hippocrates' theory before that date (against Burkert, however, the comet of 467/6 BC associated with the meteor of Aegospotamoi (1.7.344b32–4) appeared in the west (i.e., between the tropics). For other chronological arguments, Netz 1999: 272–5 and 2004: 244–8.

11 The "Pythagoreans" of Aristotle's account, Democritus, Anaxagoras, and Hippocrates of Chios (*Mete.* 1.6.342b27–343a4); cf. Diogenes of Apollonia (DK 64A15 [2.54.2] = Aët. 3.2.8 [*DG* 367.8–9].

12 Capelle 1935: 320.

13 1.3.339b6–9: "it is now clear through astronomical research" (ἤδη γὰρ ὦπται διὰ τῶν ἀστρολογικῶν θεωρημάτων); b32–4: "if they had considered what is proved sufficiently through mathematics, they would perhaps have abandoned this foolish opinion" (θεωρήσαντες ἂν τὰ νῦν δεικνύμενα διὰ τῶν μαθημάτων ἱκανῶς ἴσως ἂν ἐπαύσαντο ταύτης τῆς παιδικῆς δόξης); cf. 1.14.352a27–8 and *Cael.* 2.14.298a6–15.

14 Capelle 1919: 88–91; 1935: 316–18. Diagoras of Melos may also be added to the list: Woodbury 1965; more recently Janko 2003.

long before him.[15] Much of the difficulty in deciding the question lies in the fact that Aristotle controls most of the historical record, and there are reasons to believe that Aristotle does not represent the theories of his predecessors accurately.[16] But this difficulty is not immediately relevant (I shall return to it in due course), because the reasons why he represented his predecessors as he did were certainly related to how he conceived of the study of meteorology.

According to Aristotle's history of philosophy, Parmenides introduced the problem of change, which later Presocratic philosophers, in particular Empedocles, Anaxagoras, and Democritus, tried to solve by introducing material principles subject to locomotion but in all other respects unchanging. I suggest that in Aristotle's mind and probably in reality as well, his predecessors, in pursuing these big problems, shifted their attention away from the specific and manifest changes, the *meteora* that had so engaged the earlier Ionians, toward the general and abstract fact of change.[17]

The truth of this view, to the extent to which we can determine it from the evidence, comes in two different forms. First, though Anaxagoras was absorbed in meteorological studies, there is no evidence that his specific theories were dependent upon his general theory of matter.[18] There is reason to take the silence as a real absence, since in the case of Democritus, by contrast, there is trustworthy evidence that his atomic theory informed his specific meteorological theories. The absence of such evidence for Anaxagoras is therefore less likely to be an accidental omission of the tradition. If we may be allowed to draw this conclusion, Anaxagoras makes no use of his own elaborate theory of matter when explaining the *meteora*, in a manner reminiscent of Socrates' accusation that he introduced a useless νοῦς (*Phaedo* 97b–98). Whether Anaxagoras may have done this on purpose or by neglect, Aristotle does him no disservice in representing him in this way.

The situation is far different with Democritus. If we had only Aristotle's reports in the *Meteorologica* to go on, we might reasonably conclude that, like Anaxagoras, Democritus made no use of his general physical principles in his specific meteorological theories. However, on several occasions we

[15] Capelle 1919 is joined in his opinion by Strohm 1984: 122, who claims that "pre-Aristotelian meteorology reached its acme in Anaxagoras."

[16] Capelle 1935: 327 despaired, perhaps rightly, that a real history of pre-Aristotelian meteorology can be written.

[17] Cf. Solmsen 1960: 12–13. For a similar sentiment one step further toward the concrete, *On Ancient Medicine* 1 notes that *meteora* require general theories or postulates: "Wherefore I have deemed that [medicine] has no need of an empty postulate, as do insoluble mysteries, about which any exponent must use a postulate, for example, things in the sky (μετεώρων) or below the earth" (Jones trans.).

[18] Solmsen 1960: 405 and n50. Curd 2007: 222–5 connects his meteorological theories loosely with νοῦς, but not with the material principles. Gilbert 1907: 519 argues for a close relationship.

find Democritus invoking his atoms as indispensable explanatory principles. Our evidence comes, significantly, from Seneca: Democritus explained the wind by means of the specific behavior of more or fewer atoms confined in a narrow space.[19] Here the atomic theory is actively and essentially contributing to the explanation, however badly it may do so. Democritus' explanation of magnetism is likewise deeply dependent on atomic theory.[20] Other fragments display a lesser degree of dependence, where the atomic theory is invoked but the atoms do not as such contribute to the explanation. Thus the Hibeh papyrus describes the formation of the saline sea. Here the atoms are said to be attracted like to like (τὰ ὅμοια πρὸς τὰ ὅμοια), a controversial power active in substances such as frankincense and silphium, according to which more or less pure substances come together.[21] For the same reason, Democritus maintains, the salt of the sea gathers together into the one area (the sea) and is less present on the land. In this explanation, although atoms are mentioned, their specific nature contributes little to the like-to-like principle, which goes back to Empedocles and is not dependent on atomism.[22] Similar is Democritus' explanation of thunder and related phenomena.[23] None of these examples comes from Aristotle, who presents Democritus' explanations as if the atomic theory were wholly irrelevant, as, for example, in the case of earthquakes (2.7.365b1–7), where the excess of water in the cavities of the earth sometimes bursts out and causes the earth to tremble. There is no discernible theoretical connection here with atomism, and Aristotle makes no effort to draw one.

Is Aristotle omitting something important here? Almost certainly. Democritus probably tried to provide robust atomic explanations in most cases. But even if he tried – and this is the second way in which attention was directed away from the specific and manifest changes – the evidence

19 *QN* 5.2 = DK 68A93a (2.106.27–36): "Democritus says that when there are many little particles, which he calls atoms, in a small empty space, wind is the result; on the other hand, when there are only a few little particles in a vast void the condition of the air is quiet and calm. For example, as long as only a few people are in a marketplace or on a side-street they walk about without disturbance, but when a crowd comes running together into a narrow place, people bump into each other and start quarrelling. The same thing happens in the space surrounding us. When many bodies fill a tiny space they unavoidably knock against each other and push, are shoved back, entwined and squeezed. Winds are produced from these activities. The particles, which are struggling with one another press hard and, after being tossed about and in confusion for a long time, start to move in one direction. Whereas, when only a few bodies are moving around in plenty of room they cannot ram each other or be pushed" (Corcoran trans.). For discussion see Böker 1958a: 2227–9.

20 DK 68A165 (2.128.28–129.3). Guthrie 1962-81: II.426.

21 DK 68A99a (2.108.8–25); also 68B164 (2.176.17–177.11); Furley 1989 [1976]: 79 questions the principle as an independent power of attraction.

22 DK 31B22 (1.320.18–321.7) and KRS 307–8; 31B90 (1.344.4–5) and Guthrie 1962-81: II.165–6.

23 DK 68A93 (1.106.19–26) = Aët. 3.3.11 (*DG* 369.12–28).

suggests that Democritus was content to give an atomic interpretation of meteorological phenomena, and he was not concerned to give a decisively better, or often even different, account of the phenomena than any of his rivals.[24] Unless atoms can explain more phenomena better, then there is no reason to favor them as an explanation. Aristotle, by contrast, is everywhere eager to show that his *meteorological* principles account for the phenomena better than all other theories. But most importantly, in Aristotle's mind, anyone who reduces the *meteora* to atoms will not stop until he has reduced the soul to atoms as well, with the loss of explanatory power that comes with ignoring commensurate universals. After all, souls are different from clods of earth and have special properties, which need to be explained according to psychological principles. Why Aristotle drew the line against reduction so quickly at the *meteora* rather than waiting until the soul is a question we will address in Chapter 5.

I suggest then that for Aristotle, it was the universal reduction of all natural phenomena to general material principles that led to the withering of meteorology. Since all change could be explained by the same principles of locomotion, of combination and separation, meteorological phenomena simply became another trivial application of general physics. The later Presocratics had no specifically meteorological theory, and it is probable that the real quality of meteorological research suffered as a result.

For different but analogous reasons, Plato's pursuit of the universal and abstract principles brought about a further decline in the fortunes of meteorology. Plato sought the truth in mathematics and the intelligible Forms, of which the natural world was a paltry imitation. Even the *Timaeus*, the most physical of Plato's dialogues, has, as Solmsen has noted, "little thought to spare for" specific meteorological topics and questions, and to the extent that they are even mentioned there (59e), they are explained in terms of the unintelligent, variable cause.[25] The kinds of issues that we find in Aristotle's *de Caelo* and his *de Generatione et Corruptione*, the general structure of the universe and the nature of matter, are treated by Plato much more fully than the more specific *meteora*. Plato's neglect of

[24] For example, see below on comets, Chapter 6.

[25] Solmsen 1960: 394, also 13. In a discussion of the transformations of water, Plato offers only "if the condensation becomes very great, the water above the earth becomes hail, but on the earth, ice, and that which is congealed in a less degree and is only half solid, when above the earth is called snow, and when upon the earth and condensed from dew, hoarfrost" (59e; trans. Zeyl in Cooper and Hutchinson). Plato immediately proceeds to discuss sap. This is as close to meteorology as we get in the *Timaeus*. For the variable cause, *Ti.* 46e: "Both kinds of causes should be acknowledged by us, but a distinction should be made between those which are endowed with mind and are the workers of things fair and good, and those which are deprived of intelligence and always produce chance effects without order or design" (trans. Zeyl).

them in the *Timaeus* made clear the fate that awaited meteorology, if it was treated as a part of a single undifferentiated physical science; in this respect Plato's position is merely the inevitable endgame in a process initiated by the successors of Parmenides.

The Aristotelian renaissance

A revival of this moribund field of study had obvious attractions for Aristotle. Meteorology was, first and foremost, an integral part of his grand physical science, the comprehensive inquiry into the natural world. It had once formed the core around which the early Presocratics developed their worldviews, and, although it was no longer foundational, no consideration of the physical world could be complete without it. But Aristotle did not want merely to resuscitate a dying tradition. Whereas Plato relegated Greek meteorology to the realm of shifting appearances and material necessity, Aristotle wanted to show how the creaky Ionian apparatus could be transformed into an order of awesome beauty, one rivaling any of the works of intelligence. And as if to triumph more completely over his erstwhile master, he constructed this beauty out of the recycled and ever-changing homeostatic processes of Heraclitus. Clearly his ambitions ran far beyond merely preserving the past.

But in order to accomplish this goal Aristotle understood that he had to reverse the decline (as he saw it) caused by the reduction of meteorology to general physics, and in order to do that, the new study had to be supplied with its own domain and its own principles.[26] Aristotle signals this understanding in the first chapter with his synopsis of the physical *pragmateia*. By identifying "what all the predecessors were wont to call 'meteorology'" (ὃ πάντες οἱ πρότεροι μετεωρολογίαν ἐκάλουν 338a26) within the series of disciplines (*Physics*, *de Caelo* and *de Generatione et Corruptione*, *de Anima*, and the biological works), Aristotle shows a consciousness of the problem of integrity and reduction. Physics is not a single seamless account embracing everything from form and privation to animal reproduction; it falls into a series of discrete and integral parts related to each other in numerous and complex ways.[27] For this reason, too, Aristotle identifies

[26] Solmsen 1960: 262–4 on the importance of semiautonomous sciences within physics.

[27] Lennox 2010 emphasizes the fact that the parts of physics form a whole. He rejects (prematurely, in my mind) subordination as a model for understanding the relationship between the parts of physics, in favor of *methodos*, specifically the addition of new kinds of approaches (great genus, multiple differentiae, final cause), which distinguish terrestrial life from the cosmos as a whole and its parts.

from the very outset principles specific to meteorology, the exhalations, and he repeats them as a leitmotif throughout the treatise to remind us that they, unlike the general principles of his predecessors, are truly appropriate to the subject matter.[28] From the beginning he puts his readers on notice that meteorology will operate as an integral discipline within the array of natural science.

Aristotle has the same goal in mind in his treatment of his predecessors' opinions (*endoxa*). He engages his predecessors almost exclusively on meteorological evidence, both because his predecessors have no specifically meteorological theory to offer and because he refuses to fight again the battle concerning their general principles. These are the primary reasons for the peculiar character of the *endoxa* in the *Meteorologica* and they have escaped the notice of scholars.

It is often claimed that Aristotle's reviews of *endoxa* serve one of the functions of dialectic, to provide a way to reach the first principles.[29] Through testing and rectification, adoption and rejection of these opinions the investigator can lay down what cannot be demonstrated, the starting points and principles of a science. The *Metaphysics*, *Physics*, and *de Anima* all begin with extended endoxic reviews, which both utilize and criticize their predecessors. In the *Metaphysics* and the *Physics* Aristotle tends to be more generous and places himself at the end of a development of the science (explicitly so at *Metaphysics* 1.7.988a18–24), to which his predecessors have importantly, though defectively, contributed. The endoxic review in the *de Anima*, by contrast, is largely negative. Suzanne Mansion has argued that, in contrast to what we find in the *Metaphysics*, the predecessors of the *de Anima* have in Aristotle's opinion contributed nothing substantial. The point of the endoxic section here (and similarly in the *de Generatione et Corruptione*, as she argues) is to show instead that they failed to pose the problems at the appropriate level. She means by this that the predecessors conceived of the soul as the wrong kind of thing and consequently asked the wrong kinds of questions about it: they asked what it was made of, rather than what it did.[30]

All three treatises share both positive and negative assessments of predecessors; but most importantly all three contain major endoxic introductions, which usher in new fundamental principles. The *Meteorologica*, by

[28] In addition to the development of the theory of exhalations in 1.3, Aristotle introduces them or mentions them as his principles at 1.4.341b5–12 (as ἀρχή 341b5); 1.7.344a8–13 (ὑπόκειται 344a8); 1.9.346b32; 2.3.357b24–6 (κεῖται 357b24; ἀρχήν b25); 2.3.358a21–6 (ὥσπερ εἴπομεν 358a21); 2.4.359b27–32 (λαβόντες ἀρχήν 359b27); 360a8–10; 2.8.365b21–3 (ὥσπερ εἴπομεν 365b22); 2.9.369a11–15 (ἀρχὴν ὑπολαβεῖν 369a12); cf. 2.9.370a27–32; 3.6.378a16–19 (ὥς φαμεν 378a19).

[29] E.g., Owen 1961; Barnes 1981; Nussbaum 1982; Irwin 1988.

[30] Mansion 1961b: 43–7.

contrast, has no distinct doxographical preface, and instead this material is scattered throughout the treatise section by section, subject by subject.[31] Rather than arguing about general principles, Aristotle discusses and refutes his predecessors overwhelmingly about specific difficult questions – the nature of comets, the Milky Way, hail, the saltiness of the sea.[32] And since his refutations are mostly specific rather than general, he almost never argues that an explanation given by his predecessor is inconsistent with his own or his predecessor's principles.[33] Instead he shows in almost every case that the explanation proffered violates the meteorological appearances or violates expert technical theory.[34] When he does make reference to his own more general principles, it is usually because meteorological evidence is not decisive between theories.[35] For the most part, though, he chooses his battles carefully and discusses his predecessors explicitly, it seems, only when he thinks he has strong counter-evidence.[36]

[31] There is an endoxic section concerning the exhalations (1.3), but no opponent is named or suggested. Louis 2002: xxxvi provides a convenient catalogue of those Aristotle mentions by name: Homer (1.14.351b35; the only one to meet with even qualified approval), Anaximenes (2.7.365a18, b6–20), Pythagoreans (1.6.342b30; 1.7.345a14), Heraclitus (2.2.355a14), Empedocles (2.3.357a26; 9.369b12), Democritus (1.6.342b27; 343b25; 8.345a25; 2.3.356b10; 7.365a18, b1–6), Anaxagoras (1.3.339b22; 6.342b27; 8.345a25; 12.348b12; 2.7.365a17, 19; 9.369b14), Cleidemus (2.9.370a11), Hippocrates/Aeschylus (1.6.342b36; 343a27; 7.344b15); Plato (2.2.355b32–356a33). Anaximander is not mentioned, though he may be alluded to at 2.3.357b19. More recent contributors, such as Metrodorus, are also passed over. On one occasion he claims that his predecessors have said nothing that a man off the street could not have said (1.13.349a16; cf. 2.3.356b17–19).

[32] Cherniss 1935: 127 misrepresents the situation: "The objections Aristotle brings to previous meteorological theories are all ultimately based upon the assumption of such a generation and destruction of these four bodies. That no previous theory took this interaction into account was enough to invalidate any and all of them, for meteorological phenomena are in his system manifestations of the genesis and destruction of the simple bodies." The error is perpetuated by Coutant and Eichenlaub 1975: xxxvi.

[33] Only once does Aristotle introduce an objection from outside the meteorological subject matter: Democritus' conjunction theory of comets is refuted mathematically by claiming that point-like planets cannot conjoin to form an extended surface (1.6.343b32–344a2). Occasionally predecessors make what are probably internal errors, though never inconsistencies with their own high-level principles: Democritus says that planets are *sometimes* seen on the dissolution of comets, but this should happen always, according to his theory (1.6.343b25–8); some hold that rainwater supplies reservoirs in the earth; but condensation should happen within the earth as well as without (1.13.349b19–27); Empedocles' theory that the sea is the sweat of the earth provides no clear cause (2.3.357a24–32).

[34] For a rough catalogue of this evidence, see the Appendix at the end of this chapter.

[35] The following are some arguments based on Aristotle's peculiar commitments, rather than empirical evidence. Aristotle argues against Hippocrates perhaps, that the Milky Way is a reflection of the visual ray to the sun: the Milky Way appears reflected in water, though there cannot be double reflections (1.8.345b25–8; in this case Aristotle uses an expert argument). Some say that wind is simply a moving current of air and that all winds are one, but for Aristotle, who accepts the wind–river analogy, this is like saying that all rivers are one (1.13.349a16–26); theologians say that there are sources for the sea, but Aristotle simply accepts the principle that there are no standing natural spring-fed bodies of water (2.1.353b17–354a5).

[36] Consider two cases: shooting stars and comets. Anaxagoras had a theory for both, and in both cases the theory involved the celestial realm. But Aristotle discusses and refutes only one, the theory of

This tendency, often identified as empirical in contrast to the supposedly dialectical method of the *Physics*, has been sufficiently observed and some of its purposes have been noted.[37] As Freeland has rightly argued, Aristotle employs *endoxa* as a prompt both to gather "relevant empirical data" (e.g., Anaxagoras says that hail occurs during the summer, and this prompts Aristotle to gather data about the seasons of its occurrence) and to formulate "why-questions" based on this relevant empirical data (e.g., why does hail fall during the spring and the autumn?).[38] These are important and valid observations, but they do not account for the absence of critiques of those kinds of theories that are founded on specifically *meteorological* principles.

This absence can best be explained by the real or perceived fact that Aristotle's predecessors, to be precise, the later Presocratics, did not provide any specifically meteorological principles. This accounts for the lack of an endoxic preface at the beginning of the treatise, since there were no theories among Aristotle's immediate predecessors that covered all and only meteorological phenomena. There were only specific explanations based on the most general physical principles, and these Aristotle counted worse than the ancient tradition of the poets or the casual explanations that any man off the street could give (ὁ τυχών 1.13.349a16; cf. 2.3.356b17–19). At least the average man does not make appeal to abstract general principles.

As for arguing about the more general principles, that is a battle Aristotle has already won. By the time he arrives at the *Meteorologica*, Aristotle has long since established his highest principles. He has already engaged his predecessors in the grand disputes and feels no need to go back over that terrain and argue them again.[39] For this reason Aristotle never presents the views of his predecessors in the context of their own first principles. It is

comets. In this case he has convincing empirical evidence against the view that comets appear when two planets form a conjunction (1.6.343b14–32). By contrast, Anaxagoras' theory that shooting stars are caused by sparks shooting off from the aether, is much more difficult to refute on the basis of observation and seems no less likely than Aristotle's own theory. An exception is Democritus' theory of earthquakes, which he could refute but does not attempt to (2.7.365b1–6).

37 For some older views on the role of observation throughout the treatise, Drossaart Lulofs 1930: 9: "Aristotle's *Physics* was perhaps appropriately called by Hegel a metaphysics of physics and for that reason in many respects differs from what we understand by physics, yet he presses, specifically in the *Meteorologica*, aspects to the fore, which, free from transcendental significance, immediately speak to the physical geographer to the extent that he is interested in the history of his subject." Cherniss 1935: 327–33. Bourgey 1955: 78n3.

38 Freeland 1990: 78–9. She relates *endoxa* and signs to each phenomenon. Within the context of each specific phenomenon, she says (80), "[t]he *Meteorologica* shows that Aristotle finds reviews of previous scientific theories crucial to the process of refining why-questions." See also Viano 2006: 23–6.

39 By contrast the debate in the *de Caelo* is at a general level, where at 2.13.294b30 Aristotle says that "in general the dispute with those who talk in this manner is not about the details but about the whole notion."

a practice for which Aristotle has been frequently faulted, since it deprives his predecessors of their best justification for specific explanations, but it had its origin in Aristotle's conception of an orderly review and exposition of opinions on physical matters, which in turn depends upon the orderly descent from celestial generality to terrestrial specificity.

The conception of orderly descent goes to the very foundation of his theory of science as laid out in *Posterior Analytics* 1. Aristotle insists that scientific demonstrations proceed according to "commensurate universals," that is, that the property being proved in the demonstration be coextensive with its subject.[40] He illustrates the rule with an example from geometry: when the geometer explains why a triangle has interior angles equal to two right angles, he treats the property as belonging to *triangles* generally rather than as belonging specifically to *isosceles* triangles. While the property certainly belongs to isosceles triangles as well, it does not belong to them in virtue of their being specifically isosceles (1.5). Similarly, an isosceles triangle has two internal angles equal, because it is *isosceles*, that is, because it has two equal sides. In order to demonstrate correctly, one must prove a property of a subject coextensive with the subject. Failure to demonstrate in this way may occur within a small part of a science, for example in the study of triangles, or between independent sciences, for example geometry and arithmetic (where it is called "kind-crossing" or *metabasis*). It is also a failure within a science such as physics, when the principles of a general discipline are applied to a more specific discipline as if they were specific to that discipline and as if the specific discipline did not have principles of its own. This concern for the appropriate level of generality for a discipline, then, is not peculiar to meteorology but holds good for any study at all. The problem is more acute in meteorology only because the more recent practitioners of the discipline had been reducing it to general physics.

So Aristotle's specifically meteorological critique of his predecessors results from his desire to secure meteorology as an integral discipline within physical science, but this does not prevent him from accomplishing the two more common functions of *endoxa* as well. By showing how his predecessors have misplaced and misconceived the phenomena, he achieves the second purpose, which Mansion had identified, the refutation of theories that have asked the wrong question. At the most grievous level, there are the theories that place a phenomenon in the wrong ontological category. Those who say the sea started as fresh and has become salty, or that the

[40] Wilson 2000: 15–39; Falcon 2011.

sea has shrunk and will continue to shrink, have failed to realize what we see clearly (φανερῶς . . . ὁρῶμεν 2.2.355a25–6): all evaporated water condenses and falls again. Theories like these violate most seriously the proper boundaries of the science by denying that meteorological phenomena are continual and eternal processes. Less serious are the kinds of mistakes that Democritus and Anaxagoras make when they claim that the comet is a conjunction of planets (1.6.342b27–9). They have displaced the phenomenon into the celestial realm and alienated it from the reach of proper meteorological principles. Aristotle simply shows with evidence that comets do not behave like planets and planets do not behave like comets (343a22–35). The detection of these first two kinds of errors has the effect of ratifying Aristotle's principles and his basic distinction between the celestial and the sublunary realms, and thereby fulfills the first purpose of endoxic review, the establishment of the principles. Lastly, the phenomenon may have merely been misplaced within the discipline, as when Hippocrates of Chios claimed of the comets and perhaps the Milky Way that they are reflections (1.6.342b35–343a20; 1.8.345b10–12). The use of principles of reflection here does not remove the phenomenon from the scope of meteorological science – Aristotle himself places the aurora and rainbows under these principles (1.5.342b11–13; 3.3.372b12–17) – but it is not appropriate in this case and is a misuse of principles, a minor violation.

In this manner, Aristotle's use of his predecessors fulfills two of the functions of endoxic review. But he also exploits them for a list of problems that are in need of explanation, as Freeland has rightly noted.[41] The theories of his predecessors tried to account for various features of the phenomena, and Aristotle often agrees that these features are genuine facts, which need to be explained by any adequate theory. Hippocrates of Chios claimed that the comets lag behind the fixed stars (1.6.343a4–8). Aristotle concurs but provides a different explanation (1.7.344a33–b12). Anaxagoras argued that hail occurs in warm seasons (1.12.348a18–20); Aristotle is in qualified agreement, but rejects his reasoning (348b12–15). Conversely, the silences of the predecessors may demand a response. By noting that Anaxagoras did not say where or when earthquakes take place, Aristotle declares that he intends to do just that (2.7.365a33–5). In observing that earthquakes sometimes take place in areas that are neither droughty nor moist (contra Anaximenes 365b12–14) Aristotle signals his readiness to provide an explanation. For this reason, then, it is not enough for Aristotle merely to dismiss the predecessors at the general level and go his own way. He needs to know

[41] Freeland 1990; also Taub 2003: 93–6.

how they contradict the appearances in order that he may rectify these shortcomings in his own theory. Conversely when theories of his predecessors are wrong at a more general level (e.g., their causes are "entropic" and uni-directional rather than cyclical), his criticisms leave room to salvage many of their phenomena.[42] So in refuting Anaximenes' view that earthquakes are caused by the collapse of underground caverns, Aristotle does not merely object that this is an entropic theory; he adds that we would have noticed the earth sinking and earthquakes becoming less frequent (365b12–13). In spite of the objection, though, he accepts the theory of hollows and simply adopts a theory in accordance with his own principles that will allow the hollows to persist.

If one reads through the physical writings in their traditional order, it is easy to believe that Aristotle first isolated the celestial realm in order to explain the perfection of its movements and then imposed a strained unity on the unruly remainder.[43] But our investigation of Aristotle's use of *endoxa* in the *Meteorologica* suggests that it was also his perception of the inadequacy of his predecessors that led him to treat meteorology as an integral and irreducible discipline within natural science. Proceeding from both concerns, Aristotle provided the *Meteorologica* with an internal as well as an external unity and distinguished it from other sciences through its own peculiar principles. For this reason the common perception that Aristotle is pursuing an *empirical* investigation in the *Meteorologica* is only partly right. While Aristotle's pervasive use of empirical evidence is a sign that the sublunary phenomena are more accessible to our senses than the celestial, it is also a direct result of Aristotle's plan for the physical works, in which general physical issues, such as the shape of the universe and the nature of matter, are decided early and laid to rest. Just what the special principles of the *Meteorologica* are and how they can at the same time provide for its special domain and fix its place within the natural science are the subjects we shall take up next.

Appendix: Aristotle's criticism of his predecessors

The following is merely a rough catalogue of predecessors who violate empirical appearances (I). However theory-laden "empirical appearances" may be, Aristotle never retreats to general physical objections. At most, he

[42] By "entropic" I describe Aristotle's interpretation of theories that view the world as changing in a linear, rather than a cyclical, manner, e.g., the view that once the seas were more extensive than they are at present and are progressively drying out.

[43] Solmsen 1960: 406.

introduces expert knowledge (II), or (III) makes use of dialectical considerations appropriate to meteorology.

I. Against those who say comets are planets: many comets are seen outside the zodiac circle (1.6.343a23–5); more than one comet has appeared at the same time (343a25–6); if the tails are reflections, the planets should appear without a tail sometimes, but they do not (343a26–30); never have more than five planets appeared together (343a30–5); against Hippocrates of Chios: comets do appear in the northern sky at the summer solstice (343a35–b7), and Aristotle cites several examples including the comet that appeared during Gamelion in the archonship of Euclees (343b4–7); known fixed stars have been observed with tails; Greeks and Egyptians have observed this (343b8–14); against the conjunction theory: we have not seen comets resolve into planets, including the comet in the archonship of Asteios (343b14–25); the Egyptians have noted conjunctions of planets and stars and of two planets but have not noted a comet (343b28–32).

Against Hippocrates: more often comets appear independently of stars (1.7.344b12–17).

Against the Pythagoreans who claim that the Milky Way is the path in which the sun once travelled: the zodiac circle should be even more so affected, but it is not (1.8.345a18–25); against Anaxagoras and Democritus that the Milky Way is the appearance of stars more apparent for being in the earth's shadow from the sun: the Milky Way would appear to shift against the fixed stars (345a25–36); against Hippocrates perhaps, that the Milky Way is a reflection of the visual ray to the sun: it could not keep the same place in the mirror (345b9–25).

Against Anaxagoras' theory that hail is formed when a cloud is forced up into the upper cold air: hail is infrequent in high places (1.12.348a20–3), hail clouds are often seen very low (348a23–30), and large hail stones are rough, showing that they have not been worn down by a long descent (348a30–6).

Against those who hold that rainwater supplies reservoirs in the earth: the earth's reservoirs could not hold enough water to supply the rivers (1.13.349b15–19); water does not just gush forth as from a reservoir but trickles drop by drop from mountainsides, as the experience of irrigation workers shows (349b27–350a13).

Those who think that local desiccation is in fact a universal drying out fail to observe that the sea has encroached in many places (1.14.352a22–5).

Against those who think that the sea is the main body of water, because each of the other elements has a main body: there is no apparent main body of water (2.2.354b11–13); some think that the sun is fed by moisture, but we see the water that goes up falls again (355a21–32); moreover, terrestrial fires are not fed by the vapors boiling from cauldrons (355a15–18); against the predecessors who overlook the fact that other stars besides the sun should likewise be fed by the moisture (355a18–21); against Plato's theory that rivers flow from a network below the earth, Aristotle notes that rivers do not flow uphill (356a14–19); moreover, how does rainwater fit into this account, which seems to recognize only river water sloshing to and fro (356a19–22)? And all rivers that do not flow into one another obviously flow into the sea (356a22–33).

Against Democritus' theory that the salt of the sea comes from the entropic (i.e., linear rather than cyclical) evaporation of an originally less salty body of water: the vapor that rises falls again as rain (2.3.356b21–5; 357a6–12); against those who make the sea salty by an addition: the rivers, which carry water to the sea, are not salty (357a15–24); against Empedocles' theory that the sea is the sweat of the earth, the sea would represent a fantastic amount of sweat and the earth would still have in it an impossible amount of moisture (357b10–12).

Against Anaxagoras' theory that earthquakes are caused by air being trapped from below the disk-shaped earth by water clogging the upper pores: the earth is a sphere as shown by the horizon which changes as we move (2.7.365a25–31); moreover the earth cannot just rest on air (365a31–3); against Democritus' theory that the earthquakes are caused by excessive water falling within the earth or in droughts by water retreating to itself (365b1–6), Aristotle has no reply; against Anaximenes' theory that earthquakes are caused by earth collapsing internally during rains and droughts: there should be visible subsidences in the earth, they should not be restricted to localized areas, and they should become less frequent over time (365b6–20).

Against Anaxagoras' theory that lightning arises from aether descending into the clouds, we need an explanation for the descent of a naturally light element (2.9.369b21–2); against Empedocles' theory that lightning arises from the sun's rays trapped in clouds: water when heated by the sun then cooled does not emit lightning (370a1–5); nor does it explain why clouds are necessary (369b22–4).

II. For theories that contradict expert knowledge: against Anaxagoras' and Democritus' theory that the Milky Way is the appearance of stars more apparent for being shielded from the sun by the earth's shadow: research shows that the cone of the earth's shadow is very short (1.8.345a36–b9); against Hippocrates perhaps, that the Milky Way is a reflection of the visual ray to the sun: it could not keep the same place in the mirror (1.8.345b9–25); against those who think that local desiccation is in fact a universal drying out: they fail to observe that the earth is tiny in comparison with the universe (1.14.352a26–8); against Cleidemus' theory that lightning is a sort of appearance like the glancing of water at night: he is unaware of recent research on reflection (2.9.370a10–21).

III. For dialectical considerations appropriate to meteorology: Aristotle objects that the sea cannot have sources and adduces a division of kinds of stationary and moving water to prove it (2.1.353b18–35). Aristotle argues against earth or water being the cause of earthquakes on the grounds that only the dry exhalation has the necessary violence and power of impulse (2.8.365b28–366a5).

CHAPTER 2

From elements to exhalations

In the previous chapter I argued that Aristotle furnished meteorology with its own proper principles in an effort to save it from being absorbed into general physics. But the fact remains that meteorology is an integral part of the physical *pragmateia* and, though it behaves in many ways like an autonomous science, it also borrows and modifies principles from the more general parts of physics.[1] In the present chapter I ask what role the principles of the more general parts, especially those of the *de Caelo* and the *de Generatione et Corruptione*, play in the foundations of the *Meteorologica*. First I want to consider how Aristotle's treatment of the material cause shifts from the general theory of the elements to the exhalations; then I shall argue that the aporetic chapter 1.3 concludes by assigning an essential role to the heavens in maintaining the balance of the sublunary bodies.

From the opening lines of the treatise Aristotle emphasizes meteorology's dependence on general physics.[2] Since some physical causes are prior to others (περὶ . . . τῶν πρώτων αἰτίων τῆς φύσεως 338a20), we are led to expect that the prior ones will be applied to the posterior and more specific meteorological phenomena. Aristotle confirms this expectation in the second chapter, where he derives the material and the moving causes

[1] For discussion of the unity of the physical *pragmateia*, see Lennox 2010 and Falcon 2011. Lennox 2010: 12–13 points out that μέθοδος often refers to a sub-domain of a larger science (citing *Mete.* 1.1.338a25, *Long.* 6.467b9, and *PA* 1.4.644b16).

[2] Case 1910: 13 thought that the first chapter was a later addition: "[t]o suppose this preface, presupposing many sciences, to have been written in 356, when the *Meteorologica* had been already commenced, would be absurd." Martini 1897: 366–9 argued that Aristotle did not compose this proem on the grounds that Aristotle's predecessors did not normally use the term μετέωρα to describe only the sublunary items that Aristotle describes here (cf. 1.1.338a26), and because Aristotle could hardly have been ignorant of that fact. This argument, not plausible itself, was demolished by Capelle 1912b, who noted that Aristotle's predecessors did sometimes so use the term; that Aristotle is hardly above such misrepresentation of his philosophical opponents anyway; and that many of the features of the proem are paralleled in other introductions: the use of περὶ μὲν οὖν . . . ἔτι δὲ, λοιπὸν δέ with the new task, and run-on sentences with polysyndeton. The fact that the description of the contents of the *Meteorologica* do not include the contents of *Meteorologica* 4 is yet an additional sign that that book is not of a piece with the first three books.

of the *meteora* from the higher sciences. He begins here by describing the material cause, the general theory of the four elements, their weights and natural places, in terms familiar from the *de Caelo* and *de Generatione et Corruptione*:

> We have previously distinguished . . . four bodies arising from four principles; their motion, we say, is double: the one from the center, the other toward the center. And since there are four bodies, fire, air, water, and earth, the one that rises to the top of all these is fire, and the one that sinks to the bottom is earth; and there are two that have to each other a like relation to the former (for air is nearer to fire than the others, and water is nearer to earth). So the whole cosmos around the earth is constituted from these bodies. (1.2.339a11–20)

It is these four bodies and their consequent properties (συμβαίνοντα πάθη 339a21) that are explicitly the subject of our treatise. There is as yet no mention of the exhalations. To these simple tendencies Aristotle adds in 1.3 the theory of mutual transformation of the elements, discussed in *de Generatione et Corruptione*,[3] according to which,

> fire, air, water, and earth come to be from one another and each of them is potentially present in the other (ἕκαστον ἐν ἑκάστῳ ὑπάρχειν τούτων δυνάμει) just as is the case in other things that have one and the same substrate, into which therefore they are in the end dissolved. (339a36–b2)

The moving cause here is likewise supplied by general physics: the unending motions of the heavens (τέλος οὐκ ἔχουσα τῷ τόπῳ τῆς κινήσεως 1.2.339a25) are continuous with the lower elements and effect their motion. The elaboration we find in 1.3 is in broad conformity with the doctrine of *de Caelo* 2.7: the sun by friction on the sublunary sphere generates heat, which in turn produces all the changes here below. So far, then, the two main causes of the *Meteorologica*, material and efficient, seem to be simply adopted from general physics.

The introduction of the material and efficient causes appears to be largely complete with 1.2 and, by the beginning of 1.3, Aristotle seems ready to explain his first specific phenomenon. "Having accepted our original theses and the previously mentioned definitions," he begins, "let us speak about the appearance of the Milky Way and the comets and the other things that happen to be akin to them" (1.3.339a33–6). But instead of launching into his main task Aristotle abruptly retreats and reexamines his principles, and in particular, the question of how many elements there

[3] The elements mutually transform (*Mete.* 1.3.339a36–b2; *GC* 2.4.331a7–8); they have a common substrate (*GC* 2.1.329a24–b3); these two doctrines are mutually implying (*GC* 2.7.334a16–18).

are above the surface of the earth.[4] As if this were not strange enough, the system of four elements in their four corresponding places undergoes radical transformation and is replaced by two exhalations, hitherto ignored, which as the treatise unfolds will pair up into three different zones. Elemental fire becomes a smoky exhalation that is "potentially fire," and water, a separate element with a separate region in the *de Caelo*, is assimilated to the vaporous exhalation. Earth remains at the center. Most remarkably, although Aristotle makes repeatedly clear elsewhere that the "elements" are capable of transforming into one another, the exhalations themselves nowhere show this capacity.[5] Nowhere, that is, does vapor (ἀτμίς) change into smoky exhalation (ὑπέκκαυμα) or vice versa.[6] Given that the four elements have changed so radically, in what sense is the *Meteorologica* simply applying the material principles of the more general physics?

For the most part, commentators up to and including Ideler in the nineteenth century were content to join Aristotle in ignoring the tension between the elements and the exhalations. More recently Hans Strohm, working within the developmental research tradition of Werner Jaeger, sought and found other such discontinuities, which he explained by Aristotle's development from the abstract, deductive and speculative (i.e., Platonizing) *de Caelo* and *de Generatione et Corruptione* to the etiological and empirical *Meteorologica*, where the elements take on more specific attributes.[7]

4 Strohm 1984: 137: "unexpected."

5 *GC* 2.4.331a12–21: "Now it is clear that every simple body is disposed by nature to change into one another. For coming-to-be occurs into contraries and from contraries, and all the elements have contrariety to one another because their differences are opposites. For in some of them both of their differences are contraries, e.g., fire and water (for the one is dry and hot, the other wet and cold), in others only one is contrary, e.g., air and water (for the one is wet and warm, the other wet and cold) so that in general it is clear that they all are naturally disposed to arise from them all."

6 There are no unambiguous statements that the exhalations are mutually transformable. The most suggestive passage is found at 3.1.370b10–17, where both exhalations are said to be potentially present according their matter, and to the extent that there is a preponderance of either, so the impulse is to one or the other. Böker 1958a: 2219–20 also notes that it is unclear what happens to the dry exhalation. Alexander 134.15–27 (on 3.1.370b13 on the changes to the dominant exhalation in the cloud) thinks that the one exhalation changes into the other. Indeed the phrase ὑπάρχει μὲν οὖν ἄμφω δυνάμει ταῦτα κατὰ τὴν ὕλην suggests as much, but what follows does not bear this out: "when there is an ἀρχή of whichever δύναμις, there follows out of the material something συγκρινόμενον (a condensed mass of exhalation), which is more prevalent, and sometimes this is rain, other times it is hurricane." This does not involve change from one exhalation to the other.

7 Strohm argues that Aristotle began his meteorological investigation with a commitment to a "transcendental" principle, the celestial movements, which steer (κυβερνᾶσθαι 1.2.339a23) all things in the sublunary world. This principle worked well in explaining the kapnospheric phenomena, which are ignited by the heavenly rotation, but as he proceeded into areas where the phenomena are more readily observable (e.g., winds and earthquakes), he found it less useful and accordingly modified it. His conception of the elements had likewise to change (Strohm 1936: 8–12; 1984: 137–8).

Solmsen subsequently pressed the inconsistencies within the material cause. While noting that the *Meteorologica* bridges the gap between the elemental movements of the *de Caelo* and their mutual transformation in the *de Generatione et Corruptione*, he expressed due skepticism about how far Aristotle intended to unite their principles. "On the whole," he says, "the [*de Caelo*] left us with the impression that the elements exist as permanent and constant masses and that they keep dividing the sublunary world between themselves. This impression we are now forced to give up [in the *Meteorologica*]."[8]

In fact these innovations are not entirely new to the *Meteorologica*, and Aristotle made significant efforts to integrate his earlier treatises with the *Meteorologica* even before the exhalations were formally introduced. Thus, in *de Generatione et Corruptione* air was already described as "like vapor" (οἷον ἀτμίς 2.3.330b4) and fire as smoke (2.4.331b24–6). More importantly, Aristotle provided a way of understanding the exhalations within the basic framework of his four-element theory. While *de Generatione et Corruptione* 2.4, the *locus classicus* for elemental transformation, leaves us with the impression that there are only four simple bodies and that they are either hot or cold, wet or dry, a passage at 2.7 shows how together they can form intermediate homeomeries, tepid and damp in varying degrees. Each of the simple bodies can dull the other's excesses and a hot thing may be cold relative to a hotter thing, a cold thing hot relative to a colder thing (334b10–12). Although he never explicitly claims the exhalations are intermediate homeomeries, Aristotle provides, through the formation of the intermediate homeomeries, a bridge from the elements, understood as simple combinations of two absolute powers, to the exhalations. In fact, he extends his explanation to account for the formation of such stuffs as flesh and bone, stuffs that are the products of multiple simple bodies rather than of just one. If Aristotle thinks he can generate bone and flesh from a modification and extension of the theory of elemental transformation, then surely he can explain the exhalations, since they are merely elemental intermediaries and do not require, as the complex homeomeries do, the combination or mixture of more than one element or elemental intermediary.[9]

8 Solmsen 1960: 393–9. "Yet to tell the truth, Aristotle is not now interested in the distinction between true simple bodies and the actual elements, and we had better not press our question . . . These modifications and corrections are anything but slight" (398). "As the treatises follow one another, we look to the earlier work in which the outlines of the Cosmos are drawn in the expectation that it will provide the frame for the more specific inquiries of the *Meteorologica*. Unfortunately the frame does not entirely fit the picture which should fill it" (399).

9 There is no evidence in Aristotle for the medieval and Renaissance doctrine of the exhalations as "imperfect mixtures," found perhaps first in Albertus Magnus. See Martin 2011: 26–7.

But the difficulty with the exhalations does not lie with a theory of simple bodies that cannot account for the generation of more complex and varied stuffs – problems there may be, but they are of wider extent than our present concern. Our current problem is that Aristotle does not provide such an account in the *Meteorologica*, and he leaves us to wonder whether the exhalations have replaced the simple elements or whether they exist alongside the simple elements. As we try to answer this question and then consider the fundamental role of the sun in regulating the production and proportion of the exhalations, we will confirm the conclusion of our first chapter – that within the articulated structure of the physical *pragmateia* Aristotle intends to provide the *meteora* with their own proper principles.

Specification and *sustasis*

The problem of the transition from the four-element to the two-exhalation view admits of two genuinely Aristotelian solutions, which though different are not wholly incompatible. According to the "specification" view, the exhalations replace the simple elements, that is, the elements and the exhalations are two more and less abstract descriptions of the same material stuff. The four elements are high abstractions, ways of describing nothing but the motions, places, and basic powers of the material, and are not intended to capture other more specific properties. On this view the elements (at least fire and air) do not actually exist independently of their concrete instantiations, the exhalations. Just as there is no generic triangle that is not at the same time equilateral, isosceles, or scalene, and yet we can discuss the abstract properties of the genus triangle independent of its three species, so, according to this view, general properties of the sublunary elements must be abstracted from their actual manifestations in the specific exhalations and other sublunary materials. As an abstract entity, fire is a hot, dry, and upward-tending body and no more; and yet all real fire takes on some specific form, be it smoke or orange flickering flame.[10]

There is some evidence, not wholly conclusive, that Aristotle held the specification view both in the *de Generatione et Corruptione* and in the *Meteorologica*. As noted above, according to the *de Generatione et Corruptione*, "air is hot and moist, for air is as it were vapor" (οἷον ἀτμὶς γὰρ ὁ ἀήρ 2.3.330b4). Aristotle is trying to explain here how air can be moist, since air's place within the δύναμις theory of the elements requires that it be moist, just as fire must be dry, and so on. There is no reason to suppose that this elemental moist ἀτμίς is any different in its basic nature from the vaporous

[10] So Strohm 1936: 11.

exhalation in the *Meteorologica*. If there is no difference, we can conclude that air as such just is a sort of vapor and that there is no elemental ἀήρ that is not vaporous exhalation. Likewise at *de Generatione et Corruptione* 2.4.331b21–6, in a discussion of how a simple body can be generated by combining the powers belonging to neighboring simple bodies, Aristotle explains,

> whenever the wet of the air and the cold of the earth perish, there will be fire, because the hot of the air and the dry of the earth are left behind, which are the qualities of fire. This generation of fire is also in agreement with sense perception. For flame is fire most of all, and flame is burning smoke, and smoke arises from air and earth.

These remarks, set within a discussion of elemental transformation, suggest that this smoke is merely what elemental fire is. Just as a sort of vapor is merely what elemental air is, so smoke is the result of the combination of the dry of the earth and the hot of the air. It may be a stretch, but Aristotle is clearly trying to assimilate smoke with flame and elemental fire.

At *Meteorologica* 1.3 Aristotle seems to express himself in similar terms. After setting aside the heavy earth and water, he says,

> concerning air and what we commonly call fire – though it is not fire, for fire is an excess of heat and a sort of boiling – we must realize that of what we call air, one part surrounding the earth is wet and hot on account of the vapor and the exhalation of the earth. The part above it is hot and dry. For the nature of vapor is wet and hot, and that of exhalation is hot and dry.[11] And vapor is potentially like water and exhalation is potentially like fire. (340b21–9)

Later he calls air smoky (καπνώδης 1.4.341b10). There is no suggestion here that, apart from specific *meteora* such as the rain showers and comets, the space above the earth contains any air or fire besides vapor and dry exhalation.[12]

And yet the specification view faces an objection, namely that the species must have all the properties of the genus and some more besides, and even if

[11] There are reasons to follow Ross 1949: 109n4 and Lee 1952: 20a in altering the text so that vapor is cold and wet. Besides the parallels adduced by Ross (2.4.360a23 [definitively] and 2.8.367a34 [very obliquely]) that identify ἀτμίς as cold and wet, there is the consideration of 1.3.340b28–9, which on the received reading does not render the same sense for δυνάμει (dry exhalation is potentially fire by an excess of heat; vapor is potentially water by changing from hot to cold). Alexander 15.9–10, however, reads the received text.

[12] For a similar view Solmsen 1960: 397n17, who cites also *Mete.* 1.3.340b24–30; 1.4.341b7–18 (where Aristotle lays out the principles of the exhalation at the beginning of the chapter); 7.344a9–13 (an explicit restatement of the principles at the beginning of a section).

the genus is abstract, the species cannot have properties incompatible with it. But Aristotle makes clear both in *de Generatione et Corruptione* and in the introduction to the *Meteorologica* that elements are capable of mutual transformation, that, for example, air can turn into fire. In the body of our treatise, however, there is no material transformation between smoky and vaporous exhalations, contrary to the general principle of universal transformation. In this sense the exhalations cannot be specifications of fire and air.

According to the second interpretation, which I call the "*sustasis*" view, the four elements exist in their own right independently of the two exhalations.[13] In the *Meteorologica* each of the two exhalations gathers together in συστάσεις, concentrated masses, distinct from an ambient environment that is filled with the simple elements, especially fire, air, and water. The identity of every phenomenon in the *Meteorologica* whether it is a comet, dew, river, wind, or metal depends on its being locally distinct from the material that surrounds it. Nowhere outside 1.3 does Aristotle provide a general discussion of the nature of the ambient masses, but in some cases the ambient material is clearly an element, as, for example, in the case of the earthquake, where wind is surrounded by elemental earth. In other cases, such as comets, the surrounding material seems to be made of the same exhalation but in a different constitution or density (εὔκρατος 1.7.344a14, 20).[14] More obviously, ejected shooting stars are dry exhalation surrounded by air (1.4.341b36–342a3). While the introductory chapters of the *Meteorologica* favor the specification view, the bulk of the treatise favors *sustasis*.

The *sustasis* view is indifferent to the specific nature of the ambient material, which can be elemental or exhalational, and this indifference highlights Aristotle's concern for disciplinary unity and the flexibility of his principles. Nothing in Aristotle's account prevents the general elements from really existing, and existing alongside both the συστάσεις and whatever exhalation may be simply ambient. It is not necessary, therefore, to resolve all tensions as if the specification and *sustasis* views were exclusive. In fact, we should observe that the tension between elements and exhalations is recapitulated at the level of the exhalations themselves. The dry exhalation, especially, has many different and not wholly consistent effects (compare the passivity of the fuel (ὑπέκκαυμα) of the kapnosphere

[13] Lettinck 1999: 15 notes that most pre-modern commentators adopt the *sustasis* rather than the specification view.

[14] Düring 1966: 391n306 draws the distinction between air as the substance of wind and the ambient air, which is a mixture of stuffs. Cf. Sharples 1990: 99 on the interpretations of the commentators.

with the dynamism of the winds). That these are all the effects of a single exhalation challenges our credulity, and yet Aristotle was content to gather them all under a generic dry exhalation and never hints that he found their identity problematic.[15] Aristotle's belief in the mutability and specifiability of principles allows him to be complacent. Nothing prevents elemental fire from coexisting with the dry exhalation, just as nothing prevents various manifestations of dry exhalations, for example comet fuel and winds, from coexisting with each other. His *silence* on the subject can be attributed to methodological concerns. As he conceived of the unity and coherence of the *Meteorologica*, he preferred, as always, to pass over what is too general. If elemental air and fire exist actually in the *Meteorologica* – and for the sake of material transformation it seems that they must – they exist only *qua* ambient material, a place for the phenomena proper to the discipline to occur, no different as such from any other ambient material. It is most likely then that the sublunary world above the earth is filled with elements (including intermediate simple bodies) and exhalations in their various manifestations. Aristotle's only concern is with the exhalations.

Toward a heavenly order

The new doctrine of exhalations is introduced in chapter 1.3 against the background of the principle of the mutual transformation of the elements. The principle is invoked in order to answer the question concerning the number of simple bodies between the earth and the heavens. The introduction of the principle here is surprising in view of the fact, as we have noted, that it has very limited application in the main body of the treatise. I shall argue here that it serves as a dialectical premise, one that consistently leads to failure in producing a balanced cosmos. This premise is important for the basic structure of the chapter, which aims to establish the essential role of the sun in regulating and limiting the spontaneous and chaotic transformation of the elements, and this in turn demonstrates the need to introduce the exhalations.

Chapter 1.3, after the false start noted above, is divided into an aporetic section, which places items of specialized knowledge in conflict with common opinion (339b2–340b4), and an expository section (340b4–341a36),

[15] Olympiodorus identifies two different kinds: sooty (αἰθαλώδης) and low, and the not-sooty and high (157.12–18). The former is assigned the primary responsibility for the sea's salinity. We find something similar in the *Timaeus* (58c–d), which mentions different kinds of fire: flame, light, ash; and different kinds of air: aether, mist, and darkness. Here the general notion of the hot has a variety of species, each different from the other and useful in different explanatory circumstances.

which sets forth Aristotle's own views and explicitly looks forward to the phenomena (πρὸς τὰ λεχθησόμενα 340b5) and backward to the general principles (πρὸς τὰ νῦν εἰρημένα 340b6).[16] Within this twofold structure three questions are raised: how many elemental bodies are there above the surface of the earth (339b2–340a18)? How are these bodies arranged (340a19–341a12)? And how does the sun heat the earth (341a12–36)?

The twofold structure of the chapter and its significance have gone largely unnoticed.[17] Of foremost importance is the fact that in laying out the premises of his *aporiae* Aristotle avoids use of and commitment to his own doctrines.[18] So, though he mentions early his own first body (339b16–19), he withholds it from the argument entirely. He repeatedly uses the theory of mutual transformation of the elements (e.g., 339a36–b2), but he makes clear that this specific theory is not essential to the argument, so long as elements may be understood to maintain some proportionality of power (δύναμις) with one another (340a13–17), a proportionality he rejects elsewhere (*GC* 2.6.333a20–34).[19] This is the first of two important dialectical premises, and so long as we keep in mind this qualification concerning powers, it will be convenient for us to follow Aristotle's practice and refer simply to mutual transformation of elements. Specialized knowledge supplies the second premise. Recent research has shown, says Aristotle, that in comparison with the whole cosmos, the size of the earth is minuscule (339b6–9). Likewise, in an observation which will become important later to his positive theory of the rain cycle and the sea, Aristotle notes that the volume of water is similarly inconsequential, there being no layer of water unified and distinct (συνεστηκυῖαν καὶ ἀφωρισμένην 339b9–10) from the sea, rivers, and lakes on and under the surface of the earth. These two premises guide the

16 For the vocabulary of *aporia*: ἀπορήσειεν 339b3; διαπορήσαντες 340a19; ἠπορήσθω 340b4. On the distinction between ἀπορεῖν and διαπορεῖν see Aubenque 1961: 8, who notes that διαπορεῖν can refer to both the preliminary examination of the problems and working them through dialectically. διαπορήσαντες at 340a19 clearly means "working them through." So also Madigan 1999: xix–xxii. Strohm 1936: 10 rightly sees the aporetic section as ending without a solution.

17 Ancient commentators, Vicomercatus, Ideler, and Strohm all pass over its structure.

18 Alexander 8.12–14 believes wrongly that Aristotle's *aporia* is based on a commitment to the first body.

19 *Phys.* 3.5.204b13–29 ("There are some who make the 'boundless' this, not air or water, lest the other [materials] be destroyed by the infinite amount of it. For they have contraries to one another, e.g., the air is cold, water is wet, and fire is hot. If one of these were infinite, the others would be destroyed. As it is, they say it is different from the things that are generated from it") mentions the doctrine of overwhelming but attributes it to others or uses it dialectically to attack others (Anaximander according to Simplicius, *in Phys.* 479.33). *GC* 1.7.324a7–9 merely says "contraries and their intermediates affect and are affected by each other – for it is in these that passing-away and coming-to-be in general reside." No overwhelming here. Lee 1952: 16a notes *GC* 2.6.333a16–27, but again there is no overwhelming here.

further development of the *aporiae*: since the elements can transform into one another, if there is too much of one, there is a danger that it may overwhelm and convert the less abundant (i.e., earth and water) to its own nature.

Aristotle's first task is to show that the traditional view of the elements, especially the view that the heavens are composed of fire and air, leads to impossibilities. After leaving aside his own first element and its place in the upper cosmos (339b16–19), he turns to the corresponding bodies proposed by other philosophers, specifically Anaxagoras' aether. He stresses that men have always been speculating about the celestial material (339b19–21, 27–30), and inasmuch as they conceive of it as "always running and divine" (ἀεὶ θέον καὶ θεῖον 339b25–6) and thus different from any of the simple bodies in our region, they are on the right track. But inasmuch as they think it is fire (339b30–4; cf. *Cael.* 3.3.302b4–5), recent research has revealed their error.[20] For if the whole heaven, the celestial bodies as well as the interstices between them (τὸ περιέχον καὶ μὴ μόνον τὰ φερόμενα σώματα 339b30–1), is pure fire, and the area "between the earth and the celestial bodies" (ἄστρων 339b31–2) is air, the vast extent of heaven's fire will overwhelm the rest of the lower cosmos (340a1–2).[21]

Nor will any refinements in this first scenario yield more satisfactory results:

> [Second Scenario] But again neither are [the interstices between the heavenly bodies] filled with air alone. For it would greatly exceed the equality of shared proportion of the elemental bodies, [Third Scenario] even if the space between the earth and the heaven were filled with two elements. (340a3–6)

Aristotle has already disposed of the possibility that the heavens are made entirely of fire and the space below them is air. Here he elaborates a second and third scenario. According to the second (340a3–5), the heavenly bodies themselves are fire, and interstices between them are air; the lower realm is also composed entirely of air. In this case, one element, presumably air,

[20] According to Aristotle's criticism at *Cael.* 1.3.270b24 Anaxagoras uses the term "aether" for fire.

[21] Aristotle uses μεταξύ in five different phrases: 339b13–14 τὸ μεταξὺ τῆς γῆς τε καὶ τῶν ἐσχάτων ἄστρων; 339b31–2 τὸ μεταξὺ γῆς καὶ τῶν ἄστρων; 340a5–6 ὁ μεταξὺ γῆς τε καὶ οὐρανοῦ τόπος; 340a18 τὸν μεταξὺ τόπον; 341a11 περὶ τοῦ μεταξὺ τόπου τῶν ἄστρων καὶ τῆς γῆς. The first use has caused trouble since the time of the ancient commentators. Alexander (*pace* Philoponus 15.34–5) took the opening question of the *aporia* at 339b13–16 – how many bodies are there between the earth and the ἔσχατα ἄστρα – to be referring just to the sublunary space, where ἔσχατα ἄστρα would refer to the lunar sphere, the *remotest* from the highest heaven. Philoponus rightly objects (15.35–16.5): Aristotle is referring to the entire super-terrestrial cosmos. On the other four occasions, however, he is referring to the sublunary realm alone.

will overwhelm the others. According to the third (340a5–6), even if the space between the earth and the heaven (i.e., the atmosphere) were filled with two elements, the air would still overcome the other elements. We are to imagine in this case the heavens composed of two elements and the lower space likewise composed of the same two elements. Aristotle does not explain why this latter situation does not provide an effective balance, but probably the heavenly interstices, being so much more voluminous than the celestial bodies, will overwhelm everything including water and earth. But in neither case does there seem to be an effective balance – a fitting conclusion for the first *aporia*.

One further scenario is considered – and rejected – in the context of the second *aporia*, which asks why clouds do not appear above the tops of the mountains in spite of the fact that it is cooler up there and more prone to condensation:

> [Fourth Scenario] So either water is not naturally generated from all air or, if it is, the air around the earth is not just air but a sort of vapor, which condenses again into water. Indeed, if the air, which is of great extent, is all vapor, the air and the water would be overwhelming, since the interstices of the upper space [i.e. the celestial realm] are filled with some body – and it cannot be filled with fire lest it dry everything else out, and air and the water around the earth are the remaining candidates (since vapor is a secretion of water). (340a32–b3)

This passage elaborates the third scenario discussed above, according to which both the celestial and the sublunary realms contain mixtures of air and fire, but, since this too is an *aporia*, we are not told that an effective balance can be attained. We are told only the conditions under which a balance is not attained. For the sake of the balance, the celestial realm cannot be entirely fire, and therefore it presumably contains some air. But if the sublunary realm is entirely air, and if all air is vapor, then together with the air of the celestial realm, the totality of air will overwhelm the rest of the sublunary cosmos (i.e., the elemental earth). We may conclude minimally, as Aristotle does, that the sublunary realm above the earth and water cannot be composed strictly of vapor. So the first question ends in *aporia*.

The second question concerning the order and position of the elements above the earth begins with an *aporia* and ends with positive doctrine (340a19–341a12). Just as the previous *aporia* had introduced specialized research, the minuscule size of the earth, so here we find a specialized meteorological observation: clouds are not found above the peaks of mountains, in spite of the fact that it is colder there and more conducive to condensation (340a24–6). Aristotle's tacit answer to this difficulty is simply that the upper region is filled with a different kind of air from the lower and that the upper air does not condense. Later (340b32–341a9) he supplies two additional reasons: the celestial movement causes the clouds to disperse in the upper area; and the vapor can be raised only so high before it condenses. Neither of these is sufficient in itself or in combination to account for the absence of upper clouds – for both imply that the space above the earth is composed solely of air, a notion forbidden by the need for balance among the elemental bodies. Yet these arguments take us beyond the first *aporia*, since they show that the two sublunary elements cannot be uniformly mixed with one another but must form into layers one above the other.

In the second half of the chapter (beginning at 340b4), cutting midway through the second question, Aristotle introduces his own positive doctrine, drawing selectively on the premises of the aporetic sections. Here the doctrines of the higher sciences come explicitly and definitively into play. Here for the first time he explicitly posits the first body, different from air and fire (340b8). Generally, the *aporiae* were intended to show that no configuration of elements within the traditional four-element theory could produce cosmic stability. Aristotle's first body serves in this context as a double solution.[22] Since it is inert matter, it is not capable of overwhelming the other elements. But, insofar as it moves, it is also the main efficient cause of the sublunary world. The principle of transformation of the elements is now guided by the heavens, its movement and rest, and no longer by the chaotic battle for dominance among the elements themselves. The effect of this celestial order is most obvious in the case of the rain cycle, where instead of a struggle between air and water for cosmic supremacy, the annual cycle of the sun regulates and balances the flux. And since, of course, the solar movement is cyclical there will be no entropic and permanent overwhelming.

We find evidence for this basic interpretation when in this positive context Aristotle mentions his new doctrine of transformability of matter: the stuff of the sublunary world is a material potentially hot, cold, wet, and dry, and it becomes each of these through the *upper*, i.e., celestial, movement or lack of movement (340b14–19). Here we have a very different account of change. No longer are the elements potentially each other, subject to being overwhelmed by one another; they are now under the regulation of the first body. The celestial motion actualizes the potentiality of exhalations to undergo changes. The heavens control the change. But,

[22] See Freudenthal 1995: 140–1.

as if this were not enough to avoid the unwelcome consequences of the *aporiae*, Aristotle limits the kinds of potentialities the sublunary matter has. The dry exhalation is potentially fire (not air or earth) and the wet exhalation is potentially water (not fire) (340b28–9). The sun in its cycles determines what kinds and what quantities of materials are produced. There is now no danger that the upper fires can overwhelm the lower air.[23]

In this way, with a single celestial cause, Aristotle provides a new, more orderly sublunary material, the exhalations, together with the orderly agent of their transformation. But this order comes at a cost. Of course, the new restricted potentialities of the exhalations are difficult to square with the four elements, but they also complicate the identification of the two kinds of "air" with fire and air. For while the case can be made that the "potential fire" is the layer of elemental fire and flame is just a special form of fire, the same cannot be said for vapor, which is potentially water. For if water is just a special form of vapor, then we lose our canonical distinction between water and air.

In other important ways Aristotle distorts and changes the theory of four elements. We were accustomed to think of the heavenly realm as consisting of pure first body. But here the first body is in some places more pure and in other places less pure, and variations are particularly marked at its lower interface with the air, that is, the fiery upper air (340b6–10).[24] This passage has been taken as evidence that the heavenly region contains some dry exhalation, a doctrine that is necessitated by Aristotle's friction theory of celestial light and heat, which we shall consider more fully below. Such a solution would be welcome for mechanical reasons, but Aristotle makes no effort to clarify this argument or integrate it into the whole.

23 Strohm 1936: 8 notes that the fact of the cloud boundary contradicts the principle of the transformation of elements – the boundary is a boundary of different stuffs. Lee (on 341a8) mistakenly supposes that the transformation is general by mistranslating ἀεὶ ἄλλο καὶ ἄλλο γίγνεται ἕκαστον αὐτῶν as "each one of them [air and fire] is in constant process of transformation into the other." A comparison with 2.3.357b30 with almost precisely the same expression (which Lee translates correctly) indicates that the phrase means "constantly in a process of becoming." Lee (22a) is also probably incorrect in supposing that above the mountains there is a layer of vaporous air. Aristotle at 341a5–9 does not mean that there is churning between the vaporous and fiery layer; rather he is referring to the rain cycle, the falling precipitation (κάτω φέρεται 341a6) and rising evaporation (συναναφέρεται 341a7), separately from the rise of the smoky exhalation, which is also continually coming-to-be in the highest region. Philoponus 38.30–8 provides an interpretation similar to Lee's.

24 Aët. 2.30.6 (*DG* 362.1–4): "Aristotle says that the mixture [of the moon] is not pure on account of the terrestrial airiness of the aether, which he calls the fifth body." Cf. *GA* 3.11.761b20–2 where it shares in the fourth body, fire. Olympiodorus 28.16–26 (on 340b8, areas of the heavens being more pure and less pure) agrees with Alexander that just as fire is less dry than earth, but not because of an admixture of the wet, so the moon is less pure, not because of a mixture, but because of "difference of the body" (so also Philoponus 30.8–31.2). Strohm 1936: 35 and Louis 2002: xxiii–xxiv and n1 hold this passage to imply that the air is present among the spheres, and that the spheres rub against it.

The *aporiae* of 1.3 then, which were introduced in order to investigate the number and order of the simple bodies, seem most of all to highlight the need for the first body in maintaining stability in sublunary dynamics. The principle of chaotic mutual transformation of the elements is used to drive toward unwelcome conclusions, and in the end the cosmos is saved only by limiting its application in the two ways we have seen. The heat of the sun produces exhalations whose very untransformability suffices to establish the two non-antagonistic sublunary regions; then heaven's cycles save the sublunary elements by regulating their chaotic changes.

The heat of the sun

Finally Aristotle takes up, again on the basis of his own commitments, the question of the origin of the celestial heat.[25] The question follows naturally enough upon the role of the sun in the regulation of the sublunary elements, and completes the introduction of the principles from the more general parts of physics. Aristotle's description of his friction theory here corresponds to that of *de Caelo* 2.7 but predictably shifts the emphasis toward the sublunary world. It is worthwhile to compare the two passages (I postpone the *Meteorologica*'s second theory of celestial heat):

> The heat and light from [the heavenly bodies] is generated when the air is rubbed by their passage (παρεκτριβομένου τοῦ ἀέρος). For motion naturally ignites wood and stone and iron. So it is more reasonable that that which is closer to fire should ignite, and air is closer. This happens, for example, in the case of moving missiles. For these are ignited so as to melt lead and, when they are ignited, necessarily the air too in a circle around them undergoes the same thing. Now these things are heated by being moved in the air, which becomes fire (πῦρ) by being struck with the motion. But each of the upper bodies is moved in a sphere so that they themselves are not ignited (ἐκπυροῦσθαι), but it is necessary that the air (τοῦ ἀέρος) which is below the sphere of the revolving body be heated as the upper body moves, and there most of all where the sun happens to be fixed. (*Cael.* 2.7.289a19–32)

> We see that motion can disperse and ignite air (διακρίνειν τὸν ἀέρα καὶ ἐκπυροῦν), so that objects in motion too are often seen to melt (τηκόμενα). Now, the motion of the sun alone is sufficient to produce warmth and provide heat: for the motion must be rapid and not far off. (*Mete.* 1.3.341a17–22)

[25] For a fine discussion of the problems and a spirited defense of this doctrine, see Thorp 1982.

In both passages the sun works on the air, heating and igniting it in the *de Caelo*, and dissolving and igniting it in the *Meteorologica*. Aristotle describes a similar process when he explains the Milky Way (1.8.345b32–6), as we shall see. There are several problems in understanding the process. Both descriptions suggest that elemental air is transformed into elemental fire and that friction is the cause of the heat associated with that region. So in the case of missiles moving through air, the surrounding air becomes fire (πῦρ).[26] But, on this interpretation, there is no elemental fire in the upper place except what is created out of the air and maintained by the heavenly friction, and this seems to be in contradiction with the doctrine of the elements, which gives fire an independent existence (*Cael.* 4.4–6). But if the upper region already were fire, Aristotle's theory of celestial heat would be equally troubled, since the region would already be hot and friction would be unnecessary. Alternatively, we may suppose that air is a kind of smoke, elemental "fire," and becomes what we call fire, a ζέσις, because of friction. Though Aristotle does not identify a separate smoky layer of air here, it is clear that the air can burn, a standard characteristic of Aristotelian smoke. In this case we are left to wonder whether it is the friction or the combustion that generates the heat.

What is not explained, of course, is how the heat gets down to the surface of the earth. Although the earth has its own heat (2.4.360a15–17), it is clearly not a major causal factor for anything but the winds. The orthodox theory, whereby change occurs by contact, faces the problem that, for Aristotle as for Anaxagoras, the upper regions of the atmosphere are colder than the surface of the earth, a fact hard to explain if heat is conducted through the air to the earth.[27] The heating of the earth is also hard to explain if the heat is mechanically transported down, since Aristotle's heat causes material to become light and rise (1.10.347a30–4) and even leaves vapor and rises alone (1.9.346b26). Indeed Aristotle criticizes Anaxagoras on this account (2.9.369b21–2). In a couple of passages Aristotle mentions rays (ἀκτῖνες) that conduct heat downward, once in such a way as to suggest that they are different from the "rest of the heat from above."[28] Since Aristotle has just finished discussing the phenomena of combustion, he may be referring to

[26] The theory is a modification of Anaxagoras' view that the stars shine because of the resistance and breaking round of the aether: λάμπειν ἀντερείσει καὶ περικλάσει τοῦ αἰθέρος (DK 59A12 [2.9.28–41]).

[27] For direct contact being necessary for change: *GC* 1.6.322b21–5. For the cold upper atmosphere in Aristotle: *Mete.* 1.3.340a26–32; and in Anaxagoras: 1.12.348a14–18.

[28] 1.3.340a24–32 in a dialectical passage, but apparently a persuasive consideration for Aristotle; 1.9.346b23–6: ὑπὸ τῶν ἀκτίνων καὶ ὑπὸ τῆς ἄλλης τῆς ἄνωθεν θερμότητος.

these fires as the source of the heat from above, and the rays as generated from the celestial friction. Clarity is impossible.

Finally, the account in *Meteorologica* goes beyond that in the *de Caelo* by adding a further cause of heat:

> Another reason is that the fire which surrounds the air is frequently scattered by the motion and carried below by force. (341a30–1)

This is generally understood to mean that the sun's movement does the scattering.[29] No doubt ultimately this is true, but Aristotle explicitly says that the fire *surrounding the air* (τὸ περιέχον πῦρ τὸν ἀέρα) is scattered. It seems likely, therefore, that the turbulence arises at the interface between the fire and the air. Although this cause is presented as an additional cause (καὶ διὰ τὸ . . . διαρραίνεσθαι 341a30–1), and although it involves forced motion, it is the only explanation that succeeds on recognized Aristotelian causes. Aristotle is of course short on details, but he may have in mind something similar to the way in which ejected shooting stars occur, which are likewise subject to forced motion downwards. This form of heat is probably similar to the "other heat from above" (1.9.346b23–6) and is unlikely to be a major contributor to terrestrial heat. In the end Aristotle has no coherent theory as to how the sun can heat the earth.

But this very failure merely demonstrates that Aristotle was less concerned to provide a plausible reductive account of heat in terms of general heavenly causes, and his preference for fashioning *proper* principles for meteorology allowed him to be less concerned. By the end of chapter 1.3 Aristotle has completed the basic transition from the four-element theory characteristic of the more general physical treatises to the exhalations specific to the *Meteorologica*. He has introduced genuinely new principles with distinct properties irreducible to the simple elements from which they are formed and has fashioned them with the aid of a specific celestial principle that both amends the chaotic behavior of the elements and guides all their subsequent changes.

29 Lee's translation and Freudenthal 2009: 239.

CHAPTER 3

The exhalations

As we saw in the last chapter, the sun, when it rubs against the upper limit of the kapnosphere, generates heat, which in turn is communicated to the surface of the earth by rays. This heat generates a vapor from the water on and in the earth and a kind of smoky exhalation from the earth itself. These two exhalations are the materials from which all the phenomena of the treatise are produced.

This apparently naive theory, of which chapter 1.3 gave us only the slightest hint, begins to be revealed in the course of the treatise as the outcome of a surprising compromise between the old Ionian exhalations and the more recent theories of elements. We have seen how Aristotle thought the philosophers in the wake of Parmenides had reduced the *meteora* to the common elements of all things, and had thereby eliminated any basis for an independent study of meteorology. For this reason, Aristotle sought principles that are special and appropriate to meteorology and found them in the exhalations of early Ionian physical speculation. But for three interrelated reasons he could not adopt these exhalations without amendment. First, the Ionian exhalations had indeterminate potentialities in accordance with which through their wide-ranging changes they could explain nearly every phenomenon. We saw in the last chapter that Aristotle turned away from the indiscriminate capacity of his own four elements to change into one another and restricted his new meteorological materials to specific potentialities. Second, the Ionian exhalations change in such a way as to form lengthy chains of phenomena (e.g., the sea forms vapors, which form clouds, which form lightning, and so on), with the result that the explanatory primacy of the exhalations is lost. In place of these chains, Aristotle substituted a sort of hub-and-spoke system where the *meteora* are more immediately dependent on the exhalations than they are on each other. Third, these Ionian chains, though they start appropriately in the meteorological realm, proceed until they reach the heavens, where the exhalations feed the sun and stars. Aristotle had not blocked the Italian reduction of *meteora* to general physics merely to have meteorology turn around and usurp general physics. The heavens are governed by celestial, not terrestrial principles. In general, then, he emended the Ionian exhalations by restricting their easy and indiscriminate material transformations and modernized them by subjecting them to a dynamic mechanics characteristic of the elementalists, Empedocles, Anaxagoras, and Democritus. The result is a rectification of the Presocratic theory of matter, which makes possible a study of the *meteora* independent of the four basic elements and prefigures the patterns of change characteristic of living bodies. In this chapter I shall first describe the nature of Ionian exhalations, especially Heraclitus', and show how Aristotle emended them. Then, in reference to the peculiar case of the dry exhalation, I shall show how he limited its capacity to change. I shall end with a detailed description of Aristotle's quasi-elementalist mechanics of exhalations.

The compromise between Ionian and Italian physics just described is nowhere explicitly laid out in Aristotle's text. In fact, his very brief remarks on his predecessors' contributions ignore most of his debts:

> When the earth is heated by the sun, the exhalations (ἀναθυμιάσεις) that arise are not, as some people think (ὥς τινες οἴονται), of a single kind, but twofold. (1.4.341b6–8)[1]

The "single kind" referred to here is the wet exhalation, water vapor, which was identified first in folk thought and adapted early in the Presocratic tradition.[2] But, though water vapor was first to be identified, it can hardly have been the first to be called an "exhalation." For the word, ἀναθυμίασις, regularly translated by "exhalation," implies in its roots the rising of fumes or smoke, not vapor, and therefore must first have been applied to the dry exhalation.[3] A theory of exhalations so named presupposes, then, the discovery of the dry. The situation is sufficiently unsettled that Aristotle does not maintain a strict nomenclature but uses the term ἀναθυμίασις generically to refer to both dry and wet exhalation (e.g., 1.4.341b6–8), and specifically for the dry exhalation alone (e.g., 1.4.342a4).

[1] For the identity of "some people," Ideler 1834–6: 1.363 suggests that Plato *Ti.* 56d alone is referred to. Yet it is clear that others made use of exhalations to explain meteorological phenomena. See Solmsen 1960: 407–12. Düring 1966: 386 claims that the term, ἀναθυμίασις, is new to Aristotle. At *Sens.* 5.433a21–b1, in the context of scents Aristotle mentions both the wet and the dry exhalations more or less as we find them in the *Meteorologica* and attributes to some the theory that scents are the dry exhalation, to others that scents are the wet exhalation and to still others that scents are both.

[2] Moreover, if Aristotle were referring to the dry exhalation as the single exhalation of "some people," there would be no reason for him to mention the second, wet exhalation, since the wet exhalation is not his concern at this point in the treatise.

[3] Cf. θυμιάω: "burn so as to produce smoke," LSJ.

Vapor, then, is historically first, and throughout the early period its functions are continually extended. It is mentioned as ἀήρ, mist, by Hesiod (*Works and Days* 547–53) where it is already part of an extensive causal nexus of rivers, dew, wind, and rain.[4] Anaximander adds a rudimentary mechanics and extends its functions as far as lightning and beyond. The sun vaporizes the ἀήρ, which is composed of very tenuous vapors. These can be separated (ἀποκρινομένων) from the grosser mist (ἀτμίς) and, when they gather together and move, they form the wind. Rain arises from the grosser mist emitted from the earth under the sun.[5] So far this is a mere extension of Hesiod, but Anaximander's wind in turn accounts for thunder and lightning in its various forms and combinations.[6] Finally, this vapor explains celestial phenomena, the turnings of the sun and moon, since the sun and moon turn where there is an abundance (χορηγία) of vapor for them.[7] It is easy to conclude that the celestial bodies are fed on this vapor. The changes that Anaximander's vapor undergoes, then, are unrestricted and form lengthy chains of causation.

Xenophanes likewise uses moisture to explain a growing variety of phenomena. The sea is called the source (πηγή) and father (γενέτωρ) of the water (the clouds, winds, rains, and rivers), presumably because the sea

4 "For the morning is cold when the North Wind comes down. In the morning, from the starry sky to the earth, a mist extends over the wheat-fields of the fortunate; it draws from the ever-flowing rivers, rises high above the earth on the wind-squall, and sometimes rains toward evening, sometimes blows as the Thracian Northerly scurries the clouds on in a dense tumult" (West 1988).

5 DK 12A11.6–7 (1.84.15–20): ἐκ τῆς ἀτμίδος τῆς ἐκ γῆς ὑφ' ἥλιον ἀναδιδομένης. Hippolytus also adds that the earth is the source of the vapor. But Alexander (*In Mete.* 67.3 = DK 12A27 [1.88.10–19]) suggests that the vapors arise from the sea: on the authority of Theophrastus he attributes to Anaximander the view described by Aristotle (2.1.353b6–9), that the entire area around the earth at first was wet and then was dried out by the sun. The obvious reconciliation – that vapors arise both from the earth and from the sea – would recommend itself more strongly if Aristotle himself did not on several occasions think it important that the vapor arise primarily from the earth. Anaximenes generates the stars from the rarified moisture of the earth (13A7.5 [1.92.13–15]): γεγονέναι δὲ τὰ ἄστρα ἐκ γῆς διὰ τὸ τὴν ἰκμάδα ἐκ ταύτης ἀνίστασθαι, ἧς ἀραιουμένης τὸ πῦρ γίνεσθαι, ἐκ δὲ τοῦ πυρὸς μετεωριζομένου τοὺς ἀστέρας συνίστασθαι. Xenophanes is made to claim that πάντα τὰ τοιαῦτα [shooting stars, comets, torches] νεφῶν πεπυρωμένων συστήματα ἢ κινήματα (DK 21A44 [1.125.16–17] = Aët. 3.2.11 [*DG* 367.16–19]). For Metrodorus the vaporous clouds are still required to generate sparks (τὴν εἰς τὰ νέφη τοῦ ἡλίου βίαιον ἔμπτωσιν πολλάκις σπινθηρίζειν DK 70A14 [2.232.20–1] = Aët. 3.2.10 [*DG* 367.14–16]).

6 Seneca, *QN* 2.18: "Anaximander related all these phenomena to the wind (spiritum). Thunder, he says, is the sound of a cloud being struck. Why do the sounds of the thunder vary? Because the blow itself is also variable. Why does it thunder in a cloudless sky? Because even then the wind leaps through the thick and ripped air. But why does it sometimes not lighten when it thunders? Because a weaker wind is not strong enough for a flame, but is strong enough for noise. What then is the flash of lightning? It is the hurling of air that is dispersing and rushing together producing a languid flame that will not discharge. What is lightning? The rushing of a very intense and dense wind" (modified Corcoran trans.).

7 Alexander 67.3 reporting Theophrastus' report.

is a necessary condition for the existence of all the rest.[8] The sea may be the ultimate cause but, as with Anaximander, it is the vapor, here specifically the clouds, that is responsible for producing the burning phenomena. Lightning is formed by the ignition of the moist clouds.[9] Even St. Elmo's fire is caused by a peculiar movement of the clouds.[10] Indeed the whole celestial realm arises from this moisture. From the clouds, when they are ignited, the stars are produced.[11] The sun likewise arises from enflamed clouds, when sparks are gathered together from the moist exhalation.[12] The heavens, then, are an effect rather than a cause, as they are for Aristotle. These same clouds by being felted also produce the moon and shooting stars; the comets and beams are likewise caused by the accumulation and movement of ignited clouds.[13] For Xenophanes, then, the terrestrial water is the ultimate cause of the entire changing cosmos.

The connection of the wet exhalation with celestial phenomena became a commonplace by the second half of the fifth century: Hippocrates of Chios explains the comet by reference to the moist exhalation in a relatively offhand way (1.6.342b35–343a4).[14] Between the tropics the moisture is burned by the sun, but the comet-planet, as it travels outside the tropics, is able to draw the moisture to itself. There are even traces of this notion in Aristotle's own account (the relative absence of comets in the zodiac and the burning up of moisture and smoke by the sun, 1.7.345a5–10), but of course, for Aristotle the wet exhalation is not burned as fuel. He restricts the efficacy of his wet exhalation and thereby cuts short the catenary series of causes characteristic of the early Presocratics.

Almost certainly Aristotle owes his second, dry exhalation to Heraclitus, who alone among the Ionian predecessors recognized both.[15] The evidence here is scant but deserves careful review, because there are several features of it that have been ignored but are important for Aristotle's doctrine. We must start our discussion with Diogenes Laertius (DK 22A1.9–11

8 DK 21B30 (1.16.5–10); and 21A46 (1.125. 20–4) = Aët. 3.4.4 (*DG* 371.9–17).
9 DK 21A45 (1.125.18–19) = Aët. 3.3.6 (*DG* 368.25–7).
10 DK 21A39 (1.124.31–3) = Aët. 2.18.1 (*DG* 347.4–6).
11 DK 21A38 (1.124.28–30) = Aët. 2.13.14 (*DG* 343.1–6).
12 DK 21A40 (1.124.34–6) = Aët. 2.20.3 (*DG* 348.10–13).
13 DK 21A43 (1.125.13) = Aët. 2.25.4 (*DG* 356.3–4); 21A44 (1.125.16–17) = Aët. 3.2.11 (*DG* 367.16–18).
14 So Philolaus (Huffman 1993: 265–6). Metrodorus, to judge from the scant remains, seems not to make exhalation a major factor. As a follower of Democritus, his exhalation — and he is credited with only a wet exhalation (DK 70A18 [2.232.36–9] = Aët. 3.7.3 [*DG* 375.5–11]) — must have been reduced to atoms and the void, so it cannot have been an independent principle. That Aëtius confines the exhalation to the explanation of the winds confirms this: by his time the exhalation is no longer used as a principle to explain all phenomena above the surface of the earth.
15 For hints that Philolaus too may have had two exhalations, Huffman 1993: 265–6.

[1.141.25–142.11]), whose account goes back to Theophrastus, but whose credibility has been widely impugned.[16] Because of its complexity and importance I quote it in full:

> 9. Fire when it is thickened becomes moist, and when it gathers together (συνιστάμενον) becomes water; when it freezes, water is turned into earth. This is the road down. Again, conversely earth is dissolved, from which comes water, and from this the rest; referring pretty much everything to the exhalation from the sea. This is the road up. Exhalations arise from land and sea, some are bright and pure, others dark. Fire is increased/strengthened by the bright, and moist from the others. What sort of thing the surrounding thing is, he does not make clear, but there are in it bowls with their concave surface turned toward us, in which the bright exhalations gathered together make light, which are the stars. 10. And the light of the sun is brightest and hottest. And the other stars are further away from the earth and therefore less bright and hot, and the moon, since it is nearer to the earth does not move through the pure region. The sun is situated in translucent and pure space and has a commensurate distance from us. And it is hot and bright. The sun and moon are eclipsed when the bowls turn up, and the monthly phases of the moon arise when the bowl is turned a little in itself. Day and night arise and the months and the yearly seasons and the yearly rains and winds and suchlike according to the different exhalations. 11. The bright exhalation, shining in the circle of the sun makes day, and the opposite exhalation, when it prevails makes night. And the heat, increasing from the bright makes summer, and the moist, prevailing because of the dark exhalation, creates winter . . . (my trans.).

We have here a remarkably explicit physical account, so different in style from the familiar aphoristic fragments that we must ask whether it is Heraclitean at all. In fact, the authenticity of the bright or dry exhalation has been widely denied on the grounds that there are no fragments that attest to it, and that Diogenes Laertius' report copies Aristotelian doctrine. It must be, the argument goes, a reinterpretation of Heraclitus through a Peripatetic lens.[17] It is true that there is no explicit evidence from the

[16] E.g., Diels, *DG* 163–5 for a discussion of Theophrastus' method and reliability (poor) in this section.

[17] Kirk 1962: 270–6 in a lengthy discussion rejects Heraclitean authorship of the dual exhalation and supports a single moist exhalation; followed by Marcovich 1967: 332 and KRS 202n; on the other hand, Solmsen 1960: 409; Kahn 1979: 292–3; and Strohm 1984: 182; Menn (in an unpublished Chicago paper of 2006) supports the dual exhalation and finds Diogenes' report authentic. West 1971: 133 also believes that there must be something to the dual exhalation. He too tries to bring it into accord with continual change. An exhalation comes both from the land and sea, but the dual exhalation is bright and dark, not moist and hot. The bright exhalation is the "prester which feeds the sun and stars" (135). As for the dark exhalation, West concludes (187): "[i]n the last chapter I asked in vain how souls might turn to water after death. In the chapter before I asked in vain what were the 'dark exhalations' that rose from the earth but fell back as rain, besides being responsible

fragments, but Aëtius' doxography bears witness to exhalations that arise from two sources.[18] Moreover, the argument from the similarity of the Heraclitean and Aristotelian exhalations is weak: the similarity is superficial, and the differences are sufficiently characteristic of each philosopher as to assure us of their independence. According to Diogenes' report, Heraclitus' exhalations are bright and dark, not the Aristotelian dry and moist.[19] Diogenes' Peripatetic source, Theophrastus, was much more likely to have imposed the Peripatetic hot-cold-wet-dry analysis onto Heraclitus, if in fact he was imposing. Again, as is characteristic for Heraclitus, the two exhalations are *opposites* and have *opposite* effects: the bright exhalation causes summer and day; the dark causes winter and night. For Aristotle, in contrast, the moist and dry exhalations do not have opposite effects, though they are opposites in virtue of one of their primary qualities. Finally, Heraclitus made the bright exhalation fuel for the stars and the heavenly bodies, something Aristotle strongly rejects. There are some interesting similarities between the views, as we shall see, but these differences are sufficient, I think, to convince us that Diogenes' account is not Aristotle *réchauffé*.

If we are to credit Diogenes to this extent, we need to pay more careful attention to his text as we proceed. We start with his description of the road down, which includes fire, water, and earth, but neither air nor exhalations. On the road up, the sequence beginning with earth is broken at water, from which τὰ λοιπά arise, Heraclitus "referring pretty much everything *to the exhalation* from the sea" (ἐκ δὲ τούτου τὰ λοιπά, σχεδὸν πάντα ἐπὶ τὴν ἀναθυμίασιν ἀνάγων τὴν ἀπὸ τῆς θαλάττης). This is the road up. The absence of ἀναθυμιάσεις on the road down is striking but appropriate, since the prefix ἀνα- indicates their upward movement. To this extent the road up and the road down are not the same. The road up is different too in the introduction of the sea, which, given its special role in meteorological causation (DK 22B31), may be distinct from the ὕδωρ already mentioned twice in the passage. Now, within the broader context of Presocratic meteorology, an ἀναθυμίασις from the sea would normally suggest a moist vapor (ἀτμίς), as with Xenophanes, for whom it explains

for winds and for the diminution of the celestial fire at night and in winter. It now appears that the one problem is the solution of the other. The 'exhalations' were souls – as Aristotle told us all the time."

18 From the earth: DK 22A11 (1.146.25–6) = Aët. 2.17.4 (*DG* 346.13–15); from the sea: DK 22A12 (1.146.27–8) = Aët. 2.20.16 (*DG* 351.9–11). Also the moist exhalation as a source: "the stars, which receive their gleaming from the moist exhalation" (DK 22A12 [1.146.31–5] = Aët. 2.28.6 [*DG* 359.1–10]).

19 As has been noted by West 1971: 133; Kahn 1979: 293.

most of the *meteora*. But as we continue with Diogenes' description, we find the introduction of the two exhalations: γίνεσθαι δὲ ἀναθυμιάσεις ἀπό τε γῆς καὶ θαλάττης, ἃς μὲν λαμπρὰς καὶ καθαράς, ἃς δὲ σκοτεινάς. αὔξεσθαι δὲ τὸ μὲν πῦρ ὑπὸ τῶν λαμπρῶν, τὸ δὲ ὑγρὸν ὑπὸ τῶν ἑτέρων. The most obvious (and common) interpretation is to suppose that the bright exhalation comes from the land and the dark one comes from the sea.[20] But Diogenes does not say that, and, if he had intended it, he surely would not have singled out τὴν ἀναθυμίασιν τὴν ἀπὸ τῆς θαλάττης in the previous sentence as the source of everything; and, instead of what he says here, he would have said, γίνεσθαι δὲ ἀναθυμιάσεις τὴν μὲν ἀπὸ τῆς γῆς λαμπρὰν καὶ καθαράν, τὴν δὲ ἀπὸ τῆς θαλάττης σκοτεινήν. Moreover, the absence of the definite article before γῆς and θαλάττης gives the whole expression a much vaguer sense: "exhalations arise from land and sea. Some of them are bright and pure, others dark."[21] Diogenes' language here discourages us from contrasting the land and sea, and instead we should think of these exhalations as coming indifferently from both sources.

How do these exhalations function? First, when Diogenes says that the fire is increased or strengthened (αὔξεσθαι) by the bright exhalations, he implies that the exhalation is different from the fire and serves to feed the fire. Presumably the same holds for the wet, though how this feeding would take place is unclear. Secondly, we read much about their opposite effects, but nothing about their mutual transformation. The exhalations have influence on the same phenomena but have opposite effects, and the phenomena are paired according to these opposite effects: night–day, winter–summer. It is probably a safe bet that the months, seasons, yearly rains and winds were likewise conceived of as opposites. And yet we do not hear of the bright exhalation turning into the dark.[22]

In spite of the lack of explicit transformation, it seems hardly credible that in an ontology such as Heraclitus' the exhalations should remain unchanging and absolute.[23] Heraclitus often mentions things that have opposite effects, but that transform into one another (day and night, DK22B57; sleeping and waking, 22B88). Moreover, evidence from Aëtius suggests that

[20] Most recently Betegh 2007: 20.

[21] Gildersleeve 1900: 259: "Prepositional phrases and other formulae may dispense with the article as in the earlier language . . . Compare *wife and child . . . sun, moon, stars*. But anaphora or contrast may bring back the article at any time and there is no pedantical uniformity." Cf. *EN* 10.8.1179a4–5: δυνατὸν δὲ καὶ μὴ ἄρχοντα γῆς καὶ θαλάττης πράττειν τὰ καλά.

[22] As Böker 1958a: 2235 briefly notes.

[23] Graham 1997. Holding to a distinction between substance and accident would be odd, as is clear from DK 22B126 (1.179.7–8): τὰ ψυχρὰ θέρεται, θερμὸν ψύχεται, ὑγρὸν αὐαίνεται, καρφαλέον νοτίζεται. "Cold warms up, warm cools off, moist parches, dry dampens" (trans. Kahn 1979).

the origins or at least the functions of the two exhalations were not so sharply distinguished: the stars receive their shine from the moist exhalation (δεχομένους τὰς ἀπὸ τῆς ὑγρᾶς ἀναθυμιάσεως αὐγάς); and the sun is an intelligent ignition from the sea. Both suggest that fiery phenomena have their origin in water.[24]

If we are willing to follow Diogenes this far, a clue as to how the exhalations might have interacted comes from the *de Anima*, and precisely because it comes to us through Aristotle's interpretative lens, the passage may give us a hint as to how he viewed and used Heraclitus' exhalations. Aristotle mentions an exhalation "from which other things are composed" and identifies it with Heraclitus' conception of the soul (*de An.* 1.2.405a25–7): "Heraclitus says that the principle is the soul, if, as is the case, it is the exhalation, from which the other things are composed. And it is the most incorporeal and is always flowing."[25] The similarity here with Diogenes' claims that Heraclitus is "referring pretty much everything to the exhalation from the sea" is noteworthy, and the identification of the exhalation with the soul permits us to draw on evidence from Heraclitus' psychology.

The soul is an ἀναθυμίασις and in Aristotle's description this can only mean a smoky exhalation.[26] By turns, however, it can become both moist (associated with wine and the sea) or dry without ceasing to be a soul, though it seems that an extreme moistening will be the death of soul.[27]

[24] DK 22A12 (1.146.31–5) = Aët. 2.28.6 (*DG* 359.1–10); 22A12 (1.146.27–8) = Aët. 2.20.16 (*DG* 351.9–11). Aëtius reports that Heraclitus (and Parmenides) thought that the stars were πιλήματα of fire (DK 22A11 [1.146.25] = Aët. 2.13.8 [*DG* 342.6–7]) and that Heraclitus thought the stars were nourished from the *terrestrial* exhalation (DK 22A11 [1.146.26] = Aët. 2.17.4 [*DG* 346.13–15]).

[25] καὶ Ἡράκλειτος δὲ τὴν ἀρχὴν εἶναί φησι ψυχήν, εἴπερ τὴν ἀναθυμίασιν ἐξ ἧς τἆλλα συνίστησιν· καὶ ἀσωματώτατόν τε καὶ ῥέον ἀεί. As Colvin 2005: 266–7 points out, Galen (*de Tremore* 7) interprets the *tonos* of the Stoics in Heraclitean terms, "For inasmuch as the inborn heat is always in motion, it does not move only inward or outward, but always alternates between the one motion or the other. For the inward motion alone would soon lapse into motionlessness, or the outward motion would disperse and in this way destroy itself. But being extinguished in measures and kindled in measures, as Heraclitus said, it remains in this way always moving" (Colvin trans.).

[26] Kahn 1979: 239 holds the exhalation, viz., the soul, to be vaporous, rather than fiery, because the soul comes out of water and therefore cannot be fire. "The respectable ancient evidence is thus unanimous in confirming the natural interpretation of [DK 22B36 (1.159.8–10)], with the psyche understood as a kind of atmospheric vapor emerging from water" (1979: 240). Kahn interprets the ἀναθυμίασις as "the smoke of steam or mist," but the term ἀναθυμίασις originally refers to the burning of incense (LSJ: θυμιάζω, etc.). When Aristotle uses this word he means first the smoky exhalation, and only secondarily exhalation in general. He does not use it in the first instance to refer to the vaporous exhalation alone. If Aristotle (*de An.* 1.2.405a25–7) is using ἀναθυμίασις in his own sense when describing Heraclitus, he must be referring to the smoky exhalation; if it is not Aristotle's own expression, he must be quoting Heraclitus.

[27] DK 22B31 (1.158.6–7): "the turnings of fire: first sea, and of sea one half earth, one half firespout." 22B36 (1.159.8–10): "for souls it is death to become water, and for water death to become earth, and from earth water arises, and from water soul." 22B77 (1.168.12–13): "for souls it is delight or

We do not need to argue the details to see that this exhalation-soul is in flux between moist and dry and has tendencies in both directions, though its dry and fiery tendency is positively valorized. It is hardly a stretch of imagination to see parallels between Diogenes' descriptions of paired meteorological phenomena and the good and evil tendencies of sobriety and inebriation. On this view the exhalations are two moments of a stuff that arises from both the land and sea and that can differentiate itself into light and dark with opposite capacities.

If Heraclitus' views may be reconstructed in this way, Aristotle's debt to him is complex. First, of course, he adopts Heraclitus' two exhalations and has them both arise from the land and the sea (so Aristotle's dry exhalation also leaves the sea together with the evaporated moisture). Second, Heraclitus' bright and dark exhalations form a contrasted pair, like Aristotle's wet and dry. But while Heraclitus' exhalations are opposite in their effects, Aristotle's have analogous effects and give rise to analogous rather than opposite phenomena, like the rivers and the winds. So strong is Aristotle's impulse to analogy in fact that, after distinguishing celestial from sublunary realms, he sees fit to invent the kapnosphere as a location devoted to the phenomena of the dry exhalation, a region he makes correspond to the region of condensation, as we shall see in due course. At the same time, though, the fact that Heraclitus' two exhalations operate together in the same place and with different potentialities prefigures Aristotle's peculiar concept of the "air" in the early chapters of the *Meteorologica* as a partly fiery and partly vaporous stuff. Aristotle even talks once about exhalations that have now more fiery potential, now more vaporous potential, suggesting that the same body may have opposite potentialities (3.1.370b14–17). Indeed, it is probably no coincidence that air (as distinct from vapor and smoky exhalation) is absent both from Aristotle's account and from the "three-element" account of Heraclitus.[28]

But for all that Aristotle's dry and wet exhalations mix together to form air, they remain strictly incapable of mutual transformation. In this respect, Aristotle breaks with the early Presocratics to side with the elementalist post-Parmenideans. Whereas the early Ionians fed the sun and stars on the moist exhalation, for Aristotle fire can only arise from the dry exhalation.

death to become moist." 22B117 (1.177.1–3): "whenever a man becomes drunk, he is led tripping by a young boy, not knowing where he is going, with his soul moist." 22B118 (1.177.4–5): "dry gleam is the wisest and best soul" *or* "dry soul is wisest and best." It is reasonably clear, I think, why a soul would be moist by drinking wine, and also why this impairs a man's awareness and why that would be bad. It also seems clear why a sober man should have a dry soul and why this is good. 22B77 = 1.168.11–15 invites an identification of wine and sea water at least in their effect on the soul.

28 A conclusion arrived at as well by Betegh 2007: 21–2.

And yet his reasons were clearly not elementalist, as his doctrine of the four simple bodies and their transformations makes clear. Aristotle was first motivated by the realization that the amount of water on the earth was minuscule in comparison with the size of the heavens (1.3.339b6–9). If moisture fed the stars, the burning would result in the complete evaporation of the sea. And even if the roads up and down are equal – Heraclitus' metaphor for the cycle of material change, according to Diogenes (DK 22A1.9; cf. 22B60) – the great magnitude of the heaven, as we have seen, would overwhelm the tiny amount of moisture.[29] But, more importantly, looking forward to *Meteorologica* 4 and the biological works, Aristotle is here beginning to restrict the chaotic and unregulated changes among the four elements and replace them with a chemistry in which specific materials have more specific potentialities.

Recycling the dry exhalation

Among the many challenges we face in trying to understand Aristotle's theory of the dry exhalation, the most difficult concerns what happens to the exhalation after it rises from the earth and produces winds and the fiery phenomena. In all probability the winds ascend to the kapnosphere, but then, to judge from Aristotle's description of shooting stars and lower comets, the exhalation burns up and disappears. According to his general theory of matter, however, it ought to transform into some other form of matter.

Aristotle had compelling reasons to make the dry exhalation flow back down to the earth. He goes to great lengths to emphasize that rain and its congeners are part of a cycle, the "river of Ocean," and he rejects the contentions of his predecessors, according to whom terrestrial water is being continually consumed by the stars. He also shows himself aware elsewhere that when dry exhalations arise from earthy material, their source is diminished.[30] The earth should, therefore, however slowly, be entirely

29 Hippolytus, *Haer.* 9.10, our source for the fragment, places it in a context that suggests the unity of opposites, rather than a cycle of material change, as Diogenes understands it.

30 At *Mete.* 4.9.387a26–8 θυμιατά are said sometimes to become earth. The objection is repeated in *Pr.* 12.10.907a29–34: "Are scents smoke or [air or] vapor? . . . Also, if anything is given off (ἀπορρεῖ) by these objects, they should become smaller. And yet we see things which have the strongest smell persist longest." Again, at *Sens.* 5.443b1–2, Aristotle says, "the exhalation has the same significance as emanations (ἀπορροίαις). If the latter is mistaken, so is the former." Other evidence for the materiality of the combustion product of the dry exhalation comes from Aristotle's discussion of lightning, where the less fine pneuma that is burned generates a thunderbolt that is called "sooty" (ψολόεντα 3.1.371a21; cf. *Mu.* 4.395a26). This is, however, a traditional name, and Aristotle may not ascribe special theoretical significance to it.

transformed into exhalation and disappear. Second, all matter requires a coming-to-be and passing-away from and into opposites. So some of the upper, hot, dry fire ought to cool or moisten and fall again. Now, Aristotle rejects the moistening of the dry exhalation into vapor so as to avoid an inconvenient transformation of dry and wet exhalations, but we might still expect dry exhalation to condense into earth and to precipitate in its own cycle of change. In fact, in *Sense and Sensibilia* Aristotle recognizes a kind of condensation of the smoky exhalation (καπνώδης ἀναθυμίασις *Sens.* 5.443a23). Here he explicitly contrasts vapor (ἀτμίς) and exhalation (ἀναθυμίασις), and notes that "the former when it condenses becomes water, but the latter becomes a kind of earth" (γῆς τι εἶδος). So smoky exhalation can transform back into earth. Alternatively, even if the dry exhalation merely returned to the earth in its own form, the cosmic balance would still be maintained. But Aristotle nowhere provides an adequate mechanism for this. In fact, we see very few downward motions of dry exhalation among the *meteora*, none of them sufficient to the task. Rain absorbs dry exhalation and falls as saline water into the sea, but since this return exactly balances the exhalation that is given up in marine evaporation, it cannot explain what happens to the material that makes up the winds and the fiery phenomena (2.3.358a21–6). Again, even if a large amount of dry exhalation is churned back to earth by the celestial movement (1.3.341a30–1), Aristotle provides no account of how it gets back *into* the earth.

There are a number of ways to explain Aristotle's odd silence. Least appealing is the claim that Aristotle was content to refute his predecessors' theory that the heavens feed on water vapors and to transfer the problem of entropy to the dry exhalation where it would be less apparent. After all, the dry exhalation is much more difficult to observe than vapor, and the earth does not visibly diminish in size when the dry exhalation is released from it as some bodies of water diminish when they evaporate. A more attractive reason why Aristotle did not feel the urgency of the problem is that, as the physical lectures move toward the biological realm, the transformation of the simple elements is not complex enough to explain the material changes, which are often uni-directional. The fourth book of the *Meteorologica* is filled with processes that are not reversible in a simple manner. What is roasted, for example, cannot be unroasted. Likewise, the bodies of animals that have died are not straightforwardly recycled into other animals. No doubt the material of one human body may find its way after many transformations into another human body, but that process need not concern the biologist. Similarly Aristotle may think that there are processes by which the dry exhalation does return through many transformations to the earth, but that they result in no significant meteorological phenomena. The most compelling reason must wait until we discuss the architecture of the treatise as a whole.

Meteorological mechanics

As we have seen, Aristotle adopted the exhalations of the early Presocratics but limited their ability to transform. To these materials so conceived he added a mechanical apparatus very similar to that of the later, elementalist Presocratics but restricted it mainly to the gaseous phenomena of the kapnosphere and atmosphere. Throughout the treatise we see a great deal of combination and separation, pushing, freezing, ejection, and so on, and Aristotle describes motion and change at a level of detail largely absent in the physical works up to this point. To be sure, natural place and its effect on motion is everywhere active.[31] But the specificity of these processes, indeed their nature altogether, often escapes notice, because Aristotle's vocabulary mimics common language so closely that it masks the underlying theory.[32] However, a closer study of Aristotle's language will reveal a definite, if flexible, theory of sublunary mechanics.

The general principles: *sunkrisis* and *diakrisis*

Aristotle himself attributes combination (σύγκρισις) and separation (διάκρισις) to the later Presocratics. Empedocles is said to have introduced Love and Strife as moving causes, where Love combines (συγκρίνει) and Strife separates (διακρίνει) the roots (*Met.* 1.4.985a21–5). Elsewhere he attributes the same two notions to Anaxagoras (1.3.984a11–16) and similar ones to Democritus and Leucippus (*GC* 1.1.314a21–4).[33] Evidence from the fragments confirms the attribution: we find Empedocles using ἀποκρίνειν (DK 31B9 [1.313.1]) as well as διαφύειν (31B17 [1.315.17; 316.4]) for the process of separation. The opposite process of combination is called μίξις and διάλλαξις (31B8 [1.312.9]), and the roots are said to run through one another (δι' ἀλλήλων . . . θέοντα 31B17 [1.318.34]; cf. 31B26 [1.323.1]; 31B21

[31] By contrast, natural place is kept largely in the background in *Mete.* 4 (Furley 1989 [1983]: 134).

[32] Noted by Owen 1986 [1970]: 156.

[33] Anaximander has ἀποκρινομένων, but that expression is in Simplicius' report (DK 12A9 [1.83.11]). The general situation among the Presocratics is well summed up by Aëtius (1.24.2 [*DG* 320.20–8]): "Empedocles, Anaxagoras, Democritus, and Epicurus and all those that create the cosmos by gathering together small parts of body introduce συγκρίσεις and διακρίσεις and not generation and corruption in the primary sense."

[1.320.8]).[34] Anaxagoras explicitly uses σύγκρισις and διάκρισις: συγκρίνειν refers to that which is mixed from all of the seeds (59B4 [1.34.6]); the διακεκριμένα are the things that have been separated off, even though on Anaxagoras' peculiar theory, when they have been separated out, everything still remains in them (59B5 [2.35.8]).[35] Similarly air and aether are separated off (ἀποκρίνονται 59B2 [2.33.2]) from the large surrounding mass.[36] Direct evidence for Democritus is sketchier, but his use of the concepts is assured. Aristotle, Aëtius, and Galen all attribute the concepts σύγκρισις and διάκρισις to him.[37]

This separation and combination are the general principles of change and occur through a variety of mechanisms that are more or less distinguished from the stuffs themselves. Empedocles' Love and Strife, suffused but invisible feelings, are responsible for separation and combination but do not operate with the same kind of pushing as the roots themselves might. There is the whirl (DK 31B35.4), which causes some forced motions, and these motions in turn cause others.[38] Anaxagoras' mechanics begins with Mind, which plays a role in bringing about the revolution (περιχώρησις),

[34] ἔρχομαι and compounds are used in the Strasbourg papyrus section (a(i)6 [restored]; a(ii)17 [partly restored]; a(ii)20). There is a κρῆσις (31B21 [1.320.9]). Fragment 31B23 (1.321.10–14) introduces a pigment analogy: "As when painters adorn votive offerings, men well-learned in their craft because of cunning, and so when they take in their hands many-coloured pigments, mixing them in harmony, some more, others less, from them they prepare forms resembling all things" (trans. Inwood 2001).

[35] 59B12 (2.38.8–10): καὶ τὰ συμμισγόμενά τε καὶ ἀποκρινόμενα καὶ διακρινόμενα πάντα ἔγνω νοῦς.

[36] So also for separation: 59B4 (2.34.16 and 17); 59B6 (2.35.19); 59B7 (2.36.4). The opposition is used by Epicharmus (DK 23B9 [1.200.9]), who was active during the first quarter of the fifth century.

[37] DK 68A49 (2.97.22–4) = Galen, *de elem. sec. Hipp.* 1.2: καὶ διακρίνει δὲ καὶ συγκρίνει πάλιν εἰς ἄλληλα κατὰ τὰς τοιαύτας ὁμιλίας κἀκ τούτου τά τε ἄλλα συγκρίματα πάντα ποιεῖ καὶ τὰ ἡμέτερα σώματα καὶ τὰ παθήματα αὐτῶν καὶ τὰς αἰσθήσεις. Much the same theory is attributed to Democritus by Aristotle in his *Peri Democritou*, reported by Simplicius (*in de Caelo* 294–5 = DK 68A37 [2.93.28, 39; 2.94.1–2]): visible masses arise when the atoms συγκρίνει. Simplicius makes explicit that the atoms when they come together stay together in virtue of their shapes, until some more powerful force shakes them apart and scatters them. According to Diogenes Laertius (DK 68A1 [2.84.10–18]), "the atoms move by being whirled in the all, and in this manner all the συγκρίματα arise, fire, water, air, earth. For these things are συστήματα of certain atoms." DK 68A93 (2.106.19–20) = Aët. 3.3.11 (*DG* 369.12–15): Democritus holds that thunder arises from an irregular σύγκριμα, which forces down the surrounding cloud.

[38] On the issue of forced motion, perhaps the only direct evidence comes from the Armenian fragment in Philo's *On Providence* (DK 31A49 [1.292.15–27]), concerning which there is some dispute about translation. Inwood 2001: 106 translates, "For when aither separated and flew off from air and fire." But Kingsley 1995: 22 translates, "as the aither separated off, it was raised upwards by the wind and fire," which suggests forced motion. No doubt too, the whirl is an independent motion and a cause of mixing. In addition we find 31B56 (1.332.25–6): "salt congealed being pushed by the blows of the sun"; and 31A66 (1.295.10–15): "[the sea's] ferocious edge keeps swelling, as when swamps absorb the floating hail. For all the moisture on earth tends to be driven into its hollows, being forced by the constant whirls of the wind, by the strongest bonds as it were." For the natural or chance motions of Empedoclean fire according to Aristotle, *GC* 2.6.334a4–5.

which in turn is the reason for the ἀπόκρισις.[39] Violence (βίη) seems to have something to do with the process (59B9 [2.36.17–24]), but Democritus, like Aristotle, also works to internalize the dynamic forces in the material, making them part of the nature of the constituent atoms.[40] Finally, Plato's *Timaeus* adds to the mechanics of σύγκρισις and διάκρισις a system of mutual transformation among the elements, which anticipates Aristotle's but is based on the geometry of triangles. This system, very similar to what we find in the *Meteorologica*, is used to explain the constant churning of bodies (58b6–8; cf. 49b7–c7), but is not applied to specifically meteorological phenomena. Such in outline is the pre-Aristotelian theory of separation and combination.

In *Physics* 7 Aristotle analyzes σύγκρισις and διάκρισις as forced rather than natural motions. They are a kind of pushing, locomotions caused by something else beside the object moved. This is appropriate in a context where Aristotle is abstracting from the various kinds of motion in order to argue for the first mover, but it is quite different from what we find in the *Meteorologica*.[41] Though the sun and the heavens are the ultimate movers of sublunary mechanics, the exhalations also act through their own internal principles of change. Although Aristotle does not appear to have developed a strict classification of motions as he did for places and materials, and though his terms are subject to ambiguity and overlap, nevertheless they fall into some basic groups: gathering and dispersing, overcoming, violent motion, lateral motions, heating and cooling. Together they provide a conceptual range of internal impulses and forced motions similar to, though more sophisticated than, what he found in his predecessors. In what follows I first discuss the major terms for motions and changes, which contexts they are used in and how they are related to one another. Rarely do they rise to the level of technical terms and so I have been content to identify broad tendencies and pass over occasional exceptions. Next I turn my focus to

39 But not apparently any σύγκρισις: "What mind moves, all this is separated. And when it is moved and separated, the revolution made the separation occur even more" (59B13 [2.39.8–17]).

40 "Democritus' dynamics postulated three fundamental forces, a repulsive force which plays the role of impact in a conventional corpuscular theory and two kinds of attractive force, one which draws together atoms of the same shape and another which holds together atoms of different shapes in an atomic aggregate. There is also some rather vague evidence to suggest that atoms may have been conceived as having an internal source of mobility" (C. C. W. Taylor 1999: 194). So DK 68A38 (2.94.8–10): "For by nature like is moved by like and things of the same kind move toward one another and each of the shapes ordered for a different σύγκρισις produces a different state."

41 Many of the terms for locomotion that we find in abundance in the *Meteorologica* are treated together in *Phys.* 7.2.243a11–244b2. Here Aristotle divides forced *per se* motions into four classes: pulling (ἕλξις), pushing (ὦσις), carrying (ὄχησις), and whirling (δίνησις). Aristotle claims that all locomotions can be reduced to these four.

the dry and the wet exhalations and indicate some of the important ways in which Aristotle distinguishes their mechanics. Readers less interested in Greek semantics may pass over to the next chapter.

Aristotle's lexicon of mechanics

A large part of Aristotle's mechanical vocabulary centers around the change in internal constitution of a mass of exhalation. The noun **σύστασις** refers to a state of matter, which is often in the right or appropriate density or constitution for a certain outcome or has that appropriate constitution dissolved and destroyed by some external agency. It is used both for dry exhalations (seven times; especially comets 1.7.344a36, b11; 345a8) and wet exhalations (twelve times).[42] Dry συστάσεις are at the right density for ignition; a majority of the wet συστάσεις (eight of thirteen) concern reflection, for which the vapor has to attain the right density.[43]

The cognate verb **συνίστασθαι** can mean "gather" (1.10.347a27), and as such is similar to ἀθροίζειν (1.10.347b10 and more below).[44] More commonly, though, the verb refers to the process by which the wet exhalation turns into rain or other precipitation (though we should be wary of calling it "condensation").[45] The noun, σύστασις, by contrast, is never used to describe this process. The verb συνίστασθαι, however, can describe the whole gathering process from beginning to end (2.2.354b31), or it can refer to any part of the process: air to vapor, vapor to mist, vapor into a cloud (3.3.372b16), mist to cloud, cloud to droplet (3.4.373b20), droplet to rain. Air can συνίστασθαι into cloud and water (1.13.349b23), into small drops and then rain (1.13.349b31).[46] This συνίστασθαι can also cause a contraction of the vapor, which squeezes out an ejected shooting star (1.4.342a1). The sole cause of συνίστασθαι is the cold (1.9.346b29; 2.4.360a1; 360b35; 2.6.364b27). The process of συνίστασθαι, then, is the opposite of διαλύεσθαι (2.2.355a32) or διακρίνεσθαι, which can oppose or undo the process of συνίστασθαι (1.7.344b24; 345a8; 1.8.346a16).

42 Dry: 1.4.341b23; 1.5.342b14; 1.7.344a34; 344b18; 1.8.345b34; 346a13; 346b10. Wet: 1.10.347a35; 1.11.347b21; 1.14.352b10; 2.9.369a16, 19; 3.3.372b18, 23; 373a28, 31; 3.4.373b3; 374a12; 3.6.377b5, 32.

43 3.3.372b18, 23; 373a28, 31; 3.4.373b3; 374a12; 3.6.377b5, 32.

44 Elsewhere 1.10.347b10; 2.2.354b20; 2.3.358a10. Clouds can gather (1.3.340b30, 33; 2.6.364b9); mist can gather (3.3.373a1; 3.4.373b2).

45 32 times. (A) to mist or cloud: 1.3.340a25; 340a30; 340b30, 33; 1.12.349a3 (and rain); 2.3.358a22 (and rain); 2.4.361a10; 2.6.364b9; 2.9.369a15; 3.3.372b16, 17; 3.3.373a1; 3.4.373b2, 16, 20; 374a18. (B) to rain or water, snow: 1.3.340a34; 1.7.344b24; 1.9.346b29; 1.11.347b13; 1.12.349a3; 1.13.349b23, 31; 2.2.354b20; 355a32; 3.3.358b17, 20; 2.4.360a1, 21; 360b35; 2.6.364b27 (?). (C) dew/frost 1.10.347a27.

46 It is sometimes said to συνίστασθαι "back" (πάλιν) into water: 1.3.340a34; 2.3.358b17, 20; 2.4.360a21.

συνίστασθαι is used much less frequently with dry exhalations (eight times).[47] When upper kapnospheric phenomena συνίσταται, they seem to do little more than "occur" (1.4.342a17, 34). On one occasion the upper air is συνιστάμενος so as to ignite (1.5.342b1). Other contexts too suggest that συνίστασθαι means that the dry exhalation has a certain consistency, that it is gathered together in the appropriate density. Aristotle, however, does not associate it with πυκνότης, just as he does not associate it with vaporous condensation. For this reason the translation "con*dens*ation" is not strictly appropriate. Such a gathering of dry exhalation can be dispersed (διακρίνειν 1.7.345a8; 1.8.346a16), just as a gathering of wet exhalation can, or it can be quenched (σβέννυσθαι 1.10.347b10).

σύγκρισις and **συγκρίνεσθαι** are close synonyms of συνίστασθαι. They are used exclusively of gaseous exhalations, never of the water of rivers or of minerals under the ground. They are not found in *Meteorologica* 4. Given the important history of the -κρινειν words, as we saw above, it is disappointing that Aristotle makes so little effort to give the root a technical meaning. As with συνίστασθαι, it implies a self-gathering into one place of stuff that had been diffused throughout some mixture or that was potentially present in it. So in air, which has potentialities of wet and dry exhalation, whichever exhalation predominates will gather together in a mass (3.1.370b15). For this reason, the verb appears most frequently in the middle voice (1.3.341a4; 1.4.342a29; 2.3.358b17; 2.9.370a30; 3.1.370b15); otherwise it can be passive without explicit agent (1.5.342b17; 1.10.347a17, 19). When it is used actively, the only agents referred to are the sun (1.9.346b22) and the mountains (1.13.350a13). Though σύγκρισις and συγκρίνεσθαι are used most often for the condensation of water (1.3.341a4, 10; 1.9.346b34; 1.10.347a17, 19; 2.3.358b17; 2.9.370a30), they are less closely associated with cold than συνίστασθαι is (but cf. 1.4.342a29). There can be a σύγκρισις of the hot exhalation, strongly associated with comets and the Milky Way (1.7.344b9; 1.8.346a4, 16, 23). Though it involves internal locomotion, it does not imply motion away from, as ἔκκρισις or ἀπόκρισις do.

ἀθροίζειν and its cognates generally refer to an intense gathering of a fluid material (i.e., not earth) as occurs with violent rain and wind (1.12.348b11 [twice], 22; 3.1.370b7; 371b1, 5), freezing (1.12.348b22), earthquakes (2.8.366b21; 367a30; 368b4), mock suns (which explains why mock suns are white, 3.6.377b18). The Milky Way is an intense comet (1.7.345a9; 1.8.346a22). The term is not temperature related, so has nothing to do

[47] 1.4.342a17; 1.5.342a34, b1, 5; 1.7.344a36; 344b11; 345a8; 1.8.346a16.

with "condensation" of rain (i.e., the change of state from vapor to water). Because of its intensity, an ἄθροισις is better able to remain in its nature (2.2.355b26, 31; 2.5.361b18; 3.6.377b18). Conversely the lack of a single mass or gathering of water (2.2.354b6, 12, 13) on the earth's surface suggests the sea's instability.

πυκνοῦν and its cognates are not used for "condensation" of rain either but instead describe the contraction of wet exhalations in forced interactions (ejected shooting stars 1.4.342a13, 21; the violent Caecias 2.6.364b24; the whirlwind 3.1.371a1; thunder 2.9.369a36).[48] When applied to the dry exhalation, however, πύκνωσις makes the dry exhalation good only for burning comets (1.7.344a30, b4), not for violence like generating earthquakes.

πήγνυσθαι straightforwardly refers to the freezing of water (e.g., 1.12.348a4, 5, 13, 14) or vapor (e.g., 1.10.347a17, 26) and by extension to the congealing of metals (3.6.378a30, b2). **τήκεσθαι** is used exclusively of the underground moisture that melts and produces exhalations for the wind (2.5.362a5, 8, 18; 2.6.364a10).

διακρίνειν is used of both the dry and the wet exhalation but is less clearly defined than either συγκρίνειν or ἐκκρίνειν. Sometimes it is opposed to συνίστασθαι, as "amass" is opposed to "disperse." Thus the rays of the sun "disperse" (διακρίνει) a σύστασις of cloud (1.3.340a29; cf. 2.8.367a25) or the συνιστάμενον hot exhalation, thus preventing comets in the zodiac (1.7.345a8). It is closely related to **διαλύειν**: in droughts the vaporous wet is dispersed (διακρίνεσθαι and διαλύεσθαι) by the mass of the hot exhalation, so that it does not easily συνίστασθαι into water (1.7.344b22). διαλύειν emphasizes, however, an effective passing into non-existence, or at least causal ineffectiveness.[49] This cannot be said of any of the -κρινειν words: the hot leaves the σύστασις of the cloud as it disperses (διακρινόμενον) into the upper place (2.9.369a17; cf. 1.8.346a15). More concretely, διάκρισις accounts for changes from heavier elements to lighter elements (water to air; air to fire), perhaps as the material becomes less dense (1.3.340a10; 1.8.345b34; 2.2.354b30). The διάκρισις of the air seems to be followed by an independent ignition (1.3.340b13; 341a17). For a more violent rending of the dry exhalation **ἀποσπᾶν** (3.1.371a32) and **διασπᾶν** (2.8.367a29) are used.

With **μαραίνειν** and its compounds we begin a series of words that focus on the power of the agent to dominate an exhalation: the various

48 See Furley 1989 [1983]: 138 for the substitution of the variant παχύνω in *Mete.* 4.

49 Only at 2.2.355a31 do we see an elemental change, and the process is obscure.

compounds of μαραίνειν, when they are not simply used intransitively (e.g., 1.6.343b16), may render the active capacity of an exhalation powerless, usually through excess heat. Thus thin clouds are withered (2.8.367b11); winds are withered by the heat (2.5.361b15, 21, 27). σβεννύναι is its cold counterpart (2.5.361b26–7; see below).

Many things stop (**παύειν, καταπαύειν**) doing what they are doing, in particular the wind stops blowing or earthquakes stop shaking (2.7.365b17; 2.8.367b33; 368a7). But the sun can stop the wind from forming (associated with μάρανσις 2.5.361b22), and one wind can stop another (2.6.364b4, 8).

Similarly a force may prevent (**κωλύειν**) something. The sun, the dominant force, by its power can prevent condensation (2.3.340a29; cf. 340b32, 33) or winds (2.5.361b21, 23). The sea, too, with its great weight, can prevent the egress of earthquake winds (2.8.368b34).

Mastery (**κρατεῖν**) is associated with the violent phenomena (thunder 2.9.370a9), especially earthquakes and related phenomena (2.8.366a16, 17, b13, 28; 369a1).[50] Heat (especially the sun 2.8.366a16, 17; 3.5.376b22, 23, 26, 27 [though these passages from 3.5 may be interpolations]) and cold (3.1.371a6, 7, 9) can both have mastery.

Infrequently (**ἐγ**)**κατακλείειν** is used for closing up something in the earth (never elsewhere). It is used once for the power of the sun to confine the earthquake wind inside the earth (associated with κρατεῖν 2.8.366a16), and elsewhere of metals and minerals, and is suggestive of the pressure under which they are formed (3.6.378a15, 29).

There are a host of terms describing forced motions, starting with the -κρινειν terms. It is convenient to begin with **ἀποκρίνειν**, even though it describes a *gently* departing mass (1.3.340b20; 1.8.345b34; 1.13.349b26, 29; 2.4.360b34; 2.8.366b12 acting violently after the ἀπόκρισις; 3.3.372b32; 3.4.374a35; 3.6.378a31). Like διάκρισις, it is often accompanied by elemental change (1.8.345b34; 1.13.349b26; 3.6.378a31). It may, but need not, behave as an integrated mass (as the wind in earthquakes 2.8.366b12). The overlap with **ἐκκρίνειν** is extensive, except that in addition to gentle and natural departures (2.3.357b11; 2.5.361b18; 2.6.378a12), ἔκκρισις can describe violent ones (1.4.342a1, 15 [associated with ἔκθλιψις], 18; 2.6.364b32; 2.9.369a24, 27; 3.1.370b3, 17, 32), and there is no elemental change involved. More than ἀπόκρισις, ἔκκρισις is predominantly associated with the dry exhalation (but 369b32).

θλίβειν is the major means for forcing the dry exhalation down against its nature and is common to shooting stars (1.4.342a1, 9, 15 [associated with

[50] Elsewhere: *Phys.* 7.2.243a20–b2; *Cael.* 1.2.268b26–269a2; 2.10.291b6–10 (see Hussey 1991: 221).

πυκνοῦν and ἔκκρισις]) and lightning (2.9.369b5; 3.1.371a18) (**ῥιπτεῖν** is used only of ejected shooting stars [e.g., 1.4.342a9]; **ῥήγνυναι** is confined to analogies for thunderbolts [2.9.369a34; 3.1.371b4] and earthquakes breaking the ground [2.8.367a4]). The combination of the downward θλίψις and the upward natural tendency of the dry exhalation results in the sideways (πλάγιον) motion of many shooting stars (1.4.342a24). Aristotle might have but did not extend this dynamic to the rivers and winds. Once, oddly, θλίψις is used to expel heat upward (1.3.341a5). **κατασπᾶν** describes the firewind as it is drawn down (3.1.371a15).

πίπτειν and its compounds are used primarily for the forced motion of the dry exhalation, especially the winds (except 1.7.344b33 and 1.12.348b25). Since they are intransitive verbs, their focus is on the dynamic impulse of the exhalation itself. It is not especially important where the impulse comes from: air masses strike clouds (2.9.369a36); one wind falls upon another (2.6.364b3; 365a4). ἔκθλιψις can cause this motion (2.9.369b7).

ὠθεῖν and its compounds are likewise almost entirely restricted to the action of the air and the winds, here with an emphasis on effect. The air does most of the thrusting in the context of earthquakes (2.8.366a32; 368b4); winds (2.3.358b1; 2.4.360b21; 2.6.364a9; 364b11, 29); whirlwinds (3.1.370b20, 23); and clouds (1.12.348a15, 20 [of Anaxagoras]; 2.4.361a1). Once the sea thrusts (2.8.367a16). **ἕλκειν** and its compounds are not common. The only real standout is the celestial sphere that drags the upper air around with it (1.3.341a2). **στρέφειν** describes the violent compound motion of the whirlwind and hurricane (3.1.370b32; 371a14).

ἀντιπερίστασις, a genuine Aristotelian technical term, is used primarily in the explanation of hail and violent rains (1.12.348b2, 6, 16; 349a8; also 1.10.347b6 air masses in Pontus; even of normal rains 2.4.361a1). It is similar to ἔκθλιψις in that both involve pressure, but ἀντιπερίστασις does not result directly in locomotion, only in the concentration of hot and cold. Moreover, ἀντιπερίστασις arises usually between two masses of air, while ἔκθλιψις involves two different kinds of exhalation. A second kind of ἀντιπερίστασις involves mutual exchange and separating out of wet and dry exhalations in various districts on the earth (2.4.360b25).

σείειν, **ῥέεῖν**, and **πνέεῖν** describe the proper motions of earthquakes, rivers, and winds respectively. ῥεῖν also describes earthquake winds (2.8.365b28; 366a4, b14; 368a3), and once whirlwinds (3.1.370b21), but not normal winds (except 361a33), which blow (πνεῖν) rather than flow.

The forces of heating and cooling are relatively straightforward. **καίειν** and **πυροῦν** are close synonyms (ἐκκαίειν means to ignite; καίειν to burn; κατακαίειν to burn thoroughly). Of both roots, the active voice is causative ("to cause to burn"). Burning comes into being from non-burning and is everywhere the result of locomotion (e.g., 1.4.341b20, except where the ἀρχή of fire falls into the ὑπέκκαυμα 1.7.344a18). Boiling (**ζεῖν**) is the extreme heating of the dry exhalation into flame (1.4.341b22).

Both dry and wet exhalations are cooled (**ψύχειν, καταψύχειν**). Cooling suppresses the formation of the dry exhalation (2.4.361a2; 2.8.368b34), but it does not have a further effect on it once it is formed. Cooling serves either to condense the wet exhalation into water (1.10.347a15; 1.11.347b18) or to contract it, resulting in expulsion (1.4.342a19). Water can be cooled as well (1.12.348b32, 33); the earth can cool but cannot be cooled (1.13.350a13).

σβεννύναι and compounds never straightforwardly refer to the extinction of fire in water. Extinction occurs because of lack of fuel (1.7.344a18), but more frequently it refers to the cooling of the dry exhalation (1.9.346b28; 1.10.347b4, 10; 2.5.361b25; cf. 2.3.359b6).

The physics of rising, falling and other motions are most conveniently considered in the context of each exhalation.

The mechanics of the dry exhalation

The dry exhalation rises to the top (ἐπιπολάζειν) in virtue of its inherent heat (1.4.341b11; cf. 1.2.339a17; otherwise ἐπιπολάζειν is a surprisingly underused word). In contrast to the very frequently discussed motion of the wet exhalation, the production and ascent of the dry exhalation are largely ignored in the context of the kapnospheric phenomena. Where rising is mentioned, it is usually associated explicitly not with the dry exhalation *per se*, but rather with the "hot," which is said to be carried up (1.3.341a7 συναναφέρεται; 342a14–16). In only one passage does the dry exhalation rise with the intransitive verb ἀναβαίνειν to the area where a comet is (1.7.344a20). Unlike the wet exhalation, the dry exhalation cannot lose its heat, or at least its cooling does not lead to any further consequences.[51]

As we saw above, Aristotle regularly uses the verb συνίστασθαι to indicate that the dry exhalation is gathered together with the appropriate density (especially comets, 1.7.344a36, b11; 345a8). He associates this σύστασις with densification (πύκνωσις 1.7.344a16, 30). Conversely, a gathering of dry exhalation can be subject to dissolution (διάλυσις).

[51] Though Aristotle says the dry exhalation is cooled (2.4.361a2–3; 2.8.368b33–5), he seems to mean that the ground is cooled and is therefore unable to produce dry exhalation.

Ignition occurs in two ways that ultimately reveal the weakness of Aristotle's account. In the first, the dry exhalation is described as a fuel (ὑπέκκαυμα), which is generated from the earth and rises to the kapnosphere. There it is ignited (1.5.342b2) by a movement, however slight, of the heavens (1.4.341b20–1, 35–6) into a genuine flame. Ignition takes place where the σύστασις is well-mixed or opportune (εὐκαίρως 1.4.341b22; cf. εὔκρατος 1.7.344a14, 20; εὐφυίαν 1.7.344a28 the latter three passages describing lower comets). This explains the isolated appearance of the shooting stars, torches, and suchlike, and the lower comets, since the fire burns through the limited fuel. In order to supply the stationary comets with fuel, the dry exhalation seems to rise in ribbons from below (1.7.344a20), and this in turn accounts for why they last longer than shooting stars. The ignition of the comets involves a fiery principle being thrown in (ἐμπέσῃ 1.7.344a16–17; ἐμβάλλοι 1.7.344a27), or as it were, thrust (ὤσειε 1.7.344a26). It is not clear how this occurs.

A second process accounts for the upper comets and the Milky Way. From Aristotle's description of it, we would never know that ignition of a fuel was involved. It might best be described as a boiling, an extreme heat (1.4.341b21–2). The process is closely akin to the way in which the sun produces heat in general, and for that reason it is appropriate to the heaven-synchronous nature of these lofty phenomena. The sun, of course, generates heat indirectly by dissolving (διακρινόμενον 1.3.340b13) the potentially fiery material below it, which then becomes enflamed (ἐκπυροῦται). Similarly in the formation of comets, the motion (τῇ κινήσει) of the heavenly spheres, both planets and stars, dissolves (διακρινομένου 1.8.345b34) the air and causes a σύστασις (cf. 1.7.344a36) or a σύγκρισις (1.7.344b9; 1.8.346a4, 16, 23) to separate off (ἀποκρίνεσθαι 1.8.345b34). This process accords with Aristotle's obscure claim (1.7.344a36–b1) that the exhalation is gathered by the motion (κινήσεως) of one of the celestial bodies in order to form upper comets. Thus the celestial body is responsible for dissolving the air into a fiery substance. Again at the end of 1.7 comets are absent in the zodiac, because though the sun and the planets separate off (ἐκκρίνουσα) the exhalation, they also disperse it as it is gathering there (διακρίνουσα τὸ συνιστάμενον 1.7.345a7–9).[52] Clearly the motion of the heavenly bodies causes a separation (ἔκκρισις) of dry exhalation. I shall return to the

[52] Again at 1.8.346a6–10, where the movement of the stars in the great circle contributes to the dissolution of the air (ἐκριπίζειν), and 1.8.346a26–7, where the φορά of the stars causes the light.

problems these two views generate in the context of the kapnospheric phenomena.

The mechanics of the wet exhalation

The heat of the sun, of course, causes water to vaporize (ἀτμιδούμενον 1.9.346b25) and travel upward (φέρεται ἄνω 1.9.346b25–6). The heat transferred to the wet exhalation can then depart and cause the vapor to condense and fall. For this reason its rise is associated with transitive verbs. Especially frequent is the compound ἀνάγειν, which describes the lifting of moisture (and by metonymy, water) by the sun, by heat, by fire (1.10.347a27). There is only one unambiguous example in which the wet exhalation is carried up by the dry exhalation and this is in the introduction to the section on precipitation (ἄλλα δ' ἐν μέρει συναναφέρεται τῷ ἀναθυμιωμένῳ πυρί 1.3.341a6–7; cf. 2.4.360b34), and though the "hot," when it leaves, ascends to the upper region, this hot does not have a specified material form. Indeed there is no reason why the wet exhalation should need the dry to raise it. ἀνάγειν is not used in *Physics*, *de Caelo*, or *de Generatione et Corruptione* to describe this motion, obviously because it implies an unnatural motion (the moisture is compared to a burden, φορτίον 1.10.347a31): it is the power of the heat (τὴν τοῦ θερμοῦ δύναμιν 1.9.347a8–9) that raises it. On only one occasion do the wet vapors *rise* (with ἄνειμι 1.13.350a13); and the intransitive ἀναφέρεται is used only once (2.2.355b19). But the rain can fall with intransitive verbs, since it is now unsupported by the heat (καταβαίνειν 1.13.349b32; 2.2.355a26; 356b26; 2.4.361a15). καταφέρεται is used for precipitation repeatedly (1.10.347a15; 1.12.348a11; 2.2.354b31; 2.3.358b3). It is evident then that Aristotle prefers to make the upward motion a forced motion of a heavy vapor rather than a natural motion of a light air, and this is yet one more way in which he has modified the nature of the four elements.

CHAPTER 4

The biological method

Up to this point I have been concerned to describe the relationship of the *Meteorologica* to the more general physics of the *Physics*, *de Caelo*, and *de Generatione et Corruptione*. I have argued that Aristotle is careful to frame the *meteora* as a semi-autonomous subject matter by providing it with principles that are both appropriate to it and related, though not reducible, to the principles of the higher disciplines. I now want to turn to the lower sciences, those next in the order that Aristotle describes in the first chapter of our treatise, the study of living things, and begin to examine the relationship between meteorology and the living world.

One of the most remarkable developments of the last thirty years in scholarship on Aristotle has been the successful reintegration of his biological works into the mainstream of his corpus. The new studies of Aristotle's methods in these works have enriched our understanding of his system of classification and scientific organization, and his conceptions of essence, explanation, and cause. My aim in this chapter is to do something similar for the *Meteorologica* by arguing for two related claims:

1. In basic organization the *Meteorologica* is similar to the biological works, especially in the use of those techniques brought to light by Gotthelf, Lennox, and others.[1] This fact is important, because most scholars have supposed that those techniques, most importantly the "greatest group" system of multiple differentiae theorized in the first book of the *Parts of Animals*, are specific to the biology.[2] I shall argue that many of these techniques are present in the *Meteorologica*, and

[1] Gotthelf 1987a; 1997; Lennox 1987a; 1987b; Detel 1997.

[2] Kullmann 1997, according to whom these chapters are a specification of the rules we find in the *Posterior Analytics*: "It seems clear to me that the function of *PA* I is to give a propaedeutic to biology in fulfillment of the methodology of the *APo*, so as to secure the scientific character of the subject" (51; cf. 58); cf. Charles 2000: 312: "It is no exaggeration to say that the study of biological kinds precipitated a crisis in Aristotle's thinking about definition." Lennox 2005: 62 (and again 2010) provides a variant on Kullmann's position, namely that *PA* 1 claims to be providing methodological prescriptions for biological science "because the standards that apply to animals will include standards

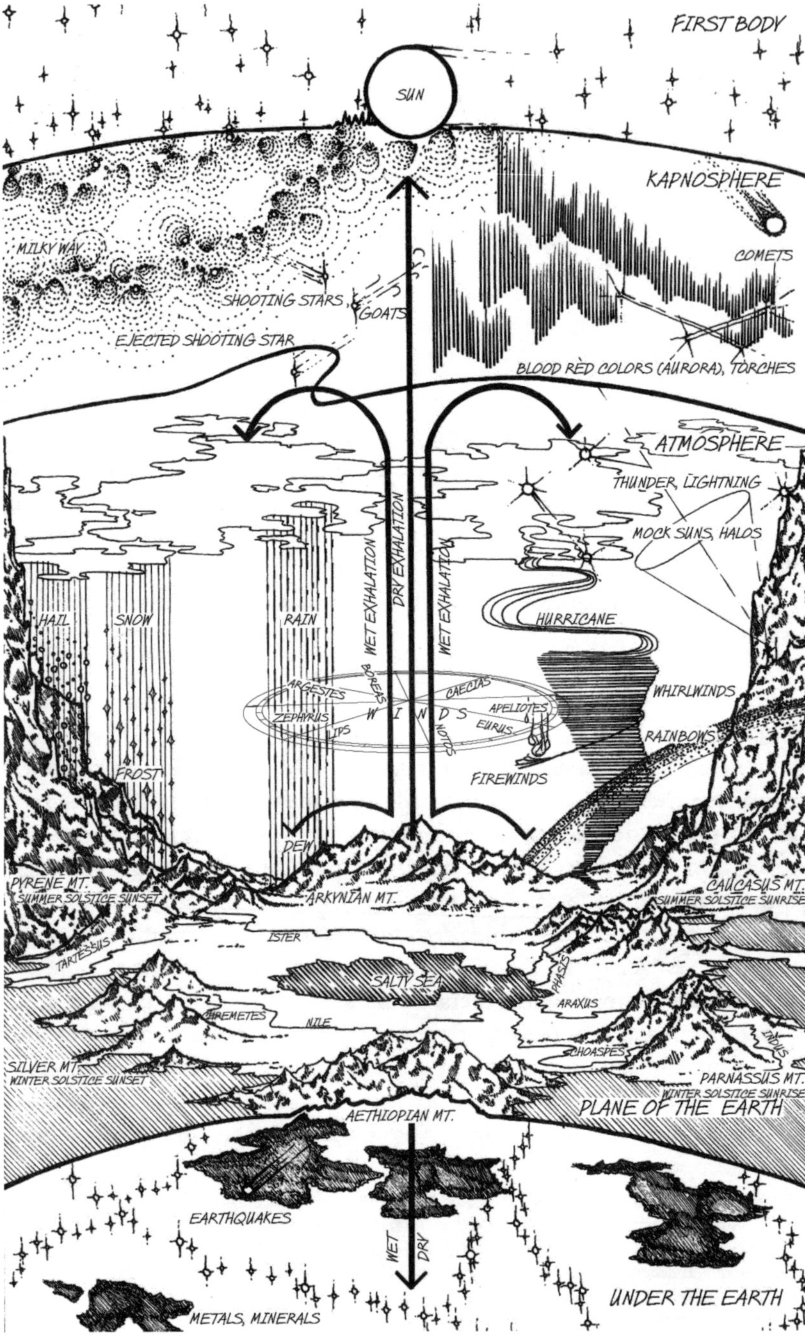

Figure 4.1 Aristotle's *meteora*

that Aristotle thought them appropriate to sublunary nature in general and not just to living things, as has universally been supposed.

2. Aristotle adapts the system of multiple differentiae familiar in the biological works in order to increase the unity of the *Meteorologica*. Here he treats his differentiae as causal, and though in a sense they are primitive, they are also joined to one another by *per se* relations. From the interactions among the causal differentiae there arise further techniques of unification – analogies, correspondences, and hierarchies – which bestow on the *Meteorologica* its remarkable coherence.

The significance of these claims lies in the way they draw the *meteora* into the orbit of the living world. In this chapter I shall argue that the *Meteorologica* bears striking methodological similarities to the biological works; in the next chapter I shall suggest that this similarity of method was appropriate, inasmuch as the underlying subjects are similar in nature. Aristotle conceives of the world of the *meteora* as in some sense lifelike.

The system of the biology and the *Meteorologica*

In the first book of the *Parts of Animals* Aristotle lays out his objections to the system of dichotomous division.[3] According to this system a genus is divided by a pair of contrary differentiae, each of which in turn is divided by contrary differentiae that are *per se* related to the first differentiae. To adapt a favorite Aristotelian example, footed animals are divided into two-footed and four-footed animals, and four-footed animals are divided in turn into cloven-footed, four-footed animals and non-cloven-footed, four-footed animals.[4] In a perfect dichotomous division, each species is defined solely by its final differentia, since this differentia implies all the previous differentiae. Aristotle objects that such a division may not in fact be possible, and that even if it is possible, the essence of a thing will have to be summed up by one single differentia, to the great impoverishment of our understanding of the object under consideration (1.2.642b5–9; 1.3.643b9–18; 644a6–8).[5]

In response, Aristotle proposes the system of multiple differentiae. Within some large group of related objects (e.g., animals) we should accept

that would be missed completely if one were to focus solely on principles that apply to natural change in general."

[3] *PA* 1.2–4. For discussion see Kullmann 1974: 53–79; Lloyd 1961.

[4] Cf. *PA* 1.3.644a4–6; 1.2.642b5–9.

[5] For recent discussions see Falcon 1997; Deslauriers 1991.

the subgroups or greatest genera that everyday language has identified (e.g., birds, fish), and define these by a number of mutually unsubordinated differentiae (e.g., having two legs, having scales; 1.3.643b9–13). Each of these generic differentiae taken singly may be of different extension both from that of the great genus and from each other; but taken together they are coextensive with that genus (1.3.643b17–26). So, for example, the distinction between viviparous and oviparous animals is not extensionally equivalent to the distinction between two-footed and four-footed animals. But all and only birds are both two-footed and oviparous, and all and only humans are two-footed and viviparous (*GA* 2.1.732b15–31). Once the generic differentiae have been determined, differences of degree may be used to define the several species within each genus. So, for example, some species of birds have long wings, others short wings (*PA* 1.4.644a19–21). According to Aristotle's programmatic passages, the differences between great genera are not characterized by differences of degree, and such similarities as exist are those of analogy (e.g., the fin is analogous to the wing; *HA* 1.1.486a14–487a1; *PA* 1.4.644a21–3; 644b7–15).

Now the *Meteorologica* considers a wide variety of phenomena, including (1) combustions, such as shooting stars, comets and the Milky Way; (2) the phenomena of condensation, namely, the rain cycle; emanations from sources, that is, (3) winds, and (4) rivers, and their related effects, (5) salinity of the seas, and (6) earthquakes; (7) violent ejections, such as lightning and hurricanes; (8) reflections, such as rainbows and mock suns; and the subterranean deposits, (9) minerals, and (10) metals. Aristotle organizes these into ten more or less distinct groups, and the chapters of the three books of the treatise are divided roughly according to these groups. The ninth and tenth groups, the minerals and metals, are barely discussed but are important for the overall structure of the treatise. These, I claim, function like the greatest groups of the biological works, and I shall consider this proposition sufficiently proved, if we find that

(a) there are several groups each with more than one member;
(b) these groups are defined by several mutually unsubordinated generic differentiae that cannot form a dichotomous division;
(c) the differences among the members of each group are differences of degree, where the differences between groups are not.

First, it is obvious that Aristotle sorts the phenomena into several groups, only the minerals and metals being so briefly treated as to be barely recognizable as such.[6] It is also obvious that there are several members within each group, though some groups are admittedly more clearly demarcated than others.[7] While the upper combustions form a distinct group, it is difficult to know whether to treat the winds and rivers separately or together, and where to place the phenomena of the sea. But for the most part Aristotle clearly signals the introduction of each distinct group.

Second, the distinctions between these groups are made on the basis of mutually unsubordinated differentiae. Among the biological works the *Historia Animalium* begins with a ὅτι inquiry according to the classification of *Posterior Analytics* 2.1–2, that is, an examination of the obvious facts, the outward appearance of the animals and their phenomenal parts.[8] Precisely because the animals are described in virtue of their many parts, it is impossible for them to be mutually subordinated in the manner of dichotomous division: one part cannot serve as a differentia for another. The meteorological phenomena, on the other hand, are divided by three causal factors: material and efficient cause, and place.[9] These causes constitute the essential features of each phenomenon. Thus at the beginning

6 1.4.341b1–5 announces the subject of flames, shooting stars, torches, and goats. And though the aurora (1.5), comets (1.6), and the Milky Way (1.8) are introduced individually, all the combustions are grouped together in the concluding formula at 1.8.346b10–15, which ends by commenting on their commonality of place (τοσαῦτα τὰ πάθη τὰ φαινόμενα περὶ τὸν τόπον τοῦτον). Of these subgroups, only the Milky Way is a unique individual. The phenomena of condensation are introduced at 1.9.346b16–20. They share a place and a set of causes and principles. The discussion is concluded with an ending formula (1.12.349a9–11) that enumerates the five members. Winds, rivers, and the sea are introduced together (1.13.349a12–14). There follows a lengthy discussion of the rivers and the sea, which is officially concluded at 2.3.359b22–6. The winds are reintroduced at 2.4.359b27–8, and wrapped up at 2.6.365a10–13, before earthquakes are introduced as a separate but related topic (2.7.365a14–15). 2.9 introduces lightning and thunder, hurricanes and firewinds (369a10–12). The discussion officially concludes at 3.1.371b14–17. Finally reflections are introduced 3.2.371b18–21. The entire discussion in the treatise so far (phenomena above the earth) is concluded at 3.6.378a12–14. The subterranean phenomena are introduced at 3.6.378a15–16 and further discussion is promised on specifics (378b5–6) but is not delivered.

7 (1) Combustions: burning flames, torches, goats, shooting stars, ejected shooting stars, aurora borealis (blood-red colors, chasms, trenches), comets (tracking and trailing), Milky Way; (2) condensations: upper (rain, snow), lower (dew, frost), hail; (3) winds (from the various points of the compass); (4) rivers (from the various mountains); (5) seas (plural, but not sharply differentiated); (6) earthquakes and their congeners, the tidal waves; (7) violent atmospheric phenomena: thunder, lightning, hurricanes, whirlwinds, firewinds, and thunderbolts; (8) reflections: haloes, rainbows, mock suns, and rods; (9) minerals: realgar, ochre, ruddle, sulphur, cinnabar; (10) metals: iron, gold, copper.

8 *PA* 1.4.644b7–11.

9 Olympiodorus 1.18–20 notes that Aristotle's purpose is to give an account of the *pathē* in the *metarsia* of the elements that have a common matter to the extent of the material cause. Place is obviously not among the canonical causes (*Phys.* 4.1.209a18–22; but 4.5.213a1–4; *Cael.* 4.3.310a31–b19; see Lang 1998: 72–4, 278–9), but it is recognized in the biological works among the other orders of difference, specific and generic. "These parts are the same and different in the aforementioned ways, either specifically or by excess or by analogy or differing in position" (*HA* 1.2.488b30–2; cf. *HA* 1.2.486b22–487a1).

of many sections throughout the treatise Aristotle asks, "what is x and what are its causes?"[10] That is, the *Meteorologica* is a τί ἐστι inquiry, and the answer to the question is the definition of the phenomenon, which is given explicitly and invariably in causal terms.[11] The causes, then, are the defining factors, and it is among these that we must seek the differentiae. And this very fact, namely, that the divisionary factors are different kinds of cause, makes it likewise impossible for them to be mutually subordinated in the manner of dichotomous division: material causes, efficient causes, and place cannot serve as differentiae to one another.[12]

Let us take up each of these three causal differentiae in turn to show how they all contribute to the nature of the phenomena, and how none is redundant or reducible to another. The *material* distinction between dry and wet exhalation is an obvious dichotomous distinction and would seem to be a good candidate for the first differentia in a dichotomous series. But if it served this function, we would expect the treatise to be divided into two halves devoted to each exhalation in turn. Instead we find the exhalations taken up as a pair in each of three successive spaces: above the earth (combustion/condensation), on the earth's surface (winds/rivers), and under the earth (minerals/metals).[13] Moreover, Aristotle makes very little effort to distinguish among species of dry or wet exhalations, even when it would be useful for him to do so. We could imagine him claiming, for example, that the shooting stars require a hotter and smokier exhalation than the winds.[14] But instead he distinguishes between them solely in virtue of their place and efficient cause.

As for the *efficient* cause, Aristotle identifies one generic mover, the circular movement of the heavens, which because of its fundamental unity

10 1.4.341b1–3 (flames); 1.5.342a36–b1 (aurora); 1.8.345a11–12 (Milky Way); 2.1.353a32–3 (sea); 2.5.363a18–20 (winds); 2.8.369a7–9 (earthquakes); 3.2.371b18–21 (haloes and rainbows).

11 *Mete.* 1.2 discusses the material and the efficient causes explicitly in those terms.

12 For Aristotle's awareness of the basic issue, see *Met.* 3.2.996a18–20: "does it belong to one or more science to study all the kinds of cause? For could it belong to one science to recognize the principles, if they are not opposites?"

13 Strohm 1936: 1 identifies three zones as well, those mentioned in 1.1: the area neighboring the motion of the heavens, the area common to water and air, and the area of the earth. This division is appropriate as a first enumeration of the commonly considered kinds but ultimately does not stand up to the critical treatment in the bulk of the treatise.

14 At 2.4.360a10–20 Aristotle says that the exhalations can be mixed as they rise from the earth, and that mixtures with a preponderance of vapor are the origin of rain, while those that are mainly dry form the winds. But he does not say that there is a difference between the kind of dry exhalation that produces combustions and those that produce winds. Again, at 1.4.341b15–18 he calls the dry exhalation "fire," and says "for we have no common name to cover all the smoky exhalation (ἐπὶ πάσης τῆς καπνώδους διακρίσεως), but because it is the most susceptible of all bodies to burn, we must make use of this name." However, Aristotle seems here to have in mind the distinction between fire and smoke, not that between different kinds of smoky exhalations.

seems unable to serve as a differentia at all. This single generic cause, however, operates in different ways in each kind, and usually in multiple ways even within a single species. It is first and foremost responsible for the production of the exhalations (1.4.341b6–10). But it operates again to produce further changes.[15] The friction of the movement of the heavens against the kapnosphere, for example, causes the combustion of the dry, smoky exhalation (1.4.341b18–22); and the withdrawal of the warming power of the sun causes condensation of vapor into water (1.9.346b20–31). For this reason it might be tempting to view the efficient causes as the first differentiae and the great groups themselves as the species: the phenomena of combustion, the phenomena of condensation, and so forth. But this cannot be so. First, the flow of the winds and rivers and their associated effects, and likewise the ejections that produce violent phenomena such as thunderbolts, can hardly be considered as species of the movement of the heavens. Thunderbolts are moved by compression in a cloud, and not by the heavens (at least not immediately, 2.9.369a12–29). Second, even the specific sources of motion do not completely divide the greatest groups: condensation, for example, is the common efficient cause of the rain cycle, of the rivers (1.13.349b27–35), and of metals (3.6.378a25–b3). Finally, the efficient cause does not adequately capture what the members of each greatest group share in common. The phenomena of combustion, while they certainly are all heated by friction from the heavens, are still irreducibly dependent on the dry exhalation for their material. Similarly too, the rain cycle requires the wet exhalation, which is an essential part of its nature.

Place, likewise, is an irreducible and essential feature of the meteorological phenomena, for all that its fundamental principles arise out of the material and the efficient causes. Admittedly the distinction in the upper places can be explained by the differences in material. Just under the circular movement of the heavens (1.4.341b13–14) and extending down to the mountain peaks (1.3.340b32–6) we find the fiery region, to which the hot, dry exhalation naturally ascends. Below this, extending to the surface of the earth, are the less buoyant phenomena common to air and water, the cycle of rain, snow, hail, dew, and frost (1.9.346b16–19). To this extent material explains place. But in the middle zone (though occupying some of the upper space), Aristotle sets another pair of phenomena dependent on wet and dry exhalations, the rivers and the winds and their related

[15] The upward movement and the condensation of the air as it contracts (to produce ejections) are explicitly called moving causes (1.4.342a27–30).

effects. Again, in a brief coda at the end of 3.6 Aristotle introduces the metals and minerals, effects again of the wet and the dry exhalations that remain entirely under the earth. Aristotle makes no effort to account for the difference between the middle and subterranean phenomena on the basis of their material or efficient cause (which they share), and they can be distinguished only in virtue of the peculiarities of their places. Place, then, is an independent defining factor. But, again, neither it nor its differences can serve as the differentiae sufficient to distinguish between all ten greatest groups, since the violent phenomena and the reflections occur in the same place as the rain cycle.

It is clear, then, that all three causal factors are necessary, and none is independently sufficient to distinguish the basic groups of phenomena Aristotle considers in the treatise. At the same time none is subordinate to another in the manner of a differentia to a genus. It follows, then, that we have a set of mutually unsubordinated and non-redundant differentiae.

It remains now only to show that among the members of each group there are distinctions of degree, of more and less:[16] differences of shape and quantity of material, short and long duration of phenomena, warm and cold, fine and coarse texture.[17] By and large they are not responsible for dividing between the greatest groups, but, as in the biological works, they serve to distinguish the species of phenomena within the greatest groups. So we find both the combustions and the condensations subdivided again, according to quantities of exhalation and longer term (upper) and shorter term (lower) phenomena.[18] Similarly, the difference between thunder and hurricanes (both ejections) is determined by the quantity of material involved (3.1.370b5–10). The generic differentiae, by contrast, are largely constructed to avoid differences of degree. Within this treatise at least, the distinction between wet and dry exhalation is absolute, and wet exhalation cannot be turned by degrees into dry. Likewise the various manifestations of the efficient cause (combustion, condensation, ejection, etc.) are not related to one another by differences of degree. Only in the case of place might we say that one place is *higher* than another, but even so each place is associated with a peculiar orientation. For the most part, differences of degree give rise to distinctions among the members of each greatest group of meteorological phenomena and do not create distinctions between the groups.

[16] 1.4.341b5: διαφέρει δὲ τῷ μᾶλλον καὶ ἧττον. [17] For fine texture: 3.1.371a15–16.

[18] Combustions differing by quantity (1.4.341b24–5); differing by altitude (1.7.344a34–5); implicitly by duration (shooting stars, comets, Milky Way 1.8.346a19–23); condensations differing by quantity (1.9.347a11–12); by time, altitude, and cold/hot (1.10.347a13–28).

It is clear, then, that the greatest group method, usually considered specific to the biological works, is a basic organizing principle in the *Meteorologica*. The reason why Aristotle used the same method is obvious: the entire sublunary world, inanimate as well as animate, manifests the same basic division of great genus, species, and individual.

Causal interactions and their hierarchical effects

So far, I have outlined the ten greatest groups into which the phenomena fall and the generic differentiae (material cause, efficient cause, and place), which, only when taken together, are adequate to distinguish them. This fact alone is interesting enough and sufficient to make us rethink the primacy of the biology in the new method outlined in the first book of the *Parts of Animals*. But now having argued that the causal differentiae are mutually irreducible, I want to show how they are dependent upon one another. In particular, I want to stress that, whereas the multiple differentiae system tends to isolate each genus from the others, the causal differentiae that underlie the genera have the opposite effect of organizing them into a single hierarchical unity. And as we shall see with increasing clarity, this alternation between the plurality of genera and a hierarchical unity characterizes the pivotal position the *Meteorologica* holds within Aristotle's natural philosophy.

Though in *Parts of Animals* 1 Aristotle explicitly introduces the system of multiple differentiae to overcome the weakness of dichotomous division, he does not discuss what is lost on account of his innovation. The great strength of the older, strict form of dichotomous division is that it prevents the kinds of slipshod divisions we find in Plato's *Sophist*, divisions in which the successive differentiae bear no essential relationship to their predecessors; that is, dichotomous division prevents the *definiendum* from falling into a heap of unrelated features having no conceptual unity. So, in the *Sophist* (220b1) swimming animals are divided into winged and water animals, in spite of the fact that swimming bears no essential relationship to having wings.[19] In dichotomous division, however, the logic of the *per se* relationship solves the problem by making clear why, for example, a cloven-footed animal must have feet.

[19] So as Aristotle complains (*PA* 1.3.643b19–23): "this is the problem that faces those who divide [animals] into wingless and winged, and winged into tame and wild or white and black; for tame and white are not differentiae of winged, but rather are the beginning of a new differentia, and are incidental in this context."

Now, the new system of *multiple* differentiae simply reintroduces the problem of slipshod division: there is no reason why a two-footed animal must be oviparous; indeed some – for example, humans – are not.[20] There is no direct *per se* relationship between having so many feet and having some specific mode of reproduction. As a result Aristotle loses the obvious (albeit sterile) unity of the dichotomous system.

In his new system of multiple differentiae, Aristotle maintains the requirement that within each generic characteristic the specific differentiae must be *per se* related both to the generic characteristic and to each other. So for example, the differentiae, "nocturnal" and "diurnal" (*HA* 1.1.488a25–6) have both a necessary relationship to "times of being active" and a necessary relationship to each other, inasmuch as they are opposites. Another generic characteristic, say "having a beak," will have other differentiae, likewise *per se* related both to that generic characteristic and to each other, for example, "having a curved beak" and "having a straight beak." Only rarely does Aristotle deviate from this practice. For example, though the differentiae of diet, "meat-eating", "grass-eating", "honey-eating", "fly-eating" (488a14–16), are necessarily forms of diet, there is no direct *per se* relation between them, that is, they are not defined in terms of one another, as day and night or curve and straight are. Aristotle generally avoids this kind of differentia.

The most usual differentiae are not merely *per se* related – they often form series, in which one differentia is prior to another. Obvious and common cases are those of form and privation: some animals have a dwelling, others do not (488a20–1); other cases involve inclusive series (mute, having voice, having articulate voice, 488a31–2). Aristotle is careful to structure his differentiae in this way in the *Meteorologica* as well. Each of the definitory factors is divided in such a way that its species form a series (e.g., place is divided into higher, middle, and lower) insuring the internal unity of each set of differentiae. So far, though, none of this compensates for the loss of unity produced by the free-floating system of multiple differentiae, since none of the *generic* differentiae are related *per se* to each other.

These problems have been frequently noted. What has been universally ignored, however, is the untheorized solution Aristotle uses in the *Meteorologica*. He shows how the generic differentiae are related to one another *causally* in addition to their more familiar role in marking difference. The role of cause in unifying the generic differences marks a clear advance in the multiple differentiae system and provides an appropriately expanded role to demonstration in the unity of the science.

[20] *GA* 2.1.732b15–30; so Charles 2000: 334 on mutual irreducibility of the generic differentiae.

Furthermore, not only are the differentiae joined together by causal relations; the series among the *differentiae* also contribute to a number of hierarchies among the *greatest groups*. So, as we shall see in greater detail, serial differences in place cause hierarchical differences in the orientation of the several greatest groups. These hierarchies among the greatest groups are invariably the result of causes working severally in each group, and never is a hierarchy itself used to explain features of a specific greatest group.

I want to take up these issues in order, first showing in more detail the nature of the causal interaction between the differentiae at a generic level, then examining the causal interactions among the species of material, place, and efficient cause. I shall then show how these species form series, and I shall argue that they cause hierarchies among the phenomena.

First, one generic characteristic can serve as the partial cause of another generic characteristic. The phenomena of the treatise occur between the source of the efficient cause (the circular movement of the heavens) and the source of the material cause (the earth and its water) that produces the exhalations. The efficient and material principles are, therefore, the causes of the extent of place in the treatise. Place is explained and therefore defined by the combination of the efficient and material causes.

Again, the exhalations, dry and wet, are principles of the science, but they are not the *first* principles. The first and only independent principle and cause is the efficient cause, the movement of the heavens (1.2.339a23–4). The exhalations are generated by the heating effect of the movement of the heavens upon the earth and upon the moisture in and on the earth (1.4.341b6–10). We have already seen how Aristotle might have chosen as material cause something independent of the efficient cause: earth, water, air, and fire are simply preexisting and independent of the movement of the heavens. But instead, he chose a special material, which has to be worked up by a specific warming cause, and which is distinct from the preexisting materials. The effect of this choice is that the exhalations are already special to the science, and their existence is not a conclusion of any of the higher sciences. Obviously, too, the material cause is not causally independent but is already a caused cause even within its own science. There are, then, not two independent first principles, matter and heavenly mover, but only one, the heavenly mover. In this way, also, the plural material causes are defined in terms of the unitary efficient cause without being species of it. At a general level, then, the efficient cause is the cause of the material cause, and the two together are the causes of place.

Turning from the general causes to the specific, we see again both how the species of each cause are caused by the other causes and how they are serialized by the other causes. Aristotle's two material causes (1.4.341b6–10), the wet and the dry, are of course, *per se* related to each other insofar as they are contraries.[21] But by calling on the other causes, Aristotle makes the dry prior to the wet in a variety of ways. First, the dry exhalation is superior, since its natural place is higher and closer to the heavens.[22] Again, the dry exhalation is superior in virtue of the efficient cause, both because the dry exhalation can carry the wet exhalation upward (1.9.346b24–31; cf. 2.9.369a13–24), and because the dry exhalation can burn and in turn heat the earth and generate both the dry and the wet exhalations (1.9.346b23–6).[23] Arguably, too, it is prior because, in the case of the Milky Way and upper comets, it is produced directly by the friction of the heavenly sphere on the upper air. Finally, Aristotle says that the wet exhalation comes from the moisture *on and in the earth* (1.4.341b9–10). Why should he mention the earth here and not merely the surface of bodies of water? The earth is essential as a source of wet exhalation, because otherwise it would be impossible to account for the origin of the rivers and the metals in the earth. The effect of this is to join causally the two material causes in their place of origin and to establish the logical priority of the dry over the wet exhalation: the wet exhalation contains in its definition the fundamental source of the dry exhalation, the earth. In these ways the serial order of the material exhalations and of the phenomena that arise from them is caused by efficient and local considerations.

The various species of place likewise form an obvious series from top to bottom beginning in the noblest location next to the heavenly movement, and ending below the earth. In view of the doctrine of the *de Caelo* and the *de Generatione et Corruptione* we should expect a simple correspondence of exhalation with its natural place, the hot, dry exhalation occupying the top half of the sublunary cosmos, and the hot, wet exhalation occupying the lower half. As such, its order would be determined merely by the matter. Instead, each of *three* places is correlated with the ability of the exhalations to attain their natural places: some of the exhalations are forced to remain in the earth, becoming minerals and metals, others emerge part way, becoming winds and rivers; others again fully escape to their natural place to become the combustions and condensations. The fact that each of the three places contains both exhalations

[21] *APo.* 1.4.73a37–b3.

[22] So Olympiodorus 36.16–18: since the dry exhalation imitates the heavens and its circular movement, it is better and should be treated first.

[23] Gilbert 1907: 483 points out that the heat of the kapnosphere resupplies the heat required for the generation of the dry exhalation, thus creating a cycle.

shows that differences in place are not determined solely by the material cause.

Moreover, although Aristotle nowhere theorizes the fact, the exhalations clearly exhibit different behavior according to their places, most strikingly in the orientation of the phenomena. These different orientations are caused only partly by the material cause; in part too they are determined by the efficient cause, and in part they are primitive and without prior cause. In the upper place both of the materials tend to ascend vertically. The phenomena of combustion and condensation take up their respective higher and lower natural places dictated by their material natures and are clearly separated from one another at the peaks of the mountains, at which point the rotary motion of the heavens translated to the upper dry exhalation peters out (1.3.340b32–6). The dry exhalation rises and does not fall again unless forced to do so.[24] But the wet cools again and descends, forming a cycle similar to the cyclical river of Ocean (1.9.347a2–8). Aristotle provides a material explanation for this cycle by calling these phenomena the common affections of air and water, indicating that they can only be understood in that combination. Their recurrent movement is also necessitated by their efficient cause, the recurrent movement of the sun in the course of the seasons (1.9.346b20–3).[25]

As we descend in the series of place, we come next to the second pair, the winds and the rivers, which are horizontally and radially, rather than vertically, oriented. Rather than rising and falling, they seem to flow. Aristotle feels no need to point out that the lateral flow of rivers is caused by their relatively heavy material nature. The lighter, dry exhalation that forms the winds is above the rivers again for obvious material reasons. What is not obvious is why the winds flow in a way similar to rivers. But all the same, Aristotle works very hard to bring out the spatial correspondences between them:[26] both winds and rivers arise from sources (2.4.360a27–33); there are

[24] Aristotle does not comment on this remarkable fact, though in *Sens.* 5 he recognizes that γῆς τι εἶδος arises from the smoky exhalation (443a29–30). At *Mete.* 4.9 the fumes exhaled into the air either dry up and vanish (ἀφανιζόμενα ξηρά) or turn into earth (387a27–8).

[25] And if there were any doubts about Aristotle's intention in characterizing this upper pair as vertically oriented, they are dispelled when we see that the subdifferentiae of the places are themselves vertically oriented. Aristotle divides the kapnosphere into upper, middle, and lower bands. The upper band, which most of all is dragged by the sphere of heaven, and which therefore tracks most closely the motion of the stars and partakes most of all in the heavenly nature, is devoted to the upper comets and the Milky Way (1.7.344a33–b1; 1.8.345b35–346a6). The middle band, which is discussed first, is populated by the ordinary torches, goats, and shooting stars (1.4.341b35–6). Finally there is the lower band, which because it is partly mixed with the cooler wet air may give rise to ejected shooting stars (1.4.341b36–342a3). Similarly too, the atmosphere is divided into an upper region, the place of rain and snow, and the lower region, the place of dew and frost (1.10.347a13–16).

[26] As Cherniss 1935: 128–30 has noted.

different winds just as there are different rivers (1.13.349a20–6); neither is merely matter in motion (1.13.349a16–20; 2.4.360a29–31); they both gather strength as they go (1.13.349b30–5; 2.4.361b1–5); when blowing, winds are situated above rivers; but they are opposite in their sources, rivers arising from the high mountains (1.13.350a2–4), winds from the low-lying areas (1.3.340b36–341a1). The differentiae of the winds and the rivers are again in both cases horizontal and radial, corresponding to the points of the compass and the solstices. In their vertical places and their differentiae, the rivers and the winds are not as sharply distinguished from one another as the upper pair: properly both winds and rivers belong above the surface of the earth, but they both have a subterranean component as well (1.13.350b30–351a18; 2.8).[27]

The orientation of the winds has a very weak etiology. Aristotle does provide an external efficient cause, though almost as an afterthought: winds move horizontally because of the movement of the heavens (2.4.361a22–5). But this cause is obviously at odds with the radial directionality of the winds and with the separation of the atmosphere from the kapnosphere, as Alexander points out.[28] Aristotle clearly felt that he needed to supply some efficient cause that was specific to the winds and that did not make their flowing dependent on the movement of the rivers.

The final pair, the metals and minerals, are indiscriminately described as being formed when the exhalations are shut up under the earth (3.6.378a15–21). Neither is said to be above the other, or in any way related to the other. In these ways, then, and to this extent, the distinctive places and orientations of the phenomena are caused by the other causes.

But it is also worth noting that as it ascends, the dry exhalation, at least, attains more and more its full material nature. Responsible for the formation of earthy deposits within the earth, it becomes the cool and incombustible wind when released. As it rises to the kapnosphere, it becomes apparently a warmer, smoky substance, finally turning into actual fire upon contact with the heavenly motions. Place itself has an effect on the material cause.

We have now seen how the three causes affect one another and how their differentiae form hierarchical series. We have seen how these hierarchies among the differentiae give rise to hierarchies among the phenomena themselves. These hierarchies among the meteorological phenomena are based on a couple of different principles, strongly reminiscent of what we

[27] Note that, like the combustions, the dry winds do not cycle back.
[28] Alexander 93.32–3. I propose a solution in Chapter 10.

find in the biological works. First, Aristotle is constructing the sublunary world as a cosmos similar to, and prefiguring, his great chain of terrestrial life. Just as we find posture among living things – the highest animals, man, being upright, and lower animals increasingly prostrate,[29] until with the plants, we find organisms that are upside down, their mouths in the ground – so here we find phenomena of different "postures": the dry combustible exhalation goes straight up and does not come back, at least within the treatise; the rain cycle explicitly forms a recurrent river (1.9.347a6–8); the winds and the rivers are radially oriented on the surface of the earth, and the metals and the minerals move back down into the earth. Aristotle does not explicitly mention this form of unity and hierarchy, but it answers the question we started to answer in the previous chapter: why does the dry exhalation in the kapnosphere not cycle back down? To have the dry exhalation cycle back down would destroy the hierarchy of posture that Aristotle is elaborating here. It also explains why Aristotle develops the elaborate analogies between winds and rivers – he is intent on developing a middle band of *meteora* analogous to the reptiles.

A second series is that of the degree of separation between the material causes, where clearer separation is better than promiscuity. The upper dry and wet exhalations have each their own clearly defined areas and their own distinct movements; the winds and rivers are more compressed in space (though still distinct) and are similar in movement; the metals and minerals are indiscriminate in space. Again, this hierarchy is familiar from the biology, where the higher living beings have highly distinct parts, while in the lower animals and plants the various organs are barely discernible.[30]

What is remarkable is that these hierarchies are produced (partly) by causes that govern the individual members of the series, but they do not in their turn supply cause, and never does Aristotle account for the nature of a particular phenomenon in virtue of its place in the hierarchy. The case of the wind shows this most clearly. Though it seems obvious to us that Aristotle's winds move horizontally because he wants them to have a certain order in his hierarchy, he never adduces this as a cause, citing instead the rotary motion of the heavens.[31] Aristotle wanted the hierarchy to be an effect of causes and not to be a cause itself.

29 The *locus classicus* is *PA* 4.10.686a24–687a2. For a discussion of this and other *scalae* in the biology, see Coles 1997.

30 *de An.* 2.1.412b1–4; *HA* 4.4.528b12–16; especially insects *HA* 4.7.532a5–8; 532b7–10; *PA* 4.5.678a28–9.

31 Note that the hierarchical cause, the *rotary* motion of the heavens corresponds to the radial symmetry of the winds, but not to their horizontal posture.

Having considered the causal interactions associated with matter and place, let us now turn to the primary *efficient* causes themselves and consider how they form a series, then look at the ways in which Aristotle uses these to strengthen and reinforce divisions in matter and place, and finally see how he uses them to build hierarchies.

Locomotion together with heating and cooling are the fundamental agents of change in the treatise. Through the repeated application and modification of these causes in the various materials and places, the characteristic "activities" of the nine greatest groups are explained. Locomotion is the first cause, and is logically prior to the other forms of change.[32] It has its origin in the heavens, and in consequence it is more independent than the other causes. Of locomotion there are several kinds arranged in an order of priority that is generated to some extent outside the causal network of the discipline. The locomotion of the heavens in general is the first within its kind, though it is a relatively weak cause: it ignites and rotates the kapnosphere (1.4.341b18–22) and prevents clouds from forming above the atmosphere (1.3.340b32–6). Next is the locomotion of the *sun* specifically, the more powerful cause, which generates heat by friction (1.4.341b6–10). Finally Aristotle adds to the sun's simple diurnal revolution its annual northern advance and southern retreat, which explain seasonal changes and climatic differences (1.9.346b20–3).

So far, these efficient causes, locomotions all, do not involve any other principles specific to the discipline. But hereafter efficient causes become increasingly involved with the material cause. The locomotions produce the alterations, which are next in the series and are responsible for every other change, including further locomotions. The friction of the sun warms the earth by some mysterious rays (ἀκτῖνες 1.9.346b24) and produces the exhalations. The heat causes the dry exhalation to *rise* (1.4.341b10–12), where it is ignited (i.e., extremely heated); the cooling of the risen wet exhalation causes the condensation and *descent* of moisture (1.9.346b31; cooling, therefore, is posterior to warming); wind is produced by simple dry exhalation, *moved* laterally by the heavens (2.4.361a22–5); rivers *flow* as a result of vapor condensing under mountains (1.13.349b30–5); ejections occur when a cooling exhalation meets a hot one and *moves* it violently either downward, or laterally, or in a circle (2.9.369a12–24); the reflective phenomena remain the odd men out, though even they double back to the sun as that to which the visual ray reflects. These secondary efficient

[32] *Phys.* 8.7.

causes are caused by the interaction of the primary locomotions with the other causes.

These causes give rise to hierarchies among the phenomena, the first of which moves from simplicity to complexity. The phenomena of combustion arise from a single efficient cause (the locomotion of the heaven) acting on a single material cause, as is appropriate for the upper phenomena. The phenomena of the rain cycle require two pairs of closely related causes. The double nature of the sun's advance and retreat acts upon the double nature of the material substrate, the common stuff that is air and water (1.9.347a8–11). Finally, violent ejections arise from contrary, natural motions (2.9.369a12–29). Since the dry exhalation must rise through the atmosphere to get to the kapnosphere, the atmosphere is the place where necessarily the dry and the wet exhalations coexist. In this case, the contrary motions do not have a single source, as with the sun's double role in the rain cycle; rather the contrary motions result from the nature of the contrary materials. Ejections, formed of complexes of material, have themselves complex efficient causes.

The second is a hierarchy of directness among the efficient causes. Since "anything has more power when it is nearer" (2.4.361a35–6), we find the agency of the heavens directly operative in the upper places, but only indirectly operative or even inoperative on and below the surface of the earth. It is fitting that the highest and noblest of the phenomena, the combustions, should be caused by locomotion, which is the primary form of change, and the more so the closer the phenomena are to the heavens.[33] Thus not only are the Milky Way and the comets ignited by the heavenly motion, but they are carried along by that motion as well. Moreover, the combustion of the dry exhalation in general warms the earth in the same way that the sun itself does, producing further exhalations (1.9.346b25). Thus the highest phenomena of our science are themselves efficient causes in the same way as the external heavenly cause, thus combining effect and imitation in one, and establishing yet again the priority of the dry exhalation over the wet. The phenomena of *condensation* are never their own efficient cause in this way, but they are their own material cause, since the moisture that returns to the earth becomes in turn the wet exhalation. And though their efficient cause comes from the heavens, it is not from the ontologically prior, celestial locomotion nor through simple translation.

[33] Olympiodorus 36.7–9, 16–18 on the kapnosphere's imitation of the celestial sphere and its priority over the atmosphere.

Rather the sun in the cycle of contrary seasons produces heat by friction dependent on the prior locomotion.

But below this level direct heavenly agency stops. The winds are simply exhalations moved by indirect horizontal translation. The rivers, earthquakes, and salty seas – apart from their material origin – owe even less to the heaven's agency; the metals and minerals owe nothing at all. As with the material causes and places, the series of efficient causes give rise to hierarchies among the phenomena. And like the other hierarchies we have seen, these are generated out of many individual causes, but perform no causal work of their own.

We can see then that the generic differentiae of the meteorological phenomena are dependent upon one another. Both at a generic and a specific level the essential causes of the phenomena interact, that is, they are causes of each other.[34] We see here in practice – though never theorized – Aristotle's solution to the problem of disunity in the system of multiple differentiae, a solution that makes demonstration the central unifying force. But what we have here is more than mere causal interaction. These causal interactions consistently fall into a peculiar pattern in which the specific causal differentiae, themselves in serial order, give rise to hierarchical arrangements among the meteorological phenomena.

But before drawing further conclusions from these observations, it is worthwhile to note another remarkable feature of the efficient causes: the way they take up "cameo roles" in alien matter-places. That is, an efficient cause that is usually associated with one kind of matter or one kind of place, becomes associated with another. Thus Aristotle describes a kind of shooting star in the lower kapnosphere that is caused, not by the usual combustion, but by ejection just as lightning is (1.4.341b35–342a13). Again, he provides alternative causes for the non-existence of frost on the mountains, one from within the appropriate causal framework for the rain cycle, namely, that the hot exhalations cannot rise so far without first dumping their moisture (1.10.347a29–34); and one from outside – the transfer of the heavenly locomotion causes the winds to blow more there, with the result that the frost cannot settle (347a34–5). Indeed, the direct transfer of heaven's locomotion, appropriate as a moving cause within the kapnosphere, is found not only dispelling clouds and frost from mountain peaks, but moving winds on their horizontal course across the earth's surface. Again, volcanoes spew fiery ejections from below the earth as if they were casting out lightning bolts (2.8.367a9–11); and wet exhalations condense

34 Cf. *Phys.* 2.3.195a8–11.

within the earth to form metal as if they were rain (3.6.378a28–31). The aurora (1.5) is a reflection from the dry exhalation in the kapnosphere, just as rainbows and haloes are reflections from vapor in the atmosphere (3.2–6).

These "cameos" serve to reinforce the rule that the causes must always come from within the genus of the science, and that we may not seek explanations from outside the general subject matter. But more importantly, they prevent the sublunary whole from breaking down into independent greatest groups and give Aristotle reason to claim that meteorology as a whole is a single science and not just the sum of kapnospherics, atmospherics, and so on. As such, "cameos" tend to the same end as the system of multiple differentiae in general. In dichotomous division the subgenera become more and more distantly removed from one another with each successive division, and since the subdifferentiae of each subgenus are unique, the subgroups have less and less to do with one another, as humans and anemones have little in common. There is little to hold the species together except the abstract genus. By contrast, the system of multiple differentiae permits "dualizing" between greatest groups, whereby major alignments of generic causes are crossed.[35] Again the biology provides the model: Aristotle noted that marine mammals, cetacea, share most of their features with viviparous quadrupeds, but, like fish, live in the water and do not have legs. For this reason, they are neither viviparous quadrupeds (land mammals) nor fish, but they participate in both or "dualize" (ἐπαμφοτερίζειν). In a like manner, the kapnospheric ejected shooting star is similar in place and matter to the upper combustions but shares its efficient cause (ejection) with the lower thunderbolts. These dualizers contribute to the unity of the science by insuring that there is some overlapping of causes among the greatest groups. The result is that, given a certain group of principles, we shall have to deal with all and only the phenomena of the subject matter; and if we try to eliminate even one of the principles, or one of the subdifferentiae, we shall not be able to explain even a whole part or section of the science. So, for example, if we eliminate reflection and ejection from the science, we will not be able to give a full account of kapnospherics, since we need those principles to account for the aurora and ejected shooting stars. If there were no "cameos," the science would risk breaking down into its constituent parts without forming a unified whole. Aristotle can have it both ways: by the limited use of dualizers he can preserve the basic divisions of the greatest groups, while preventing those greatest groups from breaking down into semi-independent sciences.

[35] For dualizing, ἐπαμφοτερίζειν, see *PA* 4.13.697b1 (and other passages cited in Bonitz [1961] 265a8).

But these "cameos" have one more function as well. They also allow Aristotle to fill out a matter-place-mover grid to the extent that the phenomena permit. The seventh and eighth greatest groups, the ejections and the reflections, are both affections of the atmosphere. Accordingly Aristotle sought and found in the kapnosphere corresponding ejections (the lower shooting stars just mentioned) as well as reflections. His discussion of the aurora borealis (1.5) uses many of the same terms of reflections, darkening and contrast (minus the geometry) that we see in his full-scale treatment of rainbows and the like. It is clear, I think, that Aristotle was interested in using efficient causality to preserve the symmetry of his pair of upper places.

We have seen abundant evidence that the definitory factors of the system of multiple differentiae in the *Meteorologica* are causal and that they serve as causes for each other both at a generic and specific level. This interdependence among the causes was a stroke of genius on Aristotle's part, since it so elegantly compensates for the weakness introduced when the system of multiple differentiae was brought in to replace the impoverished dichotomous division. It greatly strengthens the less unified classification system and sets etiology at the core of the scientific enterprise.

But we have also observed that the species of these causes are serially ordered, and that these causal series in turn give rise to a variety of hierarchies among the phenomena, though the hierarchies do not supply cause in their turn. In view of this fact, it is reasonable to surmise that the unification of the system of multiple differentiae was not Aristotle's ultimate goal in introducing this network of causal interaction. Though he certainly accomplished that, he also adapted this network to show how a cosmos, though a little less orderly than the heavens (1.1.338b20), could arise out of interactions among the multifarious sublunary principles. Why he should have chosen to do this and what that implies for the place of the *Meteorologica* in Aristotle's physical works is a subject we shall consider next.

CHAPTER 5

Teleology in the Meteorologica

In the first chapter of the *Meteorologica* Aristotle takes stock of his progress through the physical works and situates our treatise midway between cosmic and biological nature. Though his description here is not very clear, there are reasons to believe that it hints at the pivotal role the treatise plays in his physical *pragmateia*.[1] We have now seen how in certain details the *Meteorologica* is oriented relative to the *Physics*, *de Caelo*, and *de Generatione et Corruptione* on the one side and to the biological works on the other. In the last chapter we saw how Aristotle, using the material and efficient causes together with place, explained the manifold variety of meteorological phenomena. We also noticed how by means of the hierarchy among the differentiae of these causes Aristotle created a hierarchical order among the *meteora* themselves. In both ways he was using patterns of explanation prefiguring the biological works. But there is one respect in which Aristotle did not model his *meteora* on the living world: his use of the final cause.

I shall first argue that final cause in the *Meteorologica* is restricted to the natural tendencies of simple bodies, and that the heavens' role is only efficient, never final. It is a remarkable fact that while Aristotle adapted his general material cause to suit the new exhalations and rearranged the places of the four simple elements into three paired bands, he made no similar accommodation in the final cause. I shall explore reasons why Aristotle should have chosen to structure the sublunary world in this way and end by suggesting that the *meteora* act as bivalent "dualizers" between the cosmos as a whole and the world of manifold animals.

The final cause

The *Meteorologica* is the last of the treatises concerning cosmic nature before Aristotle introduces the soul in the *de Anima*. Up to this point in

[1] Cf. Lennox 2010: 6, who finds the chapter "more puzzling than illuminating."

the physical *pragmateia* the operation of the final cause has been reasonably straightforward. I do not mean to diminish the controversies but, however we understand them, simple bodies have natural tendencies to move in certain ways. The first body moves both in accordance with its own material nature and out of desire for the unmoved mover.[2] Earth, water, air, and fire have natural tendencies to move to their natural places.[3]

Since nature is a principle of order (*Phys.* 8.1.252a12), these tendencies give rise to three forms of order in the sublunary world. First, the four simple bodies in virtue of their own individual weights produce a permanently layered order of space. Second, they can transform one another by affecting each other's qualities, hot, cold, wet, and dry. These transformations in turn produce a reciprocal order of change based on the dominance of one element over another and affect the first spatial order. Third, this mutual transformation can also be guided by the sun in its power to produce heat and cold according to the days and the seasons. This will have its effects on the first and second forms of order. Such are the various kinds of order that simple bodies can produce in the sublunary world, and they are the subjects of the *de Caelo*, *de Generatione et Corruptione*, and *Meteorologica* respectively. The significance of the *Meteorologica* is that it extends the order of the sublunary world beyond these three forms toward a biological order solely through an interaction between the efficient and material causes, where the sun and the other heavenly motions explicitly provide the efficient cause, and the exhalations provide the material (1.2).

These forms of order and their relationships to one another are important in themselves and for their direct implications as we shall see, but they also touch on the programmatic discussion of teleology in *Physics* 2. Interpretation of those polemical chapters inevitably involves the winter rainfall example in 2.8 (198b16–199a5), an example that would seem to beg for a careful consideration of its role in the *Meteorologica*:

> The problem arises, what should prevent nature from not acting for the sake of something or because it is better, but rather, just as Zeus rains not in order that he may grow the grain, but by necessity (for that which is drawn up must be cooled; and that which is cooled, after it becomes water, must fall, and the growth of the grain happens when this occurs) . . . But

[2] Though *Phys.* 8.10 gives only an account of the efficiency of the unmoved mover, *Met.* 12.7 makes clear that the celestial bodies have desire, thought, and action; that the unmoved mover as the object of desire moves the celestial bodies. Here the unmoved mover functions as that for the sake of which (1072b1–3). *Cael.* 2.12.292a20–8 confirms that the celestial bodies have action and life, and this allows them to attain to the good to the extent that they are capable.

[3] *Cael.* 3.8.307b18–24 speaks of the simple sublunary bodies as discharging functions. So Johnson 2005: 140–5.

> it is impossible that it be this way... For it is not from chance or from accident that it seems to rain often in the winter, but rather if it occurs at the time of the Dog Star; nor is a heat wave accidental at the time of the Dog Star, but rather if it occurs during the winter. So if it is agreed that it occurs either accidentally or for the sake of something, if these things cannot occur accidentally or spontaneously, they would be for the sake of something.

The most recent contributors to the discussion of this passage make only the briefest mention of the *Meteorologica* and do not allow its evidence to speak for itself.[4] This negligence stems, I think, from a mistaken belief that the *Meteorologica* adds nothing relevantly new to our understanding of the final cause in the sublunary world beyond what is contained in the *Physics*, *de Caelo*, and *de Generatione et Corruptione*. My concern here is not principally with *Physics* 2 or with its enormous secondary literature, but I think that it will become clear from my understanding of the *Meteorologica* that, if in *Physics* 2.8 Aristotle is affirming *in propria persona* a final cause for the winter (or any) rain, this affirmation is contrary to the doctrine of the *Meteorologica*.[5]

[4] Johnson 2005: 149–58; Judson 2005. Scharle's 2008 account shows the deepest engagement with the *Meteorologica*. She argues that the argument of *Phys.* 2.8 is consistent internally and externally, that *winter* rain is teleological because it occurs always or for the most part, that its purpose is not to water the crops (198b18 is merely dialectical), but to imitate the circular movements of the heavens. Her argument attempts to bind the widely acknowledged teleology in the movements of simple bodies, here the rectilinear movement of rain downward to its natural place, with the heavenly cycles that the simple bodies aspire to ("Aristotle maintains that water's movement is teleologically directed toward the Prime Mover via the heavenly bodies: water's rectilinear motions imitate the circular motion of the heavenly bodies, which in turn imitates the activity of the Prime Mover," 178). She does this by assimilating the internal nature of winter rain, which falls owing to the withdrawal of the sun ("the very movement of water that imitates the divine – water's downward rectilinear motion – just is the movement by water's *own* nature – a καθ' αὑτό movement essential to water," 169) with the efficient causation of the heavenly cycles ("Although when water comes to be in the summer, it reaches its natural place, none the less its movement there is not due to its nature, which makes essential reference to its proper mover, the sun," 173). That is, the internal impulse of winter rain is just to behave in accordance with its external efficient cause. As such, she sees the relationship between winter rain (and other similar *meteora*) and the heavens as analogous to the relationship between the heavenly bodies and their movers, since the heavenly bodies have internal and external movers that are in perfect accord and produce identical circular motion. Accordingly she interprets *Met.* 12.10.1075a11–25 so that the slaves and beasts imitate (rather than serve) the free members of the household, just as the free members of the household imitate the head (160). Her interpretation undercuts the tendency of Aristotle's hierarchy to increase the unity of the causes as one ascends to the divine νοῦς and increase the conflict and separation among the causes as one descends to animals and artifacts.

[5] Furley 1989: 115–16: "... in *Physics* II.8 Aristotle *rejects* the non-teleological account of regular rainfall along with the rest of the non-teleological world view... at first sight at least, [it] seems to imply a much wider application of teleology – perhaps embracing all the workings of the whole natural world." Also Sedley's classic 1991; and Cooper 1982: "Aristotle unequivocally endorses the teleological explanation of these meteorological regularities" (217n12).

Quite simply, we find nothing in the *Meteorologica* that corresponds to the winter rain in *Physics* 2.8, which falls for the sake of crops. Aristotle never invokes the final cause in any of the hundreds of explanations he provides for meteorological phenomena. We never see the kinds of simple teleological explanations that fill the pages of the *Parts of Animals*, where, for example, the heart has many sinews *in order to* produce the contractions and relaxations that move the body (3.4.666b13–16). Nor do we find here, as we do in the biological works, hierarchies serving as causes and explanations for certain properties of phenomena. In fact, this is the significance of the observation I made in the last chapter, that the combination of causes in the *Meteorologica* gives rise to a variety of hierarchies, which, however, do not supply cause in their turn. By contrast, the biological hierarchies often do. That is, animals have causal, and therefore definitory, factors that relate them essentially to other animals within a hierarchy. For this reason, it makes sense to call a seal a deformed mammal and explain its lack of ear passages on this basis.[6] Aristotle's refusal to use this form of hierarchical explanation among the *meteora* is another indication that *meteora* are not subject to teleological explanation. The *meteora* cannot be failures, as the seal, or at least its ear, in some sense is. Least of all do we see Aristotle invoking the anthropocentric kind of teleological explanation he attributes to the Pythagoreans at *Posterior Analytics* 2.11, where thunder serves to frighten the souls in Hades (94b33–4). The absence in the *Meteorologica* of a final cause beyond the tendencies of simple bodies is a fact, but one that needs to be explained, rather than explained away.

Though there is no evidence in the *Meteorologica* for a final cause beyond that of simple bodies, a more theoretical approach to the question may be helpful in answering those who would nevertheless find one there and aid us in laying the ground for an explanation of this curious fact. Aristotle approaches natural teleology in three major ways. In *Physics* 2 it is investigated through the notion of frequency of occurrence, what Aristotle calls "always or for the most part." In the end, and very oddly, Aristotle claims that all regular and orderly processes that occur "always or for the

[6] *PA* 2.12.657a20 of seals' ear passages; 4.8.684a34–b1 of lobsters' claws. Two further examples without reference to deformation will clarify the concept sufficiently. The lack of the upper eyelid in some birds and in the oviparous quadrupeds is explained by their having hard skin (*PA* 2.13.657a25–b18). That is, they share with viviparous quadrupeds, which have the upper eyelid, the need to protect their soft eyes, and it would be good for them to have upper eyelids like the viviparous quadrupeds, but their hard skin prevents this. Viviparous quadrupeds, then, are the norm; oviparous quadrupeds are the deviation. More broadly, differences in modes of reproduction (viviparous, oviparous, larviparous) are explained by differences in heat, where decrease in heat accounts for the deviant and inferior forms of reproduction (*GA* 2.1.732b15–733a32).

most part" occur for a purpose (2.8.199a3–5).[7] This approach ill suits the phenomena of the *Meteorologica*, which display great differences in their frequency, both absolutely (e.g., the Milky Way, which exists always, and the moon-rainbow, which appears once in decades [3.2.372a24–9: ἄλλα πολλὰ δεῖ συμπεσεῖν]) and conditionally (earthquakes occur most often [μάλιστα 2.8.366b2] in spring and autumn and during rains and droughts but sometimes accompany eclipses [ἐνίοτε 2.8.367b20]). Reflective phenomena require the visual ray and cannot occur without an observer who chances to see them. Therefore the distinction between chance and "always or for the most part," central to Aristotle's analysis in *Physics* 2.8, will be of little use to us here.[8] Moreover, it is one thing to claim that every regular process occurs as a result of end-directed processes – indeed, I believe that this view is at work in the *Meteorologica* – but it is quite another to claim that every regular process is itself end-directed. In fact, Aristotle explains the *meteora* as an interaction between the end-directed movement of the heavens and the end-directed natural tendencies of sublunary simple bodies, and it is worthwhile noting Aristotle's care in adhering to his own theoretical strictures in this regard. Commentators sometimes assume that this interaction, an alternative to an end-directed causation, is a "material necessity," a temporally ordered *a tergo* efficient cause, where the heavenly cause occurs first, followed by the sublunary effect.[9] Aristotle's frequent use of ἀνάγκη and its cognates in the *Meteorologica* superficially recommends an *a tergo* causality. However, Aristotle denies that such a necessity exists (*APo.* 2.12 and *GC* 2.11), and a closer examination of the uses of ἀνάγκη reveals that he almost entirely avoids temporal sequencing in his descriptions of cause and "that of which it is the cause."[10] Rain is not the effect of a prior condensation of vapor; it just is the condensation of vapor. That

[7] εἰ οὖν ἢ ἀπὸ συμπτώματος δοκεῖ ἢ ἕνεκά του εἶναι, εἰ μὴ οἷόν τε ταῦτ' εἶναι μήτε ἀπὸ συμπτώματος μήτ' ἀπὸ ταὐτομάτου, ἕνεκά του ἂν εἴη.

[8] I assume here that all the *meteora* should be subject to a common teleological account. Scharle (unpublished paper) has argued that only the major cyclical processes of the treatise, most obviously the rain cycle, occur for the sake of something. Obviously they are in imitation of the celestial circles. Other processes, e.g., thunderstorms, which occur as a result of the interaction between wet and dry exhalation are accidents analogous to meeting one's debtor in the agora. The reflective phenomena, haloes and rainbows, are likened to the mechanism by which eyes have color, a material process that "piggybacks" on a teleological causation of the eye. I find it hard to maintain this distinction in want of any evidence in the treatise. According to Leunissen's criterion 2010: 36–40, winter rain is teleologically driven in her secondary sense, as, for example, a useful residue in the body is not made for a purpose but used because it is available. This is not a form of teleology that can be universalized to all *meteora*.

[9] E.g., Gill 1997.

[10] In almost every case (the sole exception being 1.4.360b30) causations that one might think of as temporally sequential causations (A then B) are treated as grammatically contemporaneous, and I would

is, while rejecting in the *Meteorologica* the hypothetical necessity that is associated with a temporally ordered final cause, Aristotle carefully avoids an *a tergo* material necessity that he condemns elsewhere.

A second form of teleology is deployed in the biological works and concerns the functions of organs and the directive nature of seeds.[11] This, however, is equally unsuitable to the *Meteorologica*. The directive power of living seeds and the functions of organs are obviously of no use in a treatise about inanimate nature and Aristotle never suggests that the *meteora* are developing toward some more mature form. In fact, as we have already observed, the *meteora* are invariably homeostatic *sustaseis* in constant flux, whose genesis is their being. They are not substances as animals are, and though, as I shall argue, they are in certain respects animal-like, it is misleading to suppose that they have a natural form to attain. As has been observed, *Meteorologica* 4.12 draws a distinction between those processes that can be brought about without the teleologically directed soul and those that require a final cause, and clearly places all the *Meteorologica* among the former. Moreover, the *meteora* and the exhalations they are made of do not serve in turn as material for the biological world. The exhalations are a material dead-end.

Of (τινος) and for (τινι)

In approaching the *Meteorologica* we would do better instead to begin with Aristotle's third analysis of teleology, the distinction between the two kinds of final cause, τὸ οὗ ἕνεκά τινος and τὸ οὗ ἕνεκά τινι.[12] The reason is that this is a programmatic distinction, which seems intended to be exhaustive, and with the help of Aristotle's descriptions and examples we will be able to

venture to claim physically contemporaneous as well. In the following passages efficient-material causation is associated with necessity (* indicates passages where the cause and that of which it is the cause are clearly contemporaneous; those without are unclear): 1.3.340b34; *1.4.341b7; 1.7.344b9; 344b22; 1.12.348a30; *1.14.351a36; *351b6; *353a13; 353a21; 2.1.354a10; *2.3.358a23; *2.6.364a11; *2.8.365b21; *365b23; *367a23; *367a27; *367a31; *368b3; *2.9.369a19; *369a22; *3.3.372b24; *373a3; *3.4.373b16.

11 Gotthelf 1987b and Code 1997. In this analysis, the apparatus of potentiality and actuality is central. It is worth noting, therefore, that though Aristotle assigns a new and special material to the *meteora*, he very rarely, if ever, calls the exhalations a potentiality for the phenomena that they exhibit, in the way that animal tissues and parts are potentially the whole animal (*GA* 2.5.741b7–9). The closest he comes is at 1.8.345b33, where the outermost part of the so-called air has a δύναμις for fire; 1.14.351a33: the earth takes on a differing δύναμις for moisture; 359b10: the δύναμις of fire to produce salty springs; and the clearest case at 3.1.370b13: an exhalation is δυνάμει both air and water according to its matter. The vast majority of the uses of δύναμις in *Meteorologica* refer to active powers rather than to passive capacities.

12 Mentioned in five passages: *Phys.* 2.2.194a33–6; *de An.* 2.4.415b2–3, 20–1; *EE* 8.3.1249b15–16; *Met.* 12.7.1072b1–2. See Kullmann 1985; Johnson 2005: 64–80.

consider the full range of uses and determine whether the meteorological phenomena conform to any of them. This in turn will clarify the role of the *meteora* in the cosmos and allow us to draw further observations about the dualizing and bivalent status of the *Meteorologica* between the cosmos as a whole and the manifold world of terrestrial animals.

In barest outline, τὸ οὗ ἕνεκά τινι provides an explanation by indicating the thing that is benefited by something else. The slave works for the benefit of his master, and the master is the beneficiary of the labor. The master, therefore, is the good that explains why the slave works. τὸ οὗ ἕνεκά τινος, in contrast, explains by indicating a goal that something aims at for itself. The ceaseless thought of god, for example, is the goal that the man of wisdom aims at, but cannot fully achieve.[13] It explains why a man contemplates.

The distinction, however, needs to be used with care. Though there are two senses of "that for the sake of which," the differentiae, "for" (τινι) and "of" (τινος), do not *per se* include the concept of cause. This is most evident in the case of τινι, where the beneficiary can be benefited accidentally: the debtor who happens to be in the marketplace at the same time as the creditor confers a benefit on the creditor, but the benefit to the creditor does not explain the co-presence of creditor and debtor.[14] Examples of incidental *aims* (τινος) are harder to come by, but the following might be supplied: in striving to be eternal through reproduction (*de An.* 2.4.415b2–3), a lover writes a *Symposium* to his beloved as a love-gift, something he would have done, if he had been striving to imitate the thought of the divine νοῦς. Both of these examples involve accidental goods, but even if these are regular benefits or aims, what Leunissen calls "secondary teleology," still one has not thereby given a statement of the real (i.e., non-accidental) *cause*.[15] Neither the presence of a specifically described benefit nor the presence of a specifically described aim implies that the benefit or aim so described is a cause.

Second, in order for there to be a cause, the cause must be distinct from that of which it is the cause.[16] So in *Physics* 2.2 (194a33–b9) the helmsman (the τινι) is benefited by the carpenter who makes a rudder according to the helmsman's specifications. The example makes a clear distinction

[13] *NE* 10.7–8, especially 1178b25–8.

[14] According to Lennox 2001 [1984]: 254–8, chance events are ἕνεκά του, but in a non-explanatory way. As Judson 1991b: 77 says, "E [i.e., a lucky event] actually confers some benefit."

[15] Leunissen 2010: 19–22 uses the term "secondary teleology" to describe benefits, like growing of crops, that occur regularly but are not the final cause of the winter rain, for example.

[16] Cf. *Met.* 12.10.1075b8–10, where even if medicine is in a sense health, still they are conceptually distinct.

between the benefactor and the beneficiary. Likewise τὸ οὗ ἕνεκά τινος is clearest when the goal and that which aims at it are distinct. At *de Anima* 2.4.415b2–7 the nutritive soul seeks to attain the "eternal and divine." Since the activity of the "eternal and divine" is distinct from the cycle of reproduction, we can see how the nutritive soul aims at it and how it is a cause for the nutritive soul.[17] A similar distinction can be seen in the case of the unmoved mover (or movers), τὸ οὗ ἕνεκά τινος at which the stars and planets aim: the stars and the planets aim at, but cannot attain, the full nature of their unmoved mover.[18] In these cases the good aimed at is conceptually distinct from that which strives for it.

In other passages the distinction is less marked but present nevertheless. In *de Anima* 2.4 (415b15–21) the soul is the *beneficiary* of the organs of the body and, as such, is something distinct from them.[19] This is notoriously a different way of conceiving of the relation of soul and body from that in which the soul is the form of the potentially living body. Here the parts of the body are conceived of as tools or instrumental parts for the soul, and the soul has to be understood as at least conceptually distinct from the body if it is to be benefited by the bodily organs.

In the extreme case it is possible to claim that the *being and activity* of a thing is both its end, inasmuch as it strives to maintain that being, and its benefactor, inasmuch as a thing benefits from its own existence and activity.[20] In such cases, where the aim and the striving subject are the same or where the benefactor and beneficiary are the same, the explanatory

[17] *De An.* 2.4.415a25–b7: reproduction takes place "so that [living things] may share in the eternal and divine in the way they are able. For everything strives for this and does everything it does by nature for the sake of this. "For the sake of this" is ambiguous, denoting either the aim or the beneficiary. So since it is impossible to share continually in the eternal and divine, since no corruptible thing can remain numerically one and the same . . . it shares in it in the way it is able to."

[18] The passage at *Met.* 12.7.1072b1–2 draws a clear distinction: "a distinction makes clear that that for the sake of which is present among immovables. For there is that for the sake of which as beneficiary (τινι) and as aim (τινος), and of these one is immovable and the other is not. And it moves as object of love." The cause is immovable and that of which it is the cause is moved. Which kind of motion the immovables produce seems to be described at *Cael.* 2.12.292a18–b25, where to attain the best of each thing is for everything the end. But if not this, it is always better the nearer it is to the best. It is not possible for the celestial bodies "to attain the good without activity" as it is for "the best conditioned," that is, for the immovables; the celestial bodies must attain it "through a single activity or a few, and for the things farther removed to attain it through more activity."

[19] *De An.* 2.4.415b15–21: "It is obvious that the soul is also a cause in the sense of that for the sake of which. For just as the mind acts for the sake of something, so also does nature, and this is its end. Such is the soul in animals according to nature. For all natural bodies are organs of the soul, and just as the organs of animals, so also the organs of plants, since they are for the sake of the soul. And that for the sake of which is ambiguous, the beneficiary and the aim."

[20] So Johnson 2005: 74n20 cites *GA* 1.23.731a24–b8 but misrepresents it in stating that it claims "both mere existence and even the lowest forms of life are intrinsically good and beneficial for the beings that exist or live." There is no word in the text corresponding to "beneficial."

power of the final cause is nil. This is why the good can only be a cause for changing things, in which there is a distinction between a thing and what it changes into. And even though the formal cause (what a thing is) and its final cause (its good) are sometimes identified, that is, something's nature is its aim in the process of coming-to-be, they clearly pick out different aspects of a thing's being.[21]

Finally, inasmuch as the final cause is distinct from that of which it is the cause, the final cause must be better than what benefits it or what aims at it. After all, it is the good of or for that thing, and the good must always be better than that of which it is the good. Even in the case of biological development the mature animal form is better than the fetation. Likewise, slaves benefit their masters, because that is the purpose of the relationship. No doubt a master can benefit a slave, but the benefit that the slave enjoys is not the purpose of the master–slave relationship, which must always be explained ultimately in terms of the master's good.[22] The principle is borne out by Aristotle's examples: he never suggests that the benefit a subservient thing enjoys can be a final cause.[23] Let us now apply these observations to the *meteora*.

[21] Aristotle affirms the identity of the formal and final cause at *Phys.* 2.7.198a25–6, but, of course, he is describing conceptually distinct moments of the same object. So also 198b8–9: "it is better thus (not without qualification, but with reference to the substance of each thing)."

[22] At *Pol.* 1.5.1254b10–11 tame animals are better off being ruled by man, and b16–20, "Men that differ as much as soul and body or man and beast – and they are so disposed whose function is the exercise of their body, and this is the best they can provide – they are naturally slaves and it is better for them to be ruled in this way." In the very different situation of friendship, where there is reciprocal *commonality*, a superior may benefit an inferior and do so as a purpose, as a king benefits his subjects. This is a special kind of relationship and not the kind of benefit relationship that exists between craftsman and tool or soul and body. At *EN* 8.10.1160b2–6 the king looks to the advantage of his subjects. "The man who is not autonomous and not surpassing in all goods, is not a king. For such a man has need of nothing else. And so he would not consider his own benefit but that of his subjects." On the importance of commonality, absent in the soul–body relationship, consider *EN* 8.11.1161a32–b5: "where there is nothing common to ruler and ruled, there is not friendship either, for there is no justice either, just as a craftsman has no friendship with his tool, nor soul with the body, nor master with a slave; all these things are benefited by the user, but there is no friendship or justice towards inanimate things, nor toward a horse or an ox, nor to a slave *qua* slave. For there is nothing common; the slave is a living tool and the tool an inanimate slave." Similarly, there is no commonality between helmsman and ship or between heavens and sublunary elements.

[23] The soul is obviously superior to its bodily organs (*GA* 2.1.731b28–9), but one can object that the hierarchical relationship is not always fixed. The carpenter serves the helmsman in the production of the rudder, but the helmsman may serve the carpenter by shipping timber to him. In our passage, however, the helmsman is said to be concerned with the form and to be architectonic, the carpenter is concerned with the matter (*Phys.* 2.2.194a36–b9). There is no question that Aristotle viewed the relationship among the crafts as hierarchical (e.g., *EN* 1.1.1094a6–16) and the relationships of nature even more fixedly hierarchical than those of art. *Cael.* 2.9.291a22–6 is no exception: "since [the heavenly noise shattering us] obviously does not occur, none of them moves with an animate or forced motion, as if (ὥσπερ) nature had foreseen what would happen, that if the movement did

It should be clear that, if Aristotle conceives of the final cause of the *meteora* to be explanatory in a robust sense, the final cause must be conceptually distinct from what it explains. A simple restatement of the essence of the *meteora* may be a statement of their end, but will not be a statement of cause in the sense we are seeking. It is possible, for example, to say that the final cause of rain is to aim at being a part of the cycle of evaporation and condensation and that rain is benefited by attaining this nature. This position could be argued even in the absence of textual support, but when Aristotle says that final causes are operative in nature, he means more than this. He means that the rain either provides some benefit outside itself or that it strives to be better than itself.

If we accept that the final cause must be a cause and that it must be hierarchically superior to that of which it is the cause, it is evident that the *meteora* neither are nor have a final cause in either of the two senses. First, the *meteora* do not have a final cause in the beneficiary sense. If they were a benefit to something, they would have to benefit something better than themselves. But neither the unmoved mover nor the heavenly bodies are benefited by the *meteora* either individually or as a group, and Aristotle never hints at any such benefit.

And if it were not for the *Physics* 2.8 passage, no one would argue that the *meteora* as such benefit man or any other animal or plant. No doubt the winter rain does confer a particular benefit, but the *Meteorologica* contains no mention of this, and nowhere in the treatise is there any systematic consideration of the benefit of *meteora*. Aristotle never provides the slightest hint, for example, that the kapnospheric Milky Way (the most regular and most celestially directed phenomenon in the *Meteorologica*) provides any benefit to living things. In fact, *meteora* are often harmful, as for example, the earthquake and attendant tsunami that wiped out the cities of Helike and Boura (2.8.368a34–b13).[24]

So too, although the simple sublunary bodies and the exhalations are themselves the beneficiaries of the celestial motions, they cannot be τὸ οὗ ἕνεκά τινι of the celestial motions.[25] This would invert the entire causal structure of the universe, in which the better invariably provides the final

not take place this way, nothing in our realm would be the way it is." The ὥσπερ indicates that the claim is not literally meant.

24 Martin 2010: 267 reports that Pietro Pomponazzi argued for the non-causal sense of good: "earthquakes are only good with respect to their genus if they destroy cities and provinces."

25 Bodnár 2005: 25–6 points out that the stars are similarly the beneficiaries of the movement of their spheres. The distinction between sphere and star is much more subtle than that between heavenly spheres and sublunary elements.

cause to the inferior. There is no evidence at all that Aristotle thinks the purpose of the heavens is to serve the sublunary world.[26] Just as the master may benefit a slave, but not for the sake of the slave, so the celestial motions benefit the sublunary world, but not for the sake of the sublunary world. There is a difference, of course: the master benefits the slave so that the slave can benefit the master. But, whereas the heavens benefit the sublunary world, they do not do so in order that the sublunary world can benefit them in return.

Finally, neither do the *meteora* have a final cause in the sense of a goal they are striving toward in the way that the first body strives for its unmoved mover and the nutritive soul strives for eternity through the ceaseless cycle of reproduction. Both of these strivers are ensouled and have an internal impulse toward their external good. Although having a soul is not an obvious requirement of something that aims at a good in this sense, Aristotle's explicit examples are confined to things of this kind. Minimally, the subject must have an internal impulse to strive after something that is in some sense outside, and better than, itself, for this would at least allow the natural places to act as final causes for simple bodies.[27] Examples where souls are not involved are suspect for other reasons. The simple bodies are said to imitate (μιμουμένη 337a7) the heavenly motions in *de Generatione et Corruptione* 2.10, for example, but imitation does not imply an internal principle so directed and, if any final cause is at work in this passage, it is a universal demiurge (336b30–2), unique here in the corpus.[28] In the *Meteorologica* the rain cycle is similarly said to imitate (μιμούμενος 1.9.346b36) the circle of the sun, but it is clear that it is the sun rather than an internal impulse of the water vapor that is creating this cycle.[29] The humble exhalations have their own simple tendencies closely corresponding to those of the elements, and there is no indication in the *Meteorologica* that they strive or tend to any other motions or places than these. In every case it is the efficient causality of the heavens that makes

26 Even *Met.* 12.10.1075a11–15, where the general must obviously benefit the army, does not suggest that the general is for the sake of the army.

27 Bodnár 2005: 24: "In these cases the nature of the entity which tracks the goal does this tracking on its own."

28 "God brought the whole to completion in the way that remained and fulfilled its perfection by making generation continuous." Johnson 2005: 146–8 has argued well that imitation does not require striving. In another context *Phys.* 1.9.192a16–25 matter is said to have desire for form, for which there is no support elsewhere.

29 Scharle 2008 provides the winter rain with a final cause by conceiving of the sun's efficient causality as part of the nature of the rain itself. Consequently winter rain because of this nature strives to imitate the motions of the heavens.

the simple bodies imitate their circular motions. Aristotle makes no use of striving as a final cause in the *Meteorologica*.[30]

And yet in one early passage Aristotle says that the *meteora* are steered (κυβερνᾶσθαι 1.2.339a23) by the motions of the heavens. The nautical metaphor here undeniably recommends an intelligent and purposeful direction, but, though this is precisely the place where Aristotle should describe the role of the final cause and clarify what he means by helmsmanship, no such explanation is offered. Within this chapter he makes abundantly clear that these heavenly motions are to be understood as efficient causes.[31] And though efficient causes can be and frequently are also final – as in the case of seeds – their teleological role is always made obvious and pervasive over the entire domain. In the *Meteorologica*, by contrast, we are left to wonder what this single metaphor might signify. If this passage is a programmatic promise, it is one that Aristotle never fulfilled, and in consequence our challenge is to see how this celestial guidance is good and is caused by a good cause, but it is not a final cause.

There is no reason to suppose that the good is absent. Throughout his works Aristotle discusses many materials, many motions, many forms, and many goods. Only some of these are causes; many others are caused. Some motions cause materials; some goods cause motions. It should be no surprise that some motions and materials together produce some good and do so in an orderly way that produces a regular benefit without having to suppose that the good must be the final cause.

30 The situation then is just how he describes the cosmic theories of Anaxagoras and Empedocles at *Met.* 1.7.988b8–16: "For those who claim mind or friendship place these causes among the goods, not as the final cause of anything that is or comes into being, but rather as if motion came forth from these." It is a clear indication that the final cause is not operating here that the main means of change are separation and condensation (cf. *Met.* 8.2), since these belong to non-substantial coming-to-be and passing-away. That is, there is no substance under consideration in the *Meteorologica* and there is no final cause in the manner of the biological works.

31 "Necessarily this place is continuous with the upper movements so that its whole capacity is directed (κυβερνᾶσθαι) from there. For we must suppose that the first principle is located where the principle of movement for everything is." For κυβερνᾶν in the Presocratics, Heraclitus, DK 22B41 (1.160.8) and Diogenes of Apollonia, DK 64B5 (2.61.6), both clearly teleological. Theophrastus describes how sublunary motions may be incidental (*Met.* 5b17–26): "And one might perhaps investigate first, whether these entities are or are not parts of the cosmos, and if they are parts, in what sense <they are>; for now they are (so to speak) thrust away from the thing most esteemed, not only as regards their position but also as regards their activity – assuming that the circular movement is such: for it happens to be the case that the changes <of things> – both towards their places and to one another – occur (so to speak) accidentally under the influence of the circular rotation" (trans. Van Raalte 1993). He also seems to accept that such motions cannot be attributed to a final cause (7a19–b5).

Non-causal good in the *Meteorologica*

The absence of the final cause among the *meteora* prompts one to wonder why Aristotle made use of the "biological" method we considered in the last chapter, and whether there are not more substantive similarities between biology and meteorology. In fact, there are similarities of subject matter that make the similar method appropriate, and prominent among them is the special material cause of each, the wet and dry exhalations for the *meteora*, blood and its analogue, ichor, for animals. I would like now to explore further substantive analogies between the *Meteorologica* and the biological works to determine whether Aristotle might have considered the *meteora* in some sense lifelike. I undertake this strategy in the confidence that Aristotle makes regular use of the techniques of conceptual abstraction and addition to draw comparisons between related subject matters.[32] We need only recall that artifacts and natural things are distinguished by the presence or absence of an internal principle of motion in a thing primarily and in virtue of itself, or that the addition of a certain matter makes geometry into optics. We will find that Aristotle uses similar abstractions and additions to locate the *meteora* as intermediary between the cosmos as a whole and the world of mundane animals in ways that bear on the question of teleology.

If we interpret the *Meteorologica* through the lens of biological method, we tend to see the kapnospheric and atmospheric phenomena, the winds and all the others, as genera of *meteora* analogous, say, to viviparous quadrupeds and birds. On this view, the *meteora* are a plurality corresponding to the several species and genera of animals, ranked according to the *scala naturae*. The best *meteora* – the kapnospheric fires – are most upright; the metals and minerals correspond to the plants with their heads in the ground. Importantly, on this view, the *meteora*, like the genera of animals, can be interpreted as independent expressions of a common activity analogous to the activities of the soul, sensation, locomotion, and so on; that is, all the *meteora*, with the significant exception of the lowly metals and minerals, are in one way or another an expression of the circular form that the heavenly motions (usually) confers upon them.[33] The circle that the Milky Way traces, exactly tracking the stars, the circular "river" of the rain cycle, the radial symmetry of the winds and the rivers, the forced circles of whirlwinds and firewinds and finally the perfect geometrical circularity of haloes and rainbows are all effects or imitations of the heavenly motion,

[32] Wilson 2000: 29–39. [33] On the common psychological activities, see *PA* 1.1.639a20–b3.

and they unite the *meteora* in the same kind of way as the sensitive soul is common to and unites the great variety of animal genera. In this way the *meteora* can be viewed as a plurality of genera.

But we may also conceive of the sublunary world as a *single* entity, unified as a single animal, whose best parts are uppermost, and whose worst, the sea and the metals, are at the bottom. On this conception the *meteora* relate to one another as a set of organized parts, as it were, each performing its different activity within some unified arrangement, the kapnosphere corresponding to the head, the sea to the bladder, and so on. Extreme accuracy in extending this analogy is not appropriate, but Aristotle explicitly draws correspondences between meteorological and physiological, especially digestive, processes.[34] This view is corroborated by the observation that the meteorological phenomena are made from a special material, corresponding to flesh, and form discrete, localized συστάσεις, which display a great variety of specific activities, corresponding to the discrete and spatially ordered organs of the body.

Which view influenced Aristotle? Probably both did, since Aristotle intended to mark off the sublunary world of the *Meteorologica* as transitional between the unity of the heavens and multiplicity of animals. In the heavens, the celestial spheres are made of a single kind of matter, aether or first body, and move with a single kind of motion, caused ultimately by a single mover. They operate as a single genus differing from one another only by more and less, and each sphere is arguably the single member of its species.[35] Similarly the various *meteora* of the sublunary world are all produced by the same celestial motions and are located within a single vertically organized place. There is only one sublunary world. To this extent the *meteora* participate in the singularity and unity of the heavens. But, like animals, the *meteora* come in many different genera formed of two different materials (dry and wet exhalation corresponding to the blood and its analogue) and produced by a variety of causes. And, with the important exception of the Milky Way, there are many members of each species (i.e., many comets, many rivers, etc.). Finally, among animals the unity of cosmic place breaks up into a plurality of microcosms, each animal's

[34] Earthquakes and bodily tremors, throbbings (2.8.366b14–19), and shivers attendant on urination (b19–23); the sea and stomachs (2.2.355b6–15; 356b1–3) and bladders (2.3.358a3–11).

[35] Aristotle's discussion of this issue in *Met.* 12.8 is notoriously obscure and this is hardly the place to take up the question in any detail. While at 1073a23–b3 he leads us to believe that the prime unmoved mover moves the first sphere of heaven and other unmoved movers move the other spheres, the passage at 1074a31–8 argues that there can only be one heaven because there can only be one unmoved mover. Whatever reconciliation is possible between these positions, it is certain that the cosmos must have a single cause.

body being a spatially ordered unity unto itself. The unity of the heaven's efficient and final cause is replaced by a plurality of souls, one for each individual animal.

Two models

Within this framework of unity and multiplicity we can explore how Aristotle uses two biological models to draw the inanimate *meteora* into the orbit of the living world. The first, appropriate to a unified cosmos, is that of the soul as the chief mover of the body. The second, appropriate to the multiplicity of animal genera, is that of the adult reproducing offspring, which on closer examination yields to that of spontaneous generation. It is through these models that Aristotle moves from unity to plurality in the absence of the final cause.

Body and soul

The structure of the sublunary world, as we just noted, is in some ways similar to that of a single animal body. As with animals, the *meteora* have their own special material, the exhalations, out of which they are formed, distinct from the ambient environment. Like an animal in its normative posture, the sublunary world has a proper vertical orientation. The various *meteora*, like the organs of the body, engage in different activities that frequently interact with one another. Like the animal body and its organs, too, the *meteora* do not move themselves but require some principle of motion. In the case of the animal body it is the soul; for the *meteora*, it is the heavens in their diurnal and annual movements.

But this principle of motion is not internal to the sublunary sphere as the soul is to the body, and if we wish to press the analogy between the sublunary world and the animal body, we would do better to treat the entire cosmos as an animal rather than just the sublunary world. This is by no means farfetched.[36] The model is already familiar from the *Timaeus*, where the world soul is located primarily in the heavens and placed in the intersecting diurnal and annual circles (36b–d). It is by no means un-Aristotelian that the soul should have some specific habitation distinct from the rest of the body. Although according to Aristotle's hylomorphic psychology the soul is the form of the body and therefore not material in

[36] *Cael.* 1.9.278b11–21 distinguishes three senses of οὐρανός: the stars, the stars and planets, and, importantly for our purpose here, the whole cosmos.

the same way as the body is, nevertheless its primary functions are located in the heart. On the cosmic level, the heavens are moved by desire for the divine mind through the final cause and in turn move the sublunary world mechanically, just as the object of desire moves the soul, and the soul, through the efficient causality of the σύμφυτον πνεῦμα in the heart, moves the limbs.[37]

Aristotle goes to considerable lengths in the *de Motu Animalium* to draw such comparisons between the heavens and individual animals. In both cases the final and efficient causes are treated as distinct moments, and the pushing and pulling of the σύμφυτον πνεῦμα in the animal's body corresponds to the efficient actions of the heavens on the sublunary world. In neither case is the movement of the limbs or the movement of the sublunary world given a final cause.

We may anticipate two objections to this analogy. First, though the heavens with their diurnal and annual motions serve as a kind of bipartite soul to the sublunary world, the heavenly spheres themselves are treated as having separate souls and bodies of their own (*Cael.* 2.12; *Met.* 12.8). From this point of view it is difficult to see the sublunary world as the *body* of the heavenly soul. In response, we should hardly be surprised that much of the soul of heaven has no effect on the lowest part of the cosmos, since the intellectual parts of the human soul are likewise not directed toward lower functions such as the processes of digestion. Indeed, Aristotle goes so far as to liken different centers of impulse in a single animal to separate animals (*MA* 11.703b20–3), and this in turn reveals the extent to which Aristotle was willing to view the same part both as a part of a whole and as an independent whole in its own right. In two different contexts Aristotle uses household and political comparisons to indicate this middle ground between unity and plurality. In the *de Motu Animalium* the commanding part of the soul is likened to a monarch who does not meddle in every little detail in the well-governed city:

> One must suppose that the animal is constituted like a well-governed city. For in the city, when once order is established, there is no need for a distinct monarch to preside over everything that happens, but each person looks after his own business, and one thing follows on another by habit. In animals this same thing occurs by nature and because each of their parts is so naturally constituted to do its own task, so that there is no need for the soul to be in

37 *MA* 7–10; *PA* 3.4.666b16–7: "the heart is like an animal in that which contains it"; see Theiler 1965: 96–7; Bodnár 2004: 144–6, who highlights Aristotle's discussion in the *de Motu Animalium*. *MA* 4.700a16–17: "For all inanimate things are moved by something else, and the principle of all the things that are thus moved is the things that move themselves."

> each part. But because the soul is in some one part, the others live because they are attached to it, and do their own task by nature. (*MA* 10.703a31–b2)

In this manner the soul and the body can be treated as a political unity composed of different individuals. The higher part provides minimal direction to the lower parts, which act largely according to their own endowed nature. Again, in explaining the order and unity of the cosmos in the *Metaphysics*, Aristotle uses the image of the household, in which

> the freemen have the least ability to act at random, but all or most of their actions are dictated, while for the slaves and the beasts, their common contribution is small, and the majority of what they do is random. For the nature of each of them is such a principle. (*Met.* 12.10.1075a19–22)

The household is a unity composed of a plurality of freemen (corresponding to the heavenly spheres) whose actions are closely regulated, and of slaves and beasts (corresponding to the sublunary sphere), which are largely left to their own natures. In these analogies Aristotle envisions a cosmic and biological unity that is consistent with a surprising degree of plurality and autonomy. It is worth noting also that while these analogies are germane to the *meteora* as parts of the cosmic whole, they rely on a beneficiary (τινος) teleology that is alien to the *meteora*, which do not serve or benefit the heavens.

The second objection concerns how on this analogy we are to understand the role of the unmoved mover, which, though it is certainly super-physical, is nevertheless alive (*Met.* 12.8). Should it not be considered part of the world soul on this analogy? No, we should understand it to be outside the physical cosmos and no part of the world it inspires, as Aristotle makes clear.[38] But in one respect the analogy between cosmos and animal is strained: *de Caelo* 2.12 treats the heavenly bodies as animals requiring more or fewer motions to accomplish their end (292a14–b25). There is no suggestion here that the bodies or their spheres work together as a composite world soul. I would respond first that Aristotle is developing here in answer to a specific problem a *scala* different from the one present in the *Meteorologica*. The multiplicity of *scalae* in the biological works (e.g., posture and heat) show that Aristotle never felt he had to restrict himself to only one. Moreover, as we have just seen, Aristotle recognizes that the animal soul can itself function as a collection of souls rather than as a simple unity (*MA* 11.703b20–2).

[38] *Phys.* 8.10; *MA* 3.699a25–7; 4.699b32–670a6.

In other ways the analogy is deliberately inappropriate. Heavenly bodies need no instrumental parts, as the sublunary souls do, to perform their activities, and the motions of the sublunary world are not beneficial to the heavens as the motions of the body are beneficial to the soul. The good of the sublunary world rests, then, not in the functions it performs, but in the celestial order it is made to imitate. The good is present in the sublunary world, but the final cause has no place, since the *meteora* are merely the mechanical responses of the heavenly desire.[39]

Seed, body, and spontaneous generation

As we saw, the various *meteora* can also resemble the genera and species of animals, not in any particular features, of course, but rather by analogy. The genera of animals display very great differences among one another both in their bodily parts and in their shape as a whole, while the functions of their souls are identical or very nearly so. Almost all animals feed and reproduce, perceive and move locally, that is, their basic activities are very similar. Aristotle is making use here of the familiar Academic association of the form with the universal, and matter with the particular. In a similar way, the heavenly motions confer a unified form and order on the *meteora*, whose basic material and spatial causes differ tremendously. The predominant form that *meteora* take is some kind of cycle, whether they are merely activities at one moment of the cycle of transformation, or whether they themselves take on a circular appearance. The differing material components together with the various places in which the *meteora* occur give rise to their different appearances.

This model, in which a common form is impressed on differing materials, emphasizes a rather different set of relations from the first one we just considered. Here we conceive of the cosmos not as a single whole animal but rather as divided into two parts, celestial and sublunary, where the celestial realm acts as an external source of motion and form similar to a male parent. And just as in sexual reproduction the parent produces in a new material something similar to and imitative of itself, so the heavens move in circles and produce circular motion of a variety of kinds in the sublunary world.

39 It is worth quoting Owens' 1978: 453 comment on *Met.* 12.10.1075b8–10: "The formal cause 'health' is in the body as in matter, and in the soul (as medical art) without matter. In the soul it is a secondary instance, and functions as an efficient cause. In the body it is the final cause. In both cases it is a form, but as primary and secondary instances. Efficient causality, then, seems to belong only to secondary instances of forms. The unmoved movers cannot be efficient causes, and correspondingly, cannot have final causes. But they can be the final cause of all the movements in the universe."

This model allows us to understand how the biological world ranks lower than the inanimate world in the *scala naturae*. The heavenly motions provide a single cause for all the circularity we find among the *meteora*, and this singleness of cause is a mark of high status. There is no corresponding single cause for all souls and their functions. Each species of soul is a first principle unto itself. That is, the biological realm makes plural the principle of change that had been singular and common to all *meteora*. The *meteora* are, as it were, the last moment in the hypostasis of celestial unity, and the introduction of the soul brings with it a radically new and inferior plurality of cause.

Of course, there are important respects in which Aristotle does not intend the analogy to hold: in sexual reproduction the seed works on its specifically appropriate matter and produces not an inferior token of itself, but an equal embodiment of its own form. Here instead the heavens, perfect instantiations of their form, produce the same form in a variety of different materials, some more and others less able to receive the form fully.

So if we want to press the analogy, we would do better to treat the order generated among the *meteora* as corresponding to that of animals generated spontaneously. There is an extensive group of simple animals that are not reproduced from seed according to their kind. Similarly it is obvious that the comets do not engender comets, nor do winds give birth to winds. Among spontaneously generated animals and among the *meteora* the sun both works on the material to make it suitable for further elaboration and provides the moving cause to elaborate that matter further. We have already seen how the sun both produces the exhalations and affects them further. In the case of spontaneously generated animals,

> what the heat in the animals works up from the nutriment, this the warmth of the season in the environment combines and produces from the sea and land by concoction. The part of the psychic principle that is enclosed and separated in the *pneuma* creates a fetation and instills movement. (*GA* 3.11.762b13–18) [40]

Thus out of earth and water the sun's heat concocts a material from which the animal is then formed. And while it is obvious that *meteora* are inanimate, it is important to note that Aristotle also refrains from calling the *pneuma*, which has generative heat, soul:

[40] There is apparently a different doctrine at *Met.* 7.9.1034b3–5: "spontaneously generated things are generated just as [some sexually generated things], since their matter can undergo by its own agency the same motion as the seed confers" – no *pneuma* here at all.

> Animals and plants are generated in earth and in moisture because water is present in the earth and *pneuma* is present in water and in all of these psychic heat (θερμότητα ψυχικήν) is present, so that in a certain way (τρόπον τινά) all things are full of soul. For this reason they are formed quickly whenever [the *pneuma*] has been enclosed. (*GA* 3.11.762a19–23)

The very fact that this generation is distinguished as αὐτόματον makes clear that it is not for a purpose in the way that the whole sexual apparatus of other animals is clearly made for the purpose of reproduction.[41] Aristotle's denial of the final cause to spontaneously generated animals is explicitly corroborated in the *de Anima* (2.4.415a26–b1), where animals that are *not maimed or spontaneously generated* assimilate their food to themselves "in order that they may participate in the eternal and the divine to the extent they can." Aristotle is clearly denying here the τινι form of the final cause to spontaneously generated animals. Moreover, while the generative organs, indeed the whole soul, are specific and different in each kind of animal, there is, as far as we can see, only one kind of *pneuma*. The generic heat of the sun works on the earth and the sea to produce the appropriate matter for animals in a process, which, though more complex than that which generates the *meteora*, does not require a specific seed. The differences among spontaneous animals are explained by differences in their matter and the places (τόποι) where the generation takes place.[42] In a similar manner the heavenly motions are numerically unique, but indiscriminately available to the sublunary world, and the exhalations and the *meteora* arise spontaneously as a result of the sublunary world's interaction with them. In the *Meteorologica* the sun works up a material, the exhalations, common to all the *meteora* and then generates a broad variety of simple phenomena depending on the matter and the places where the *meteora* occur. Importantly, the heavenly mover is not determinative of the specific forms that the *meteora* or the spontaneous animals become: in each case the matter determines. If the specific *meteora* as such, the rain and the rivers,

[41] The tensions between "spontaneous" generation and τὸ αὐτόματον of *Phys.* 2 have been noted by Balme 1962. See Lennox 1982 for a defense of the consistency of spontaneous generation with Aristotle's account of αὐτόματον in *Physics* 2.

[42] This depends on the interpretation of *GA* 3.11.762a24–7: αἱ μὲν οὖν διαφοραὶ τοῦ τιμιώτερον εἶναι τὸ γένος καὶ ἀτιμότερον τὸ συνιστάμενον ἐν τῇ περιλήψει τῆς ἀρχῆς τῆς ψυχικῆς εἰσιν. Peck translates "The object which thus takes shape may be more valuable in kind or less valuable; and the differences herein depend upon the envelope which encloses the soul-principle." Lennox 2001 [1982]: 233 translates "The differentiae which determine whether the kind is more or less honorable are determined by the organization of the vital principle in the enclosure." *HA* 5.15.547b18–23 (where differences of mud account for differences in testacea) favors Peck's interpretation. *HA* 5.19.550b31–551a27 points less strongly in the same direction. Compare *GA* 2.4.738b34–6 on the role of the district in determining the form of the organism.

and so on, are to be considered ends and governed by the final cause, what they are specifically is not the result of the heavens but of their own material. By contrast, the soul of the animal is in all cases, determinative of the matter and the form.

It is a highly significant corollary of this interpretation that the assimilation of the *meteora* to spontaneously generated animals allows us to understand better Aristotle's motivation in introducing the peculiar and puzzling theory of spontaneous generation into his biology, inasmuch as it provides a much-needed step on the *scala naturae* between the *meteora* and sexually generated animals. That spontaneously generated animals are found only at the lowest rungs of the living *scala* rather than at the highest is no argument against the connection: Aristotle did not want seed-based souls to be merely the most etiolated and exhausted expression of the heavenly motions, but rather fresh starts and, in the case of humans, co-workers in the divine. It is enough for the continuity of the *scala* that *meteora* can behave like animals.

Dualizing

These similarities suggest that spontaneously generated animals act as dualizers between the *meteora* and sexually generated animals. They are similar to the *meteora* in that the generic heat of the sun rather than a specific kind of seed generates them. But spontaneously generated animals differ in requiring an individually enclosed *pneuma*, a numerically different portion for each animal, similar to a seed. By contrast, the *meteora* require no such individual agent, and the sun acts directly on the material in the sublunary world without ever being enclosed in it. They obviously differ too in that spontaneous animals have bodies and move like sexually generated animals; *meteora* do not.[43]

At the same time the *meteora* can be seen as dualizers partaking at once in the unity of the cosmos and the plurality of the animate world. I hazard the following as Aristotle's architectonic vision of the physical world and the place of the *Meteorologica* within it. (1) The cosmos as a whole, which consists of simple bodies, forms a unified macrocosm, inside which the plurality of ensouled living things are the corresponding microcosms. The cosmos is unified by the unmoved mover, the οὗ ἕνεκά τινος of the heavenly

[43] In another sense the *meteora* serve as the humble correlates of the ensouled heavenly bodies, just as spontaneously generated animals are the inferior counterparts of the sexually generated animals. For each of the heavenly bodies and each ensouled animal has its own individual mover, while the *meteora* rely on a common external mover, as do the spontaneously generated animals.

bodies, which in turn serve as efficient causes for the sublunary world. The sublunary world corresponds to the organic body of the animal. (2) At this point the absence of the final cause is critical, because it allows Aristotle now to shift the *meteora* away from their place in the unified world toward their role in prefiguring the plurality of the ensouled world: though the sublunary world corresponds to the organic body of the animal, that body neither strives toward nor serves the cosmos and so cannot be completely absorbed into that unity; it manifests rather the common heavenly form in various ways and to varying degrees according to specific material and local factors. These manifestations now correspond to the several great genera of living things, on their way to being independent substances, their mode of generation similar to spontaneous generation (i.e., generation without a final cause). The *meteora* are now treated as a plurality and subject to the "biological" method. They remain, however, homeostatic processes depending on an external mover without an internal nature of their own. They need only the addition of individual portions of *pneuma* or individual souls to complete their descent to the bottom of the *scala naturae*.

PART II

The meteora

CHAPTER 6

Kapnosphere (1.4–8)

Aristotle's theory of the upper region or "kapnosphere" as I shall call it – the shooting stars (ἀστέρες διαθέοντες), torches (δαλοί), and "goats" (αἶγες), the comets (κομῆται) and Milky Way (γάλα) – combines elements of early Ionian, Italian, and Platonic philosophy. The early Ionians had explained the upper fires as ignitions of terrestrial moisture. Aristotle likewise finds their fuel source in the earth but attributes it to the dry rather than the wet exhalation. His reason for doing so is that the kapnospheric phenomena are ontologically and causally prior to the other *meteora* and require a prior cause – while water vapor requires several changes before it can burn, dry exhalation ignites immediately.

The kapnosphere also corresponds to the firesphere, familiar from the *de Caelo*, that Aristotle had inherited from the Italian tradition of Empedocles. It thus provides a concrete interpretation for the transformation of the sublunary simple bodies and a means to unite the lowest and the highest levels of this region through the change of earth into smoke and its subsequent ascent.

Finally, like the Platonic forms, Aristotle's celestial substances are permanent objects and uninfluenced by the sublunary sphere, in spite of their having a causal effect upon it. The phenomena of the sublunary sphere, like the Platonic particulars, are in constant flux. Aristotle redeems this unpromising realm by granting it its own intelligible, even orderly, nature, which is dependent on, though distinct from, the heavens. The kapnosphere occupies the upper zone of the sublunary world and serves as an analogue and close imitation of the heavens above it. Its highest member, the Milky Way, in its light and its motion is almost indistinguishable from the stars. In an imaginative *tour de force* Aristotle fuses these three elements of the philosophical tradition, unifies this great group, and augments it with the aurora.

My purpose throughout this chapter is to show how the kapnospheric phenomena dualize between the heavens and the lower sublunary realm.

After adding a few more words in order to complete my investigation of the material cause, I shall argue that Aristotle develops his classification of meteors (used here in the modern sense, as distinct from *meteora*) in implicit opposition to Anaxagoras' theory of the descent of fiery aether. I shall argue that Aristotle correctly identified the aurora and that by examining his conception of it we can solve an interpretative problem among meteors. Finally, my discussion of comets and the Milky Way will focus on their quasi-celestial nature and end with Aristotle's meteorological solution to his own metaphysical problem of defining a unique object.

The kapnospheric exhalation

Chapter 1.3 left us with only the vaguest hints concerning the origin of the exhalations. Aristotle called the dry and hot material an exhalation and potentially fire, but his only purpose there was to distinguish it from vapor (ἀτμίς 340b23–9), and within the confines of the chapter it appears merely as the product of the friction of the sun upon the upper air (ὁρῶμεν δὴ τὴν κίνησιν ὅτι δύναται διακρίνειν τὸν ἀέρα καὶ ἐκπυροῦν 341a17–18). It is only in 1.4 in the context of meteors, which require a combustible fuel (ὑπέκκαυμα), that the exhalation's *terrestrial* origin is introduced (τὴν δ' αὐτῆς τῆς γῆς οὔσης ξηρᾶς καπνώδη 341b10). No doubt the terrestrial exhalation is necessary for the treatise as a whole, for otherwise the mixing of simple bodies would be confined to water and air, and no lower phenomena could arise from the dry exhalation. But pressing as this consideration is in the lower region, it is less exigent in the kapnosphere, and less so the higher we ascend.

This is because the fires of the kapnosphere can be produced either directly by the friction of the celestial bodies on the air immediately below them or indirectly by the ignition of the smoky dry exhalation previously generated from the earth. Indeed, to the extent that the nature of fire (dry-hot) is situated between the powers of earth (*dry*-cold) and air (moist-*hot*), there should be no reason why fire could not be generated from either earth or air. Obviously Aristotle does form the dry exhalation by warming the cold earth, and we have already seen how the sun generates fire from air through friction. But it is remarkable that we invariably find the solar friction heating the air, rather than drying it, and producing fire rather than exhalation. For reasons we have already considered, Aristotle wanted to avoid an explicit transformation of vapor into smoke and was willing to create considerable confusion in order to prevent it. Throughout the treatise, the ambiguity in the term "air" allows for an uneasy transition between the two views: as dry exhalation, the air is naturally hot and can be further heated or ignited into fire, while as non-smoky air – for we dare not call it "vapor" – it can be transformed by friction into fire.

As Aristotle ascends from meteors and comets to the Milky Way, he abandons the terrestrial exhalation in favor of the immediate product of friction. Meteors require a fuel (ὑπέκκαυμα 1.4.341b19, 24, 25, 29), which in general is spread throughout the kapnosphere but is capable of localized concentrations. These concentrations ignite when subject to movements, however slight (1.4.341b18–24).[1] The account of comets, which follows, introduces some innovations. Their longer duration and star-tracking speed require a different mode of burning and a kapnosphere that rotates with the heavens (1.7.344a8–22). But still a burning fuel (ἐκκαίειν 344a18; ὑπέκκαυμα 344a29, 31, b14) supplies their material.[2] The Milky Way, however, though it is explicitly "the *comet* of the greatest circle" (1.8.346b6), is never said to be the effect of a burning fuel. Rather,

> the outermost part of the so-called air has the capacity to become fire, so that when the air is dissolved (διακρινομένου τοῦ ἀέρος) by motion, a mass separates off which we call comets [and the Milky Way]. (1.8.345b32–5)

Though the similarity to comets is explicit, no exhalation, no fuel, no combustion is mentioned in this chapter. Instead the air is subject to διάκρισις, not here the "dispersal" or "scattering" we often find elsewhere, but the process of generating a lighter element from a heavier.[3]

We could reconcile the processes of combustion and friction by supposing that the heavenly sphere is simply dissolving and burning the ὑπέκκαυμα that has arisen from the earth. But Aristotle's silence on this

[1] καίειν 1.4.341b2, 26, 27, 30; ἐκκαίειν 1.4.341b16, 20, 23, 36; 342a2, 15 (ἔκκαυσις), 17. On the ignition 1.4.341b18–24: "we must suppose that this 'fuel' as it were, what we just called fire, is spread out at the remotest part of the sphere around the earth, so that when it meets with a slight movement it often catches fire like smoke. For flame is a boiling of dry *pneuma*. So wherever such a concentration is most opportune, whenever it is moved somehow by the celestial circle, it ignites." Cf. *de An.* 2.5.417a6–9, where the sensitive faculty is compared to combustible material (καυστόν) which "does not burn just by itself without something to ignite it. For in that case, it would burn itself and have no need for some actual fire."

[2] "It is laid down by us that, of the part of the cosmos surrounding the earth which is below the circular motion, the first part is dry and hot exhalation. This itself and the greater part of the air which is next to it is carried along around the earth by the circular motion. And being carried in this way, wherever it happens to be well mixed, it often ignites. For this reason, we say sporadic shooting stars occur. So whenever a fiery principle falls into this sort of dense body on account of the motion from above, neither so great as to burn quickly and extensively, nor so weak as to be quickly extinguished, but strong enough and widespread enough, and at the same time whenever there happens to arise from below a well mixed exhalation, this becomes a comet star, in whatever shape the exhausted material happens to be arranged" (1.7.344a8–22).

[3] See Chapter 3, p. 67 above.

score, his stress on διάκρισις, and the absence of the dry exhalation suggest that he was looking for a different cause for the Milky Way. His reasons for doing so are not hard to see. His treatment of comets concludes with a discussion of the kinds of periods and years that are particularly prone to them (hot and dry periods; 1.7.344b18–345a10), but such variations in the production of the material cause are difficult to square with the unchanging permanence of the Milky Way.[4] In view of this consideration, it would be better to rely on a cause that can generate an unvarying quantity of the necessary material, and one that, consistent with the dignity of the Milky Way, does not require a terrestrial phase or the consumption of a fuel.

And yet in important ways the material that forms the Milky Way is similar to a terrestrial exhalation. Like the terrestrial exhalation, it need not remain where it was originally produced. It forms a concentrated mass (σύστασις), and gathers (ἀθροίζειν 1.8.346a22) around the brightest stars, because on each rotation its material has been and is being separated off (ἀποκεκρίσθαι καὶ ἀποκρίνεσθαι 346b8) and moves there from elsewhere across the heavens.[5] Like an exhalation, too, this σύστασις can be dispersed by the intense heat of the sun and planets in the zodiac (346a11–16). For this reason it gathers into the areas outside the tropics (ἀθροίζεσθαι τὴν σύγκρισιν 346a22–3).[6] In sum, then, the Milky Way partakes, albeit awkwardly, in both material natures. We shall consider its reflective nature below.

Meteors (1.4): polemic with Anaxagoras

Aristotle does not discuss his predecessors' opinions about meteors, and little can be attributed to anyone before Anaxagoras.[7] But by the time he was writing, the species of meteors had names, though some, like "torches" and "goats," seem technical and not in common parlance (οἱ καλούμενοι ὑπό τινων δαλοὶ καὶ αἶγες 1.4.341b3).[8] We may safely conclude that the

[4] So Olympiodorus 75.24–9 attributed to Ammonius.

[5] In fact, other compounds of κρίνειν are abundant here, though less frequent for the meteors, and only present there in the context of ejected shooting stars. It is hard to interpret the evidence clearly. Milky Way: ἀποκριν- 1.8.345b34; 346b8 (twice); διακριν- 345b34; 346a9a, 15; ἐκκριν- 346a1, b6; συγκριν- 346a4, 16, 23. Meteors: διακριν- 1.4.341b15; ἐκκριν- 342a1, 15, 18; συγκριν- 342a29.

[6] It is strange, but not an insurmountable difficulty, that the sun should both produce and in the same moment dispel this fiery material. One also wonders what happens when the sun passes through the Milky Way (cf. Philoponus 114.20–30).

[7] Aëtius may credit Xenophanes with some thoughts on the issue, but the evidence is unclear (3.2.11 [*DG* 367.17–19]): πάντα τὰ τοιαῦτα are merely the gathering and movement of burning clouds.

[8] Ideler 1834–6: 1.362–3 alone argues that "goat" does not derive from the animal, but from the verb ἀίσσω, "to dart." He is not supported by Frisk 1960: 1.41–2, who takes it as an extended use from the

distinction he draws between them, at least in their phenomenal descriptions, predates him, for example, the goats have sparks shooting off their sides and torches do not. These sparks go back at least to Anaxagoras whose "so-called shooting stars descend from the aether like sparks and are immediately extinguished."[9] Now, for Anaxagoras the spark-like descent and its extinction are undoubtedly the causes of a wider group of such phenomena, including lightning (2.9.369b15–18), and Aristotle probably rejects his explanation for reasons that will become clear when we consider his theory of thunder. However, he adopts Anaxagoras' "sparks" (σπινθήρων) and applies it to one of his species, the so-called "goats" (ἀποσπινθηρίζῃ 341b30).

But Aristotle's dispute with Anaxagoras is not confined to this minor detail. In what follows I want to argue that his critique of Anaxagoras is visible in his entire description and classification of meteors and that only by understanding this critique correctly can we resolve the problems in his system of classification:

> These things [meteors] differ according to the orientation and quantity of the fuel. If the fuel has breadth and length, a burning flame (φλόξ) is often observed just as when stubble is being burnt in a field; if it has length only, then the so-called torches (δαλοί) and goats (αἶγες) and shooting stars (οἱ δοκοῦντες ἀστέρες διάττειν) are observed. If the fuel is greater in length than in breadth, whenever it throws off sparks as it burns – and this occurs because small portions burn alongside it but at its head[10] – it is called a goat; but when it occurs without this characteristic it is called a torch; and if the parts of the exhalation are scattered into little bits all over in breadth and depth, then the apparently shooting stars are produced. (341b24–35)

It would be easy to suppose that Aristotle is providing a straightforward classification of what we today call meteors, and with a little ingenuity all these phenomena can be made to fit modern categories.[11] Thus, the

mammal (so also Chantraine 1968: 1.36). Seneca understood it as a goat: *Aristoteles quoddam genus horum capram vocat* (*QN* 1.1–2). He goes on to mock the name and those who try to figure out why Aristotle used that term. Philoponus 60.5–6 and Olympiodorus 38.1–2 explain the name from the flocks of a goat's wool that hang down and are separated from one another, perhaps inspired by Homer, *Il.* 2.446–9.

9 Gilbert 1907: 688n3 notes the sparks coming off stars in Homer, *Il.* 4.75 and *Hymn. Hom. Ap.* 440–3. Anaxagoras (Aët. 3.2.9 [*DG* 367.10–13]): τοὺς καλουμένους διᾴττοντας ἀπὸ τοῦ αἰθέρος σπινθήρων δίκην καταφέρεσθαι, διὸ καὶ παραυτίκα σβέννυσθαι. So also Metrodorus of Chios, DK 70A14 (2.232.20–1) = Aët. 3.2.10 (*DG* 367.14–16), for whom the sun forcibly descends into the clouds and sparks (σπινθηρίζειν).

10 Philoponus interprets ἐπ' ἀρχήν as "continuously with the beginning" (συνημμένως τῇ ἀρχῇ 60.3–4).

11 Littmann 1998: 36.

φλόξ corresponds to our fireball, a pea-sized piece of debris that produces a light brighter than Venus and of discernible breadth.[12] The δαλός, a simple, single fire, corresponds to the sporadic meteor and produces a thin bright line.[13] Upon entry into our atmosphere, meteors can break up into fragments, causing the appearance of sparks commonly, though not invariably, associated with fireballs (αἶξ).[14] And finally the apparently darting stars (οἱ δοκοῦντες ἀστέρες διάττειν) that are scattered in small parts and in many directions would seem to be meteor showers.

On this account kapnospheric phenomena in general are distinguished by the position and quantity of their material (κατὰ τὴν τοῦ ὑπεκκαύματος θέσιν ἢ τὸ πλῆθος 341b24–5). Shooting stars and such like are differentiated again by length and breadth; so too, by extension are the aurorae of chapter 1.5. Comets (1.6–7), besides differing in position and quantity, differ in shape, some being similar in all directions (the hairy "comets" specifically speaking), others being elongated, like beards (πωγωνίαι). The fact that the Milky Way is long and thin (unlike the point-like celestial bodies) may have persuaded Aristotle to include it in this group.[15]

But a more careful examination of the text suggests that Aristotle was less concerned to provide a taxonomy based on shape and appearance than he was to account for the strange direction and movements these phenomena display, which were so naively accepted by Anaxagoras and at odds with the principles of Aristotle's own physics. He begins with a description of the burning flame (φλόξ), whose fuel has both length (μῆκος) and breadth (πλάτος), and which he likens to the burning of stubble in a field (341b25–7). The fuel of torches, goats, and shooting stars have length only (341b27–8). The obvious interpretation is to suppose that the φλόξ makes a thick streak and the others a thin. But properly speaking the two-dimensionality of the field (ἄρουρα) in the analogy serves to describe not the φλόξ itself, but the fuel through which the φλόξ burns, that is, the analogy seems mainly directed at showing how the ὑπέκκαυμα can burn in two directions at the same time. But even this is strange, since in a field a fire burns in a straight or curved line that moves across the field, an appearance that corresponds better to the aurora than to the fireball of the modern interpretation (we shall consider this issue again in the context of the aurora). Oddly too, shooting stars (οἱ δοκοῦντες ἀστέρες διάττειν) are said to be scattered in breadth and *depth* (βάθος) (341b32–5).

12 Bone 1993: 41, 48–9. 13 Bone 1993: 7. 14 Bone 1993: 41.
15 This issue may lie at the basis of Aristotle's argument against Democritus' planetary conjunction theory of comets, that two points in conjunction still just make a point (1.6.343b32–344a2).

Now, it is possible that Aristotle is stressing the distinction between meteors that have thin trails and those that have deep trails, but in the following section, where he discusses meteors that descend (341b36–342a16), explanations for their unnatural motion become the focal issue. According to the first and more pervasive mode of ignition, a flame passes through and ignites the dry exhalation on a downward path, a motion Aristotle likens to the swift descent of a flame that is placed in the smoke of a freshly extinguished lamp.[16] The second kind of descending meteor results from ejection (ἔκθλιψις 342a15), when moist air surrounding some dry exhalation contracts and forces the exhalation down and out, just as when one squeezes a pip from between one's fingers. It is probable that Aristotle intended to contrast these latter forms of burning, which take up the bulk of his discussion here, with the form illustrated by the burning field of stubble.

In light of these other forms of meteor, then, it is likely that when Aristotle says the fuel has length and breadth, he is not concerned about the shape of the flame (φλόξ). He is trying instead to make clear how a burning light can seem to move in a transverse direction in spite of its naturally upward motion. His answer: it is like a fire sweeping across a field. And this in turn suggests that the primary purpose of his three analogies (stubble, extinguished lamp, and pip) is to illustrate the unexpected directionality of the meteors' apparent motion. It seems, then, that in this section (from 341b36) Aristotle intends to distinguish the torches (δαλοί), flames (φλόγες), and goats (αἶγες), which move more transversely, from the ἀστέρες that descend in the manner of either the lamp or the pip (most ἀστέρες move crosswise as a result of the combination of natural and unnatural motions).[17] Philoponus clearly understood the text in this way, and a further parallel between Aristotle's analogies confirms the interpretation. In comparison with burning fields, the ignition of the lamp smoke and the ejection of the pip suggest small objects, as is appropriate for masses being moved contrary to nature.

There are objections to this interpretation. The flame (φλόξ) burns in length and breadth while torches and goats burn in length only, and this distinction does not seem to correspond to different kinds of

[16] Philoponus 61.14–25, conscious that Aristotle is explaining motion contrary to nature, says that the ἔκκαυσις in descending meteors travels through a rare exhalation and this rareness causes the flame to burn out above as it descends.

[17] διὸ καὶ τῶν διαθεόντων ἀστέρων ἡ πλείστη λοξὴ γίγνεται φορά (1.4.342a26–7). In the following chapter we hear of torches (δαλοί) too being ejected by their surrounding air (1.5.342b16–17).

natural or unnatural motions.[18] According to Philoponus, some people have interpreted length (μῆκος) and breadth (πλάτος) as the east–west and north–south direction respectively; βάθος is up and down (59.12–15). But, he continues, since neither Aristotle nor Alexander supports this usage, we should suppose that μῆκος means the longer direction, however the object is oriented. Philoponus is clearly wrestling with the same kind of interpretative problem we are, and he is certainly correct to conclude that the flame (φλόξ) has breadth that the torch and the goat lack. On his interpretation some meteors move transversely in one direction, others move in two transverse directions ("like when a field of straw or a heap of wood burns from all sides" [πανταχόθεν 59.32–3]), and others again move in all. In one passage Aristotle suggests that "breadth" (πλάτος) denotes an axis of direction: though ἀστέρες are classed with torches and goats as phenomena that burn in length only (341b27–8), five lines later they fly in all directions.[19] Here shooting stars are associated with breadth and depth, and the significance of "breadth" (πλάτος) is clearly not the thickness of the light, but rather that thin shooting stars move on *both axes* across the heavens.[20] This is as far as we can go within the confines of this chapter to explain the appearance of the φλόξ, and I shall propose a new solution later in my discussion of the aurora. For now though, while it is unclear to what extent Aristotle intended a distinction of shape (and there seems little escaping this interpretation in the case of the flame), there is no doubt that directionality was uppermost in his mind. Both considerations are certainly captured in the official differentia stated at the beginning of the section: "in accordance with the position of the fuel" (κατὰ τὴν τοῦ ὑπεκκαύματος θέσιν 341b24).

18 Alexander, Olympiodorus, Ideler, and Strohm are of no help in distinguishing the kinds of meteor. There is no evidence that it is a concern of Anaxagoras. In 1.5 on the aurora, Aristotle seems to be hinting at a distinction between the δαλός and ἀστέρες that move and the φλόξ that "seems to burn" (1.5.342b3–4).

19 ἐὰν δὲ τὰ μήκη τῆς ἀναθυμιάσεως κατὰ μικρά τε καὶ πολλαχῇ διεσπαρμένα ᾖ καὶ ὁμοίως κατὰ πλάτος καὶ βάθος, οἱ δοκοῦντες ἀστέρες διάττειν γίγνονται (1.4.341b32–5). "If the lengths of the exhalation are scattered in small [bits] and in various directions, and likewise in breadth and depth, shooting stars seem to appear." Lee, following Ideler, reads μέρη (with $E_{corr.m.1}$ and Parisinus 314), in which case we are reduced to two axes from three; Fobes reads μήκη with E, J, and the lemma of Philoponus and Alexander's interpretation (21.26–7).

20 In describing how the shape of the wet exhalation can affect the direction of the ejection, Aristotle again uses πλάτος to denote an axis of direction rather than thickness: διὰ δὲ τὴν θέσιν τῆς ἀναθυμιάσεως, ὅπως ἂν τύχῃ κειμένη τοῦ πλάτους καὶ τοῦ βάθους, οὕτω φέρεται ἢ ἄνω ἢ κάτω ἢ εἰς τὸ πλάγιον (1.4.342a21–4). Cf. Alexander 22.30–1. Philoponus 65.20–3 concurs, explaining that if the condensation occurs to the north, the ejection will occur towards the south, etc. (65.25–7); he adds (65.27–32) that the θέσις can also be interpreted as the position of the ὑπέκκαυμα: if it is scattered along the east–west axis, the flame will travel along that axis. This latter, however, makes a very poor explanation for an ejected shooting star.

Aristotle's concern with the direction of motion is probably a response to the theory of Anaxagoras. He is eager to show that the apparent movement of the shooting stars is really the progress of a flame through the more or less stationary fuel, and not the locomotion of the fuel itself, as Anaxagoras had claimed.[21] He structures his account of the meteors so as to make apparent motion the rule and real motion (ejection) the exception. In this way he can account for the apparently unnatural motion as Anaxagoras could not, by showing that it is only apparent. The analogy of a flame placed above the smoke of a freshly extinguished lamp makes this clear (342a3–8). Anaxagoras cannot explain why the sparks should be carried down by the aether, and Aristotle explicitly criticizes him on this account later when he discusses his theory of lightning (2.9.369b21–2). As for the really unnatural motion of the ejected shooting star Aristotle shows the kind of physics Anaxagoras ought to have used, and in so doing he preserves a remnant of the Ionian tradition in which the wet exhalation is involved in the production of fiery phenomena. In this case though, Aristotle has transformed the Ionian material cause into a purely efficient cause, since the wet exhalation merely ejects the dry. Here again, Aristotle emends a predecessor's general theory and transforms it into a species, thereby preserving some truth.

Aristotle's account also seems to be directed against Anaxagoras' apparent failure to provide sufficient conditions for the phenomena, when he says that shooting stars and other such phenomena occur where there is a *well-formed concentration* of dry exhalation (341b22; cf. 344a14, 20). This is assuredly not a very definite condition, but it shows a clear awareness that conditions must be right for the production of the phenomena, and that it is not simply a matter of chance. Now, we have no evidence that Anaxagoras' theory of shooting stars was vulnerable to this criticism, or whether the aether generated sparks under any particular conditions or at any particular times, but Aristotle brings the same criticism against Anaxagoras' theory of lightning (2.9.369b22–4), and so in Aristotle's mind the criticism was probably justified here as well.

Aristotle had other reasons to distinguish the meteors from celestial phenomena, which have nothing to do with Anaxagoras and become clear only when he discusses comets and the Milky Way. The comets and especially the Milky Way move closely with the fixed stars, whose rotary motion both drives and limits the speed of the kapnosphere below it. But the shooting

[21] Cf. 2.3.357b30–2, where Aristotle compares the flux of salinity of the sea to both the flowing of rivers and the streaming of the flame.

stars do not move at the same speed as the stars and appear to move much faster.[22] It is evident, then, that their apparent motion cannot be caused by the celestial sphere, and Aristotle uses their speed as proof that they are sublunary: "their apparent speed is similar to that of things thrown by us" (1.4.342a30–3). A moving baseball, for example, viewed against the background of the sky, appears to move much faster than the stars. Aristotle concludes – in a way that shows he is aware of parallactic effects for low altitude phenomena – that the real speed of meteors is relatively slow, and this confirms their sublunary nature.

But inasmuch as Aristotle sees the kapnosphere as "dualizing" between the celestial and sublunary realms, he draws a contrast between these kapnospheric phenomena, which occur when the dry exhalation has already arrived at its natural place, and the phenomena of precipitation, which occur as a result of matter moving from place to place. True generation and corruption can only come about with the introduction of the motion of the sun along the ecliptic.[23] In this respect the kapnospheric phenomena are activities (in contrast to the atmospheric changes), actualizations of already fiery capacities, and therefore partake more fully in the celestial nature they are imitating.[24] And this in turn is another reason to avoid discussing the cyclical transformation of the dry exhalation.

Hierarchy and the rotation of the kapnosphere

Aristotle's commitment to hierarchical order in the cosmos and to the continuity between members of the hierarchy led him to conceive of the kapnosphere in close imitation of the heavens, and this generated a problem recognized by Xenarchus in the first century BC and revisited regularly throughout late antiquity. For those who thought that the heavens were made of fire (not Aristotle's view), it was common to claim that fire has a natural motion upward when it is out of its natural place, but that, when it reaches its natural place, it naturally rotates.[25] Aristotle, of course, denied

[22] That meteors, in fact, track the fixed stars was noted by von Humboldt and by Denison Olmsted, who observed during the Leonid storm of November 13, 1833 that the radiant (the area from which the meteors of a meteor shower seem to emanate) moves across the sky together with the fixed stars (Bone 1993: 51).

[23] *GC* 2.10.336a31–4: "for this reason too the first motion is not the cause of coming-to-be and passing-away, but rather the motion in the ecliptic circle. For in this motion there is both continuity and a combination of two motions."

[24] *Met.* 9.6.1048b29–34; 11.9.1065b14–1066a7. Hankinson 2003: 34.

[25] Philoponus 37.14–23; 97.4–16. Simplicius, *in de Cael.* 20.11–15 attributes the opinion to Ptolemy and Plotinus. My thanks to István Baksa for clarifying many of these issues.

that rotary motion is the natural motion of fire. But Xenarchus noted that in the *Meteorologica* Aristotle made the motion of the kapnosphere, albeit moved by the heavens, rotary and eternal. In the *de Caelo*, by contrast, a natural rotary motion for fire is impossible, since one simple body has only one simple natural motion, and fire moves naturally upwards; nor can fire move unnaturally in a rotary motion, since an unnatural motion cannot be eternal.[26] Especially troublesome is Aristotle's argument that makes the rotary motion of fire neither natural nor unnatural (1.3.269a9–17), since it is in clear contradiction with premises and conclusions of other arguments that admit that fire can rotate albeit unnaturally.[27] But in late antiquity a compromise was sought between the *de Caelo* and the *Meteorologica*, first by Damascius, according to which the rotary motion of the kapnosphere was "preternatural" (ὑπὲρ φύσιν).[28] Since according to Aristotle neither natural nor unnatural rotary motion is possible, commentators argued that this rotary motion is caused by a third, preternatural and beneficial force, which is compared, at least by Simplicius, to the kinds of motions the souls of animals produce in their bodies in addition to the tendency of the various material constituents of their bodies to seek their natural places.[29]

There is no evidence of a preternatural motion for fire in Aristotle's text and Xenarchus' arguments against the aether take us beyond our present concerns. But this is a suitable opportunity for observing how strange the doctrine of rotary fire is. Its physical motivations arise strictly from the problems of the *Meteorologica*, specifically the motion of the comets, the Milky Way, and later the winds, and are curiously at odds with the friction theory of heat with which they are so intimately related. On the one hand, in order to generate the friction necessary for combustion or dissolution, the kapnosphere must be stationary relative to the earth while the celestial sphere rubs quickly past it. On the other, the motion of the comets and the Milky Way show that the upper kapnosphere moves at

26 "There is a single natural movement for each of the simple bodies" (1.2.269a8–9); "it is fantastic and completely unreasonable that this motion alone is continuous and eternal, since it is contrary to nature" (269b7–8).

27 "So it is necessary that motion in a circle, since it is unnatural to these bodies [fire and earth], be natural to some body" (1.2.269b1). "It is possible for a simple body to be forcibly moved in accordance with the natural motion of a different simple body; but it cannot do so naturally" (269a8). "If, as some say, the body that is moved [in a circle] is fire, this movement is no less unnatural to it than downward movement" (269b10–11).

28 For Damascius, see Philoponus, *in Mete.* 97.20–1. Discussions of Xenarchus are found in Simplicius, *in de Cael.* 13.22–14.29; 20.10–25.21; 50.18–24; 55.25–56.25; 286.2–27; there is one in Julian, *Orat.* 8 (5) 3. For modern discussions, see Sambursky 1962: 122–32; Moraux 1973: 198–206; Falcon 2005: 62–9; 2012; Hankinson 2003; 2009; Baksa.

29 Simplicius, *in de Cael.* 21.24 calls it beneficial; Olympiodorus 2.26–9.

the same rate as the celestial sphere, and consequently cannot be subject to friction.[30] This is not a problem for the meteors, since they have only apparent motion and do not track the stars. It is, however, a problem for the comets and Milky Way, and would require for a solution the addition of *ad hoc* principles Aristotle does not see fit to introduce. We shall return to this issue below. These facts suggest that Aristotle first developed his theory of meteors with a geo-stationary kapnosphere in mind (in the same order as his presentation). Its rotation with the celestial sphere is a subsequent refinement whose only physical justification is the tracking motion of the comets and the Milky Way. Its later involvement in the production of the winds is, as we shall see, highly problematic. There is no reason to believe that the rotating kapnosphere was a deep physical commitment on Aristotle's part (it is but a secondary consideration in the distinction between the vapor and the smoky layer in 1.3.340b32–341a2). It served a narrow purpose in the upper sublunary, and the embarrassments it generated for cosmology and meteorology were tolerated because of the orderliness of its hierarchical effects in creating a region in which the comets and the Milky Way could imitate, however imperfectly, the celestial nature.

Chasms, trenches, and bloody colors (1.5)

The chasms, trenches, and bloody colors are the secondary effects of lower kapnospheric fires, dependent on the flames, torches, and shooting stars of the previous chapter (1.5.342b3–5) and for that reason treated next in order. Here, as before, our first challenge is identification. Thereafter we shall consider their mutual differences and finally their similarity with other phenomena and especially with the elusive flame (φλόξ).

The common view, for the most part uncritically adopted, is that Aristotle is talking about various forms of the aurora borealis. None of the ancients seem to have known exactly what Aristotle was referring to, and they tend to be very casual in their terminology. When interest in these phenomena resumed in the early modern period among northern Europeans, scientists adopted Aristotelian terms to refer to phenomena of which they had abundant first-hand experience.

In favor of the identification with the aurora are the following considerations. (1) Aristotle clearly distinguishes this group of phenomena from the torches, goats, and shooting stars on the one side and the comets on the other. He thought that they were different in appearance and in

[30] So Thorp 1982: 118.

etiology. (2) These phenomena are appearances (φάσματα), some of them reflections, whereas the torches and the like are real combustions. The term φάσματα seems to capture well the shimmering appearance of the aurora. (3) Aristotle mentions that they alone of the upper phenomena have color. The bloody colors would correspond best to the kinds of aurorae that are most frequently visible at mid-latitudes.[31] The terms "chasms" (χάσματα) and "trenches" (βόθυνοι) suggest the appearance of depth that the classical northern-latitude aurora presents. (4) It is clear from Aristotle's opening description that many of these φάσματα appear in the sky at the same time (φαίνεται δέ ποτε . . . πολλὰ φάσματα 342a34–5); they do not remain long (342b13–14) and their appearance is rapid (ταχείας τὰς φαντασίας 342b23). Such a description might accurately capture the disappearing and reappearing shimmers of the aurora. (5) He is explicit that these appear only at night (νύκτωρ 342a34), whereas bright fireballs can sometimes be seen during the day.[32] (6) It is not geographically impossible that Aristotle is describing the aurora. Though Athens is not favored for such observations (being at 38°N in eastern Europe, it is relatively far away from geomagnetic north), the aurora has been seen in Greece more than very rarely.[33] There are reports of aurora-like phenomena in antiquity.[34] (7) There is no other really suitable candidate for these descriptions.

The identification has been denied on the grounds that (1) Aristotle does not mention specific instances of a phenomenon that is very rare in Greece. It has been suggested instead that he is describing "phenomena of cloud coloration."[35] Aristotle is explicit, however, that the phenomenon occurs at night in a clear sky (αἰθρίας 342a34).[36] Ideler, who supports the aurora

[31] Bone 1996: 110–25; Brekke and Egeland 1983: 37.

[32] Bone 1993: 41; Seneca also reports that more than one has been seen (*QN* 1.1.11–12).

[33] Carapiperis 1956: 141 relates a report of an aurora: "On Saturday the 22nd January 1872 at about two o'clock in the morning the sky became red and appeared everywhere and within it was something like white columns and over Mavrovouni it shone to day in such a way but in a lesser intensity than in the autumn of 1870." Aurorae have been observed even to within a couple of degrees of the equator in extraordinary circumstances (the aurora of February 4, 1872 was visible in Aden and Egypt; the aurora of September 25, 1909 was visible from Singapore (1° 25' N)). Bone 1996: 18, 21, 108–9. See a fine red aurora photographed over Athens November 20, 2003, www.perseus.gr/Astro-Aurorae-20031120-000.htm.

[34] Pliny, *HN* 2.26.97 (a red chasma in 349 BC associated with Philip II's aggression against Greece) and Seneca, *QN* 2.15.5 (in the time of Tiberius).

[35] Gilbert 1907: 594–7; Webster (1.5 note 1); Strohm 1984: 142–3. Vicomercatus suggests St. Elmo's fire (26H–27C).

[36] Olympiodorus 46.34–5 interprets αἰθρίας as "without wind" (presumably because otherwise the dry exhalation would be dispersed). This is not a condition, however, that Aristotle sets for the appearance of other kapnospheric phenomena. Moreover there is no indication that αἰθρίας means anything but "clear sky" for Aristotle (1.4.342a12; 2.9.369b23; at 1.10.347a26 αἰθρίας is clearly different from νηνεμίας). As for cloud coloration, there are a couple of possibilities. Noctilucent

view, points out another drawback: (2) Aristotle does not mention that the phenomenon appears only in the northern sky. Although it can suffuse the entire sky, it is seen predominantly in the north.[37] (3) What do we make of the torches (δαλοί), which we thought were very short-lived phenomena, a kind of shooting star, if they are associated with the somewhat longer-lived aurora?

It is especially frustrating that the ancients are of so little help. In fact, there is not a single unambiguous eyewitness report that is informed by Aristotelian terminology. Aëtius has no heading for χάσματα, βόθυνοι, or αἱματώδη χρώματα, and the ancient commentators on the *Meteorologica* are interested only in parsing Aristotle's words. Aristotle mentions no predecessors on this issue, and there is no other evidence.

A short glance at the tradition is worthwhile to reveal the extent of the confusion and to honor Aristotle's powers of classification, if these phenomena are indeed aurorae. Difficulties begin already in the Peripatetic *de Mundo* with the introduction of a host of unAristotelian terms:

> In this so-called fiery and disordered substance, flashes (σέλα) shoot and fires (φλόγες) are hurled along with beams (δοκίδες) and trenches (βόθυνοι), and so-called comets (κομῆται) are often fixed in place and extinguished. (2.392b1–5)

The distinction between moving and fixed phenomena is still observed here. Again,

> many other forms of appearance are seen, the so-called torches (λαμπάδες) and beams (δοκίδες) and jars (πίθοι), and trenches (βόθυνοι), receiving their names from their likeness to these objects. Some of them are seen in the west, others in the east, some in both places, but rarely in the north and south. They are all unstable. For none of them has ever been reported as seen fixed in place. (395b10–15)

These are still described as appearances and distinguished from comets and shooting stars. But later these reflective appearances become simply assimilated to other upper fires, and their specifically reflective nature

clouds are predominantly a northern phenomenon but can appear at mid-latitudes. They do not, however, display the variety of colors Aristotle describes, and they occur during twilight before night (i.e., before the sun is 18° below the horizon). Nacreous clouds display a range of iridescent colors, but they are also a twilight phenomenon. Moreover nacreous clouds "stand majestically in almost the same place" (www.atoptics.co.uk/highsky/nacr1.htm). Contrast Aristotle's description μὴ πολὺν χρόνον μένειν 1.5.342b13. My thanks to István Baksa for pointing these possibilities out to me.

[37] Ideler 1832: 50.

disappears.[38] The tendency to treat these phenomena as forms of comets becomes the general practice even later.[39]

We return to firmer ground only in the early modern period when authors are no longer relying merely on book descriptions. Tycho Brahe in his *Meteorological Journal* is clearly describing the aurora, to which he applies the Aristotelian term "chasmata."[40] In southern latitudes the phenomenon is called *capre saltanti* ("leaping goats," obviously a variation on the Aristotelian αἴξ) or *tizzoni ardenti* ("flaming trees"); alternatively they are called "comets."[41]

38 There is no evidence that Posidonius discussed these phenomena under the names χάσματα or βόθυνοι. Seneca (*QN* 1.14) groups the celestial fires by their appearance rather than by their etiology. He discusses among the other celestial fires *bothunoi* or *putei*, "wells" (both are emendations for a thoroughly corrupt *ut ei*), *pithiae*, *pithoi*, and *chasmata*. The list is similar to what we find in the *de Mundo*. Seneca's description of the "well" suggests an aurora: *cum velut corona cingente introrsus ingens caeli recessus est similis effossae in orbem specus* ("within a surrounding corona there is a great gap in the sky like a hole dug in a circle," Corcoran trans.). There is no clear indication that he ever saw these things himself. The *pitheus* is also mentioned by Pliny, *HN* 2.22.90, but here it is listed among comets; his general description of comets (2.22.89) evokes Aristotle's αἱματώδη χρώματα: "the Greeks call them 'comets,' in our language 'long-haired stars,' because they have a blood-red shock of what looks like shaggy hair at their top" (*cometas Graeci vocant, nostri crinitas horrentis crine sanguineo et comarum modo in vertice hispidas*) (Rackham trans.). Neither Pliny nor Seneca mentions reflection as a cause of the appearance and in both cases they are considered fires.

39 Ptolemy, *Tetrabiblos* 2.9.18.5 (followed by Hephaistion, *Apotelesmatica* 50.24) mentions χάσματα together with βρονταί, πρηστῆρες, σεισμοί, and ἀστραπαί but does not define the terms further. Suidas K1996.7 places the βόθυνος among four kinds of comet, among which is the very common πωγωνίας; similarly John Lydus, *Ostentis* 10b, who also places goats (now called τράγοι) into the class of comets. He and Joannes Camaterus, *Introductio in Astronomiam* (1908: 3381–94), discuss χάσματα among fiery phenomena, following Pliny's account (or some other source) concerning Philip II, merely saying that χάσματα are like a fiery phenomenon, and not specifying times or locations. The latter does, however, mention a στέφανος, which appears in the north and appears as a comet (3367–9).

40 Brahe 1876. For a list of aurora sightings, see the Résumé, lxxii.

41 Brekke and Egeland 1983: 38–9 for details. Negri 1929: 188–90 used the terms, though he did not believe that the two phenomena were the same in cause as the aurora. The *King's Mirror* (2.2) from around AD 1250 describes the aurora independently of Aristotle but with striking parallels (Brekke and Egeland 1983: 37): "the darker the night is, the brighter they seem, and they always appear at night but never by day, — most frequently in the densest darkness and rarely by moonlight. In appearance they resemble a vast flame of fire viewed from a great distance. It also looks as if sharp points were shot from this flame up into the sky, these are of uneven height and in constant motion, now one, now another darting highest; and the light appears to blaze like a living flame. While these rays are at their highest and brightest they give forth so much light that people out of doors can easily find their way about . . . The light is very changeable. Sometimes it appears to grow dim, as if a black smoke or a dark fog were blown up among the rays; and then it looks very much as if the light were overcome by this smoke and about to be quenched. But as soon as the smoke begins to grow thinner, the light begins to brighten again; and it happens at times that people think they see large sparks shooting out of it as from glowing iron which has just been taken from the forge. But as night declines and day approaches, the light begins to fade; and when daylight appears, it seems to vanish entirely." Nordic folk theories often treat it as a reflection (Brekke and Egeland 1983: 2).

The ancient successors to Aristotle, then, did not have a clear idea of what he was referring to and were content to confuse these appearances with the rest of the burning phenomena. In such circumstances, we can only return to Aristotle's description. He says that these phenomena all share the same cause (342a36–b1), and to a limited extent this is true. All arise from an interaction between an unignited smoky exhalation and an ignited one, where the unignited mass somehow alters the appearance of the burning light source. Nevertheless, the three kinds of appearance (φάσματα) have distinct causes according to how the unignited mass affects their appearance. Moreover, since they share in the same range of combustions (φλόξ, δαλός, and ἀστέρες 342b3–4), they probably also differ according to which form of combustion they are appearances (φάσματα) of. On this issue, however, Aristotle offers few specifics.

Color is a universal property. The first kind of appearance occurs when a light seen through a smoky cloud produces a red or purple color (342b5–11).[42] This is formed through a mixture of fiery and white colors superimposed, and the effect is similar to when a rising or setting star is seen through a smoky haze. In such a case, evidently the form of the light source is preserved and there is no reflection. The phenomenon is similar to cloud coloration, though Aristotle is careful to explain that the light must come through *smoke* and not vapor.

Secondly, an aurora may be produced by the reflection of the optical ray from dense air to a light source (342b11–13).[43] This reflection causes a change of color similar to the first, but the details must await Aristotle's later discussion of rainbows.[44] Unlike the first kind of aurora, the form and shape of the light source are not preserved and may produce a more diffuse appearance. There should be observable differences between these first two kinds of aurorae, but it is difficult to assign them to known phenomena. It is probable that these first two causes account for the bloody colors.

Finally, the chasms and trenches arise when a torch (δαλός) is being ejected by a condensing sack of dry exhalation:

> Chasms cause an appearance of some depth when the light breaks forth from blue and black. And often torches are extruded (ἐκπίπτουσι) from

42 Aristotle's presentation is confusing. He introduces filtering and reflections together (διά τε γὰρ πυκνοτέρου διαφαινόμενον ἔλαττον φῶς καὶ ἀνάκλασιν δεχόμενος ὁ ἀήρ 1.5.342b5–7), and though he says in the context of rainbows that the effect is the same (3.4.374a4), the process is clearly different.

43 This ray emanates from the eye and strikes the object of vision. For more details, see Chapter 12.

44 Philoponus 70.20–7 says that in this case the cloud must not be continuous as with the first cause but rather dispersed into mirror-like drops.

> such formations whenever it gets rather condensed (συγκριθῇ), and as it is coming together, a chasm appears. (342b14–18)[45]

The torches (δαλοί) are treated here as phenomena of ejection (in the previous chapter only shooting stars were ejected). They are concentrated masses of ignited dry exhalation (συνιστάμενος ὁ ἄνω ἀήρ 342b1) surrounded by unignited exhalation. Aristotle draws a distinction, therefore, between the torch, which is ejected, and the chasm, in which the surrounding air is coming together. It is clear, then, that the chasm does not require a torch to have already separated. That is, the pocket has not yet fully condensed and has not yet fully come together so as to eject the fiery part as a δαλός. We see the light directly in this case, unlike in the first two kinds of φάσματα. But this phenomenon is still a φάσμα because the dark background of the pocket provides only the appearance of depth, and because the white of the torch on the black background produces a variety of colors (342b18–19). It is not, then, as Philoponus says, that the cloud pocket blocks some of the light, but rather the light of the torch shows up against the darkness of condensed clouds of dry exhalation, which in consequence appear blue and black.[46] But why is the black background of the cloud of exhalation necessary when it is already dark at night? The reason is to be sought among the rainbows, another reflective phenomenon, in which colors appear against the dark background of the water drops (3.4.374a3–4). Since the theory is much more appropriate to rainbows than to aurorae, it is clear that it had its origin among the former and was adapted, somewhat inappropriately, to the latter.

In all these cases aurorae are modifications of meteors, and this fact returns us to the problem we faced above in identifying the meteoric phenomenon that produces the field-like burning of the φλόξ. For this form of combustion there are no easy candidates among the familiar kinds of meteors. Mid-latitude aurorae, however, regularly appear as glowing red

[45] τὰ δὲ χάσματα ἀναρρηγνυμένου τοῦ φωτὸς ἐκ κυανοῦ καὶ μέλανος ποιεῖ τι βάθος ἔχειν δοκεῖν. πολλάκις δ᾽ ἐκ τῶν τοιούτων καὶ δαλοὶ ἐκπίπτουσιν, ὅταν συγκριθῇ μᾶλλον. συνιὸν δ᾽ ἔτι χάσμα δοκεῖ.

[46] Philoponus (73.26–8) says: "the light is broken off by the cloud and its continuity is interrupted by the imposition of the black and the blue in the middle." That is, the chasm appears because of the partial eclipse of the light source by a dark cloud of dry exhalation. The light is broken off by a cloud that blocks the light and creates the appearance of the chasm. Philoponus' interpretation seems incorrect, because Aristotle expressly says that the chasm appears while the cloud is coming together. That is, after it comes together the δαλός is either ejected or obscured and the chasm φάσμα no longer appears. ἀναρρηγνυμένου is probably being used in the middle voice. ἀναρρήγνυμι is regularly used for breaking forth or eruption: breaking through a wall (transitive), eruptions of volcanoes, floods or sores on the body (intransitive) (LSJ). We should compare our passage with 2.8.368a26 ὕδατα ἀνερράγη γιγνομένων σεισμῶν, where the aorist passive is used intransitively.

flames and take on the appearance of burning fields. For this reason we may speculate that for Aristotle the φλόξ does not really exist independently of its auroral form, and that its place among the simple combustions in 1.4 is merely Aristotle's attempt to complete the classification there. On this interpretation, Aristotle supposed that the widely diffuse field-shaped burning of the aurora was caused by a similarly shaped φλόξ seen through a smoky haze, even if he never saw the φλόξ by itself without the haze. This interpretation is corroborated by the statement at 1.5.342b1–4 that the φλόξ is immobile while the δαλός and the ἀστέρες are mobile (φέρεσθαι), a distinction that fits well with the types of aurorae discussed above, but nowhere operative in 1.4.

It is worth considering too that in using the notions of reflection and flashing forth against dark backgrounds, Aristotle may be adapting explanations his predecessors had used to explain the Milky Way. We know that the comets and the Milky Way both had been treated as reflections (probably both by Hippocrates of Chios), but that for Aristotle this causation implied that the image must move when the source of light or the mirror moves. By joining in his account of the aurora both his theory of shooting stars and the physics of reflection, he provides two reasons for their remarkable instability: they are *reflections* of *moving* flames. Finally, Aristotle tells us that according to Anaxagoras and Democritus, the Milky Way is the appearance of stars shielded from the sun by the shadow of the earth (1.8.345a25–31), and this may have inspired his own theory of the chasm, which is a light appearing against a darkened background.

Comets and the Milky Way (1.6–8)

Aristotle caps his account of the kapnospheric phenomena with a lengthy discussion of comets (1.6–7) and the Milky Way (1.8). He treats them together, because the Milky Way is merely a special form of comet and is subject to a similar explanation.[47] His predecessors, by contrast, whom he describes and criticizes here for the first time in the treatise, tend to explain comets and the Milky Way by different causes, and this for Aristotle is a sign of their failure. The predecessors all agreed that comets are a planetary phenomenon, but their opinions about the Milky Way are quite disparate. First, both Anaxagoras and Democritus hold that a comet is a conjunction of two planets, but their Milky Way is an effect produced by the many

[47] Explicitly at 1.8.346b1–6.

stars that shine when they are in the earth's shadow and therefore shielded from the extinguishing brightness of the sun (1.6.342b27–9; 1.8.345a25–31). Anaxagoras is probably extending his interest in eclipses here. Next, for the Pythagoreans the comet is a rarely seen planet, whereas the Milky Way is the old path of the sun (1.6.342b29–35; 1.8.345a13–18).[48] Finally, the theory of Hippocrates of Chios and his associate, Aeschylus, claims that there is a single comet, caused by a planet that drags vaporous exhalation behind it when it is outside the zodiacal band, and that our visual ray reflects off this exhalation to the sun below the horizon (1.6.342b35–343a20).[49] The reflection theory of the Milky Way mentioned in 1.8 is without ascription but may plausibly be attributed to Hippocrates (345b9–12). If this is so, it is only with him that the comet and Milky Way receive similar explanations.

Aristotle presents the theories of comets in order of increasing sophistication, those of Anaxagoras–Democritus and Hippocrates of Chios being quite recent and exploiting the then current fashion of celestial research.[50] Because they share a lot of common ground, he outlines them first before criticizing them and introduces his general objections before taking up the specific. All three explanations are presented as empirically motivated, and none is so embedded in a general theory as to be irrefutable at the level of the meteorological phenomena.

It is remarkable that Aristotle never explicitly asks whether comets and the Milky Way are celestial or sublunary, but his answer is implicit in his predecessors' inability to account for the phenomena, and since the assumptions they started with are in conflict with the facts, their assumptions must be wrong. His critique draws on a whole range of evidentiary techniques largely but not exclusively meteorological. He cites many comets that have been observed by himself and others in places and times inconsistent with

48 Metrodorus picks up this opinion: Aët. 3.1.3 (*DG* 365.7–9).

49 For details and an interpretation of the theory, see Wilson 2008.

50 Aristotle chose to ignore the theory of Xenophanes, which is most similar to his own, that comets, shooting stars, and the like (including the sun and the stars) are all condensations or movements of burning clouds (DK 21A44 [1.125.16–17] = Aët. 3.2.11 [*DG* 367.17–19]). Xenophanes failed to provide different explanations for the different phenomena, and he instead explained everything by reference to the sea: "the great sea is the father of the clouds and the winds and the rivers" (DK 21B30 [1.136.9–10]). Other theories were more celestial in their tenor but were so embedded in their peculiar principles as to make criticism of them pointless. Parmenides, for example, says that "the sun and the moon were separated off from the circle of the Milky Way; the sun from the rarer mixture, since it is hot; the moon from the denser, which is cold" (DK 28A43 [1.225.8–10] = Aët. 2.20.8a [*DG* 349.12–16]). This is clearly related to the Pythagorean view that the Milky Way was the old path of the sun, a theory that Aristotle explicitly criticizes (1.8.345a13–25). Elsewhere we learn that "the mixture of the dense and the rare produced the milky color," the rare and the dense being the foundations of Parmenidean physics (DK 28A43a [1.225.11–12] = Aët. 3.1.4 [*DG* 365.10–12]).

his predecessors' theories.[51] But he also calls on expert evidence from astronomical research in his refutation of Anaxagoras' explanation of the Milky Way to show that the vertex of the earth's conical shadow does not fall far from the earth and therefore cannot shield the stars from the sun's intense light (345a36–b9). His aim here is partly rhetorical. He might have noted the obvious and empirical fact that the cord-like shape of the Milky Way is inconsistent with the shadow cast by a sphere or even a disk (Aristotle assumes a spherical earth, Anaxagoras a disk). In choosing the expert argument instead, he implicitly objects that this celestial theory of the Milky Way is not even consistent with recognized celestial theorems. At the same time he reinforces his basic view of the cosmos (i.e., that the earth is tiny in proportion to the whole) and shows his audience that while he champions an old-fashioned exhalation theory, he remains abreast of the latest astronomical discoveries.

Not all of his evidence hits the mark. His most general objection, that the planets are all in the circle of the zodiac while comets are seen outside the zodiac (1.6.343a22–5), seems aptly directed against the Pythagoreans, who argued that the comet is a planet that always remains close to the sun, but for Hippocrates of Chios, this objection is precisely the question in dispute. Hippocrates' comet appears only north or south of the tropical band and has an orbit outside the ecliptic. Since Aristotle is in no position to refute the existence of such a planet except on *a priori* grounds, he can only deny it. On one occasion he even adds an abstract argument from outside the subject matter (343b32–344a2): against Democritus' theory that comets are formed by the conjunction of planets, he objects that planets are like mathematical points, and a conjunction of points only makes a point. Through all this, Aristotle directs his arguments against the planetary rather than generally *celestial* nature of comets and the Milky Way.

It is this context that allows us to interpret correctly the general methodological remark that introduces his own theory at the beginning of 1.7:

> Since concerning things unapparent to the senses (ἀφανῶν τῇ αἰσθήσει) we suppose that we have provided a sufficient demonstration in accordance with reason (κατὰ τὸν λόγον) if we refer to what is possible, it is from the

[51] Aristotle uses the following kinds of temporal and circumstantial markers: the great earthquake in Achaia (1.6.343b1–3); the year by Athenian archon and the month (343b4–6); and contemporaneous weather conditions (343b19) such as was common in the so-called parapegmata tradition (Lehoux 2007; Taub 2003: 15–69). A comet's first appearance in the evening sky is noted (343b20–1). Some evidence he himself vouches for (343b9–11, 30–2); for other evidence he refers to the Egyptians (343b9–11, 28–30): both he and they have observed that some fixed stars have tails, and both have observed without the appearance of a tail both conjunctions of planets with each other and of planets with fixed stars.

present phenomena that one might suppose that this is most likely to come about concerning these things. (344a5–8)[52]

Though comets and the Milky Way are the most remote from us, they are not so remote that the phenomena cannot be decisive between some theories, and Aristotle has just shown that the theories of his predecessors are in contradiction with the phenomena. The phenomena are not, however, so accessible that we can actually see the cause, which instead must be inferred from or tested by the empirical signs we are given. The phrase κατὰ τὸν λόγον is doing a lot of work in the prescription: when we deduce from our principles (as the predecessors have done and Aristotle immediately goes on to do), the conclusion must be possible, that is, must be consistent with the empirical evidence.[53] It is not sufficient to have a merely logically consistent theory. In this respect the predecessors have failed.

Of course, Aristotle's prescription here is directed immediately against his predecessors and should not be taken as a full description of an adequate meteorological theory.[54] He nowhere provides that description, and yet in this context it would have to mention the elegance of causal structure, which is such a powerful consideration that it compensates in Aristotle's mind for the empirical weakness of his own theory. In fact, Aristotle avoids direct engagement in the question concerning the celestial–sublunary demarcation, precisely because he is aware that his own solution is just as open to empirical refutation as those of his predecessors and is perhaps even more dependent on controversial physical principles. Though it is important for reasons of general physics that he set a boundary between celestial and sublunary realms, it is not clear that he must set it where he does. Comets, unlike meteors, regularly follow the diurnal rotation of the heavens and, like the planets, they wander. His predecessors' claim, then, that comets are planets is quite strong, much stronger than Aristotle is willing to admit.

52 ἐπεὶ δὲ περὶ τῶν ἀφανῶν τῇ αἰσθήσει νομίζομεν ἱκανῶς ἀποδεδεῖχθαι κατὰ τὸν λόγον, ἐὰν εἰς τὸ δυνατὸν ἀναγάγωμεν, ἔκ τε τῶν νῦν φαινομένων ὑπολάβοι τις ἂν ὧδε περὶ τούτων μάλιστα συμβαίνειν.

53 So Philoponus 90.4: τοῖς φαινομένοις μαχόμενον. Philoponus says that we must avoid the impossibilities of our predecessors and give our own account in harmony with the facts (πράγμασι) (90.5–10). Aristotle may also be responding in a general way to *On Ancient Medicine* 1, where a student of μετέωρα needs postulates, since these things are ἀφανέα τε καὶ ἀπορεόμενα. "If a man were to learn and declare the state of these, neither to the speaker himself nor to his audience would it be clear whether his statement were true or not. For there is no test the application of which would give certainty" (οὐ γὰρ ἔστι πρὸς ὅτι χρὴ ἀνενέγκαντα εἰδέναι τὸ σαφές [Jones trans.]). Cf. Aristotle's νομίζομεν ἱκανῶς ἀποδεδεῖχθαι κατὰ τὸν λόγον, ἐὰν εἰς τὸ δυνατὸν ἀναγάγωμεν 344a5–7.

54 Bolton 2009, especially 67, assimilates the epistemic concerns of the *de Caelo* to our passage and consequently discounts the degree to which empirical evidence figures in Aristotle's treatment of the comets.

For similar reasons making the Milky Way into a sublunary phenomenon is a hard sell, and Aristotle makes it harder still by assigning to the stars an important causal role in its nature.

It is clear that he did have legitimate objections to the planetary nature of comets. He explicitly objects that many comets appear outside the zodiac, and many fade before they set (343a24–5; 343b14–16). Their behavior is different from that of planets and different even from each other. To invent a special orbit for them, as Hippocrates of Chios did, though logically consistent, seems to be special pleading and *ad hoc*. And if comets are not celestial, neither can the Milky Way be.

But the greatest support for Aristotle's sublunary theory comes from architectonic features of his own account. He first turns to his own advantage the fact that his predecessors had treated comets as celestial. While denying their celestial nature, he takes this opportunity to bind the rift between his celestial and sublunary realms by constructing a *scala naturae*, in which comets and especially the Milky Way are made to imitate the highest realm: they follow the heavens and their motion is partly (and obscurely) caused by the fixed stars or planets. They are the most celestial of the sublunary realm and their orderly place within the structure of the cosmic whole is a sign of this truth.[55]

To aid in this new conception Aristotle adopts the partly scientific and partly rhetorical method of displacing his treatment of comets and the Milky Way from their natural position as first in the kapnosphere and postponing his first endoxic review until after describing the shooting stars and the aurora.[56] By treating the shooting stars first he can confidently display his basic causes among the lower and uncontroversial cases. It is especially significant that, though Anaxagoras was a spur to Aristotle's theory of meteors, Anaxagoras is never mentioned in that discussion. Aristotle did not want to engage Anaxagoras' celestial hypothesis at that early stage of his investigation. But after Aristotle has laid his foundation, he can extend his principles to the more problematic phenomena. His audience is then more likely to view the comets and Milky Way as situated in a series of short-lived and long-lived fiery phenomena, where meteors are very brief and passing, comets are more persistent, and the Milky Way is permanent. As we shall see, this is the same technique as he uses later when he explains hail, and it serves the same purpose.

[55] Olympiodorus 36.7–9 notes that the meteorology is the greatest study of things under the moon, since they imitate the heavens: the ὑπέκκαυμα unceasingly follows the motion of the sphere.

[56] Note that the vertical order, rather than the order of presentation, is given at 1.1.338b21–4 and 1.3.339a34–6.

No doubt, comets and the Milky Way receive such elaborate treatment because their causes are remote from the senses, but more importantly they are resistant to classification. Their inclusion in or exclusion from a given genus is, according to Aristotle's theory of science, a matter of definition and admits of no direct proof.[57] Ultimately, as we learn from *Posterior Analytics* 2.4–5, classification and definition are matters of intuition that cannot be demonstrated. But by fitting comets and the Milky Way appropriately and convincingly into the genus of kapnospheric phenomena, Aristotle can lend credibility to his definitions. In such cases the elegance of the classification itself serves as an informal criterion and sign of its own veracity, and this in turn makes the ordering of the phenomena relative to one another of utmost importance, for without it the essence of these recalcitrant phenomena would elude us. Etiology, which we saw in Chapter 4 to be in the service of order and beauty, is now being supported in turn by that very order. Order is a guarantor of essence.

In general, then, Aristotle supports his own theory metaphysically, while refuting his opponents empirically. In more specific ways too he raises implicit objections concerning methodology that recall concerns discussed in Chapter 1. By refuting Anaxagoras and Democritus together, for example, Aristotle is implying that, though their basic theories of material reality differ quite markedly, those theories make no difference when it comes to explaining comets and the Milky Way. Moreover, Democritus, copying from the elder Anaxagoras, is especially guilty of having said nothing better than his predecessor. Aristotle for his part is constantly pointing out that his predecessors' theories cannot adequately explain the phenomena, and that his theory is better in these respects. The indictment implicit in both objections is that his predecessors proclaim important-sounding first principles but cannot show how they are superior in explaining the appearances.

Closely related in purpose are his objections to Democritus (but not apparently Anaxagoras) concerning his conjunction theory of comets. Democritus notes that stars have been seen at the dissolution of *some* comets (343b25–8). Aristotle objects that on this theory stars should *always* appear on the dissolution of comets, and that he has seen conjunctions (albeit of planet and star) with no comet forming. Democritus does not even understand that an explanation, in order to be adequate, must explain every relevant case.

[57] *APo.* 2.3–7.

Aristotle's theory of comets and the Milky Way

Aristotle's own theory is an awkward amalgam of three features, combustion, friction, and reflection. First, combustion plays a fundamental role among the lower comets, where the fiery principle (ἀρχὴ πυρώδης 344a17), on account of the movement from above (διὰ τὴν ἄνωθεν κίνησιν 344a16–17), falls into a dense concentration of exhalation rising from below (ἀναβαίνειν εὔκρατον ἀναθυμίασιν a20). What this principle is (is it actual fire or the cause of fire?), and from where it descends, is nowhere stated. Aristotle's interest here is not so much the source of combustion as accounting for the combustion itself, the comet's bright center, and the persistence of its burning. There is clearly, however, a burning exhalation.

In the case of upper comets, though the material is still an exhalation (ἀναθυμίασις 344b1), no specific mode of combustion is described and the bright center is provided by a star (344a35–b8), which by its motion gathers the exhalation (συνιστῆται 344a36). But even if the principle of the comet's condensation arises from the stars, the comet, being sublunary, cannot be in actual contact with the stars (344b2). Though Aristotle does not explicitly say that these upper comets are ignited, he makes no distinction between upper and lower when, in his discussion of signs and seasons (344b18–26), he calls all comets fiery.

When we get to the Milky Way, however, there is no evidence that its light is the product of the combustion of exhalation. And yet in many ways the material of the Milky Way seems to behave as an exhalation as well, being capable of mixture (346a5–6), motion (346b8–10), and gathering (346a22–3) around the brightest stars.

Secondly, we have already seen how friction plays a role in the production of actual fire through the agency of the sun. But this process seems to be extended to the stars as well (ὑπό τινος τῶν ἄστρων 346a1–2). At least, the reason why comets do not appear around the sun is that the sun dispels the hot material as quickly as it creates it (1.8.346a11–14; cf. 1.8.346a8–9a), and this implies that the stars create the material too but do not dispel it. On this view the upper comets and the Milky Way glow for the same reason as the sun does – direct friction on the air –– and there is no need for a combustible exhalation to gather around the brightest stars from elsewhere.

Finally, the upper comets and Milky Way are both compared with the reflective halo, which is introduced here precisely to avoid a direct causation by the stars. There is no reason to suppose that this is a vague metaphor and that the comet does not arise somehow by reflection.[58] The

[58] Strohm 1984: 146 and Taub 2003: 108 have noted some of the tensions here.

only qualifications Aristotle adds are that reflection does not cause its *color* (344b8) and that the appearance is not caused by reflection to the sun, but rather to the star itself (b13–18). These really are reflective phenomena. On this view the upper comets and the Milky Way are the diffuse images of stars as seen through concentrations of dry exhalation (similar to the first kind of aurora), and for this reason they would be especially associated with bright stars (346a10–11), whose light could penetrate the *sustasis*. Still, Aristotle does not explain why the *sustasis* should gather directly below a star or planet. In the case of the vaporous halo, which appears irregularly, this is a matter of chance. But chance is inadequate to explain the permanence of the Milky Way. He might have suggested that the exhalation gathers in grooves in the celestial sphere, indeed even drifts up into the heavenly realm as the famous passage in 1.3 may be interpreted as suggesting (340b6–10). That there should be irregularities in the celestial surface immediately next to the kapnosphere, as there are among the other layers, would not be entirely unexpected.

Though these three causes, combustion, friction, and reflection, may "refer to what is possible" (1.7.344a6–7), their reconciliation is highly problematic and Aristotle's motivation in introducing them was less empirical than it was architectonic. Combustion provides continuity with the meteors and lower comets, friction provides continuity with the heavens, and reflection aims to mitigate the problems with the friction theory, which arise because of the great distance of the stars.

Aristotle's theory borrows significantly from his predecessors. They had stressed the celestial aspects of these phenomena, and though he rejects their heavenly essence, he does not want to deny a heavenly influence. Accordingly, he has the kapnosphere rotated by the celestial sphere and thereby makes the comets more or less track the celestial bodies. As a result too his comets appear for a long time, unlike the short-lived meteors and aurorae, and for this reason their form of ignition must be different. Aristotle's curious obsession with the apparent movement of the meteors discussed above becomes intelligible now in light of the ignition of comets. He already knew then that he would need to distinguish the ignition of comets from the ignition of meteors, and that need in turn suggests that he had decided on the precise demarcation between celestial and sublunary before becoming too entangled in the cause of ignition.

Again, all three of his predecessors, and especially Hippocrates, stressed the infrequency of the comet and provided planetary reasons for it. Aristotle admits the fact but claims that comets are infrequent because the Milky Way collects most of the hot substance that the heat of the sun does not disperse (345a5–10). Again, Hippocrates places the comets outside the tropics, and Aristotle, believing that he is partly correct, feels compelled to explain: the sun and the planets dispel most of the inter-tropical exhalation (1.7.345a5–10). Most remarkably, he takes over and modifies Hippocrates' theory of ὑπόλειψις, a motion in which the comet "falls behind" the stars more slowly than any of the other planets (1.6.343a4–7). But for Aristotle "being left behind" is no longer a description of the cometary and planetary motion relative to the sidereal, but rather the comet's falling behind the star *or* planet with which it is associated. In fact, Aristotle would probably have been better off abandoning the connection of comets with celestial motions, but since the comets' relationship to celestial bodies was the dominant position of his predecessors (implicitly for the Pythagoreans, Anaxagoras, and Democritus; explicitly for Hippocrates and Aeschylus) and since the similar Milky Way permanently tracks the fixed stars, he feels obliged to provide a celestial cause.

In the end Aristotle can explain the difference between a long-haired comet and a bearded comet better than Hippocrates and Aeschylus can, for whom comets indifferently drag a trail of vapor. By attributing it instead to a difference in the shape of the *sustasis* of the exhalation, to a difference in length (πλάτος) and breadth (μῆκος), he draws attention to a similarity with meteors and so draws the whole genus into closer unity (344a22–3). Similarly, his tracking theory, borrowed from Hippocrates, gains credibility by being used twice in closely related phenomena: like the tracking comets, the Milky Way permanently tracks the brightest and most numerous stars in the heavens.[59] Aristotle is also careful in other ways to make the explanations appropriate to the genus. Because of the kapnosphere's proximity to the celestial sphere and because of the permanence of the Milky Way, seasonal variations play no role.[60] Only in the case of comets is there discussion of times, and even here they are related to wind and drought; there are years, but not seasons, when comets are frequent (344b26–8).

Phenomena of his predecessors that Aristotle cannot fit into subordinate species, he sometimes turns into attendant circumstances. He mentions the rock that fell at Aigospotamoi at the same time as a comet appeared (344b28–34).[61] Anaxagoras famously thought that the heavenly bodies were

59 "There is more than one reason to believe one has the genuine scientific account if the same causal story can explain several *pathē*" (Freeland 1990: 92).

60 1.7.344b34–6 merely states that the winter was dry, not that comets occur more frequently in the winter. It is precisely because of the lack of seasonal variation that Olympiodorus rejects the sublunary Milky Way (75.24–9); so also Philoponus 114–16 (especially 116.18–23).

61 According to Plutarch, *Lys.* 12, citing Daimachus, a νέφος φλογοειδές was seen for five days before the fall of the rock (DK 59A12 [2.9.37–40]).

fiery rocks, and Aristotle here alludes to that theory.[62] Without directly refuting it, he claims that the stone and the comet are concomitant effects of a common cause, the dry exhalation, which burns as a comet and which as wind can pick up a stone by force and transport it through the air. This meteorite, perhaps the most challenging evidence against Aristotle's whole theory, is safely deposited here as a footnote at the end of his discussion.

Aristotle's account of the Milky Way was widely rejected in antiquity. Philoponus objects that evidence from parallax and eclipse is not in Aristotle's favor, and he is particularly savage in pointing out that, whereas all sublunary phenomena are transient, the Milky Way is a permanent feature of the world.[63] On this last point at least, Philoponus failed to understand Aristotle's intent. Aristotle wanted the Milky Way to be a permanent homeostatic feature of the world, and he was eager to place it in immediate proximity to the celestial sphere.

The essence of the Milky Way

I would like to conclude with a brief discussion of some problems of definition and essence that arise in the context of the Milky Way. The Milky Way is the only species in the *Meteorologica* of which there is only one member; of the comets and all the other species of phenomena there are many individual instances. It is unvarying in its homeostatic flux and is as eternal a thing as the sublunary realm permits. Because it is unique, fixing its essence is difficult. *Metaphysics* 7.15 (cf. *APo*. 2.13.97b26–8) argues that there cannot be definitions or demonstrations concerning individuals that are either (1) sensible, destructible substances or (2) eternal objects, like the sun.[64] Of course, the Milky Way is not a substance, and so presumably *a fortiori* cannot be defined. But if we think of it like a substance, and as falling under the first class, we must ask whether it is liable to Aristotle's

62 These are unseen bodies, which revolve below the sun and the moon, are made to shine by the friction of the aether, and on falling are extinguished (DK 59A12 [2.9.31–7]; 59A42.6 [2.16.16–19]).

63 Philoponus also objects that the Milky Way should appear in Orion and around other bright and numerous stars (114.17–20); that the sun and the planets as they move north and south through the zodiac ought to have some effect on the appearance of the Milky Way (114.20–30); that we should see eclipses by the Milky Way (114.30–115.6) and parallactic effects (115.6–16); he also adds the delightful myth (115.16–21) that Athena placed Heracles at the breast of the sleeping Hera so that he might become immortal. But when he started to nurse too forcefully, Hera suddenly drew back her breast and the spraying milk created the Milky Way. Philoponus does not rate Aristotle's theory any higher than this myth (115.24–5). Olympiodorus, following Ammonius, likewise rejects the sublunary Milky Way (66.17–20; 75.24–76.5). In general see Gilbert 1907: 661.

64 He also considers the impossibility of defining Platonic forms, but that issue need not concern us.

objections (1039b20–1040a7): inasmuch as it is composed of matter, and matter is destructible, its existence is contingent, and it can cease to be. But definition and demonstration are of things that are necessary and necessarily exist. The Milky Way, of course, maintains an unvarying constancy and, if it is fueled by the dry exhalation, it may be conceived of as a sort of serial plurality of generation and corruption, where at each moment it is a different instance of the same form. Insofar as it is the same at every instant, it can be defined as a species form and have essential features and properties. Alternatively, if the Milky Way shines because heavenly friction actualizes the capacity of air for becoming fire, its material is as indestructible as the heavens, and the Milky Way needs to be understood as an eternal object, like the sun.

If we are to treat the Milky Way as an eternal object, Aristotle offers two arguments against its definability (1040a27–b4).[65] First, it is difficult to know whether a feature present in the definition is truly essential. If the sun does not move around the earth, for example, is it still the sun? Aristotle clearly thinks that this is a relational attribute and not part of the essence of the sun, which is a substance. Second, even if we do provide a definition without such superfluous additions, it is conceivable that there may be another instance that fits the definition. Would we be willing to call that thing another sun? Within the dynamic of *Metaphysics* 7, Aristotle seems to imply that we would have to, and that having two suns would be a problem.

His treatment of the Milky Way, however, suggests a different approach. The problem of additional accidental features and the answer to the question of duplication are addressed by defining the Milky Way within a great genus. Since it is a kapnospheric phenomenon, it shares in the basic common features of that genus, the efficient and material causes. None of these generic features or the differences that define the species, whether it be shooting star, comet, or Milky Way, can be absent from the definitions. For this reason the Milky Way must be made of a certain material and must burn by the agency of the heavenly rotation. This is a fact not noted in the *Metaphysics* passage, where the sun is treated as unrelated to other

[65] "So as we said, we fail to notice that it is impossible to provide a definition in the case of eternal objects, especially those that are unique, such as the sun and the moon. For people make mistakes by adding things – and even if they were subtracted the sun would still be the sun – like 'going around the earth' or 'night-hidden' (for if it rests or appears at night, it will no longer be the sun. But it is strange, if it should not be. For the sun refers to a substance). Furthermore, they make mistakes by adding things that can belong to something else, for example, if another such should come to be, clearly it will be a sun. So the definition will be common to them. But the sun is an individual, like Cleon or Socrates" (7.15.1040a27–b2).

objects in its genus. Moreover, Aristotle is especially careful to show how the Milky Way differs from the upper comets only by more and less and does not introduce any new differentia however minor, as he often does: its essential similarity to comets safeguards its place in the genus and therefore its definition.[66] By proceeding by the greatest genus method, there is no real danger that superfluous features will be added to the definition. There remains the fact that a putative definition is a universal account, which could conceivably be shared by another individual. But the object may be defined in terms that make it unique and, though Aristotle's exact meaning is not clear, he does say that the circle of the Milky Way is the *greatest* and lies far outside the zodiac (346a17–19) among the *brightest* stars (346a28).[67] The significance of "the greatest" is not clear (it is not merely a "great circle," a circle concentric with the earth's center), but it is a unique designation and makes it impossible for another such Milky Way to exist, since there can only be one greatest. In both ways the Milky Way, whatever its precise definition may finally be, can overcome the challenges to defining unique, eternal objects laid out in the *Metaphysics*.

66 The comet associated with the earthquake in Achaia was called ὁδός (1.6.343b23), suggesting a phenomenal similarity between some comets and the Milky Way.

67 Alternatively, Aristotle might simply say that the Milky Way is a plurality, either twofold (δίπλωμα 346a24) or as numerous as the stars it forms around.

CHAPTER 7

Condensation and precipitation (1.9–12)

From the fires of the kapnosphere Aristotle now descends to the atmosphere, the place between the surface of the earth and the mountain peaks where the rain cycle occurs. By locating the atmosphere below the kapnosphere, he confines the wet exhalation within narrower spatial and causal limits than had been the case among many of his predecessors, for whom vapor had extended to and fed the stars.[1] The first and controlling efficient cause here (κινοῦσα καὶ κυρία καὶ πρώτη 1.9.346b20–1) remains celestial, but in place of the constant and undifferentiated sidereal rotation that ignited the upper smoky exhalation, Aristotle employs here the sun to evaporate water and condense vapor by its warming approach and cooling withdrawal. He deliberately obscures at the beginning of chapter 1.9 the distinction between the daily passage of the sun and the annual cycle that produces the seasons.[2] The sun is responsible for both, of course, and each has its own effects on temperature and the cycle of precipitation, but for now Aristotle wants to keep his account as general as possible – the sun simply causes the σύγκρισις and διάκρισις in cycles of generation and corruption. Again, whereas in the upper region the kapnospheric fires were generated by immediate contact with the celestial motion, here below the sun produces rays (ἀκτῖνες), which together with other heat generated in the kapnosphere (346b23–6), operate from a distance on the moisture on and in the earth. Aristotle never tries to explain how these rays can warm the distant earth, but clearly their less immediate agency served his architectonic purposes by showing that the atmosphere does not enjoy the kapnosphere's immediate proximity with the celestial. In service of the same structural ends, the simple unity of the kapnospheric efficient cause

[1] Gilbert 1907: 475–6.

[2] It is worth noting that at *GC* 2.10.336a31–3 Aristotle denies that it is the "primary motion," i.e., the diurnal cycle that causes generation and corruption, whereas here it is responsible for the generation of dew and frost, and elsewhere of the winds. No doubt Aristotle meant by "primary motion" the *sidereal* motion rather than the sun's daily course.

(the rotation of the stars) is contrasted with the twofold duality of the sun, which heats and cools both daily and annually. The form that the motion takes in the atmosphere is different as well. While the dry exhalation had certainly ascended to the kapnosphere, the phenomena themselves were not characterized by that upward motion but in every case by a motion at least apparently contrary to nature. In the atmosphere, however, the vapor rises and, when it condenses, it actually descends again as its nature requires. Aristotle is remarkably coy about whether the dry exhalation helps to raise the wet.[3] We are told only that heat (θερμότης or πῦρ) causes the vapor to rise and then disperses to a yet higher place or is quenched (1.9.346b26–31; 1.10.347a14, 25, b26–9). What becomes of this fire or what its relation is to the dry exhalation remains a mystery, but we must suppose that Aristotle wants to establish an efficient cause of the rain cycle independent of the dry exhalation.

Aristotle chooses to ignore most of the well-trodden territory of his philosophical predecessors and reach back over the centuries to the ancient wisdom of the poets. Homer and earlier writers (οἱ πρότερον) had identified Ocean as the river that flows in a circle (1.9.347a6–8).[4] On Aristotle's interpretation they were alluding to the rain cycle in a riddling manner (ᾐνίττοντο). No doubt the cycle of Ocean caused Aristotle to draw the connection with the cycle of the rise and decline of human civilization, which generates such riddling truths.[5] But more properly, the metaphor of the river of Ocean is related in his mind to the notion of the "common place of air and water" (κοινὸς ὕδατός τε τόπος καὶ ἀέρος 346b17–18), an expression itself in need of interpretation. On the most likely interpretation, this "common place" applies to the rain cycle narrowly and does not extend to the rivers, winds, or earthquakes, as is sometimes supposed.[6] At least, Aristotle does not use the expression outside the immediate context of the rain cycle, and though he says it is the second place after the kapnosphere and the first above the earth (an area appropriate to the *de Caelo*'s conception of the middle two elements), he singles out the *upper* effects of generation there (τῶν συμβαινόντων περὶ τὴν ἄνω γένεσιν αὐτοῦ 346b16–21), and

[3] Olympiodorus 80.3–5 thinks it does.

[4] Gilbert 1907: 393, on the basis of weak evidence (citing Homer, *Od.* 10.508ff.; 11.1–22; 12.1–6), claims that Oceanus was for Homer a river of fresh water.

[5] On Aristotelian treatment of myth and allegorization, see Brisson 2004: 38–40; Johansen 1999. See also 1.14 and my discussion of land–sea exchange in Chapter 8.

[6] Philoponus understands this region as the entire area where air and water are found together: there is water in the air in the form of rain and there is air in the water as bubbles (119.24–27). Alexander concurs, claiming that Aristotle now separates off τὴν ἄνω γένεσιν, but will treat the lower part, common to air and water, later (44.2–6).

later he explicitly refers to the cyclical river of precipitation as common to air and water (κοινὸν ἀέρος καὶ ὕδατος 347a3). The "common place," then, refers just to the rain cycle. Later, as we shall see, he works hard to isolate the rivers from a causal origin in precipitation. What happens to rain after it falls (except that it eventually returns as vapor) is a subject he studiously postpones until later chapters.

The isolation of this cycle serves two purposes. First, though Aristotle was attracted to the notion of the cycle because of the cyclical course of the sun, he did not want to extend the cycle to include all the other watery and airy phenomena, as Anaximenes and Xenophanes had done, since that inclusion would destroy the independence of each great genus and make each a cause and effect of the other. Second, in introducing a cycle here, his purpose is not to show respect for the law of universal change or conservation of matter, but rather to indicate that the very cycle is an essential part of this particular meteorological phenomenon, and in this respect he is drawing a sharp contrast with the kapnosphere, where in apparent contradiction to his highest general physical theory, the dry exhalation ascends but is never seen to return to earth.

The common place, then, is important for these reasons, but its precise meaning requires further clarification. "It is necessary," Aristotle says, "to suppose that this is a river flowing in a circle up and down, sharing in air and water" (κοινὸν ἀέρος καὶ ὕδατος 347a3). The point here is not that the "river" is composed of water and air, as mere material constituents (as is the case at *Sens.* 5.443a21–2 ὀσμὴ οὖσα κοινὴ γῆς τε καὶ ἀέρος), for the rain cycle is not unique in that respect. Nor does κοινόν mean here what it most commonly means, that some universal feature is found equally in several subjects.[7] Instead Aristotle is trying to make clear that the circular flow of the "river" depends on there being both air and water. The best commentary on the relevant sense of "common" comes from the *de Anima* (1.1.403a4–b19), where the question concerns whether the affections of the soul can be studied apart from the bodies in which they are found.[8] If they cannot, then those affections are common to the soul and the body (τὰ τῆς ψυχῆς πάθη μετὰ σώματος 403b16–17; cf. τῶν συμβαινόντων in our passage, 346b18). Aristotle contrasts such common affections with the separable and abstractable properties studied by mathematics, which, though not

[7] E.g., *Sens.* 5.445a11–13: "So the object of smell is an affection of both [air and water] (τὸ ὀσφραντὸν κοινόν τι τούτων ἀμφοτέρων), and it belongs to the tangible and the audible and the transparent."

[8] See also passages at *de An.* 1.4.408b25–9, where old age seems to be an affection of the body only and not the soul; *Sens.* 1.436a7–8, b2–3.

actually separated from matter, can be separated in conception (403a10–16). An account of the common affections of the body and soul must include an account of both the body and the soul and their interactions. Both are essential explanatory factors.

Similarly in the rain cycle both air and water and their mutual transformations are essential components of the explanation. The air rises higher or less high, and this accounts for the difference between dew and rain. The water freezes or does not freeze and this accounts for the difference between rain and snow. Air and water, then, are not just two adventitious material causes, whose interaction happens to give rise to these effects (such as is the case with the violent phenomena). They are essential phases of the same phenomena and, by calling these properties common, Aristotle stresses the fact that the cycle itself forms a single process. And this returns us to our earlier observation that the common properties are restricted to the rain cycle narrowly conceived. Aristotle is not interested here in a cycle of material change that would extend to the rivers and the sea. Air is the first product of evaporation and the precipitations are the first products of condensation. For this reason, Aristotle provides a detailed analysis of the states of vapor immediately before its condensation into water: the exhalation from water is vapor (ἀτμίς 346b33), but water arises from ἀήρ through the intermediary of cloud (346b32–3). Rivers, by contrast, arise from sources in mountains and the sea from rivers. Neither is an immediate product of vapor. Moreover, rivers and the sea, and perhaps also the winds, though they are water and air, are strictly speaking unnecessary for the rain cycle and do not form part of the κοινόν. As far as the cycle is concerned, what evaporates is just water on and in the earth (μενούσης δὲ τῆς γῆς, τὸ περὶ αὐτὴν ὑγρόν 1.9.346b23–4), and which specific containers hold the water are irrelevant.

The sea had been the source of atmospheric vapor for Xenophanes and Heraclitus, as we have seen. Aristotle, by treating the rain cycle first before other phenomena, inverts this order and makes the rain's nature prior to that of the sea. Salinity loses its position as the original or natural state of water and becomes instead the result of an addition, which must be explained within the prior constraints of the rain cycle, and, in the end, Aristotle's solution to the problem of salinity has only the slightest effect on the rain. The rain cycle is likewise prior to drought and abundant rainfall, which are postponed until the discussion of climate change (1.14) and the winds (2.4), since they involve terrestrial factors beyond the rain cycle itself.

This conceptual order is remarkable, because precipitation had long been associated in the popular imagination with wind and storms, as Hesiod and Homer's storm descriptions bear witness.[9] For Anaximander winds and rain are both products of mists that arise from the earth.[10] Xenophanes' wind and rain have a similar origin. Aristotle minimizes this connection for now, first because wind has a different material origin and also because wind and rain are not invariably concomitant and cannot be the universal cause or effect of one another. Aristotle mentions wind only as a negative cause in preventing the settling of dew and frost and in their association with temperatures necessary to generate moisture (347a34–7).

One of the great advantages of treating the rain cycle in isolation is that Aristotle is better able to display its internal structure, and this structure is best viewed by considering how the differentiae define the phenomena and how they are related to the differentiae of the kapnospheric phenomena that precede. I have already outlined at the beginning of this section some of the general ways in which the kapnosphere and the atmosphere are related. It remains to consider the specific differences.

Aristotle is careful to juxtapose the atmosphere and kapnosphere by emphasizing three major sets of common differentiae. Within both great genera phenomena are distinguished by altitude. Kapnospheric phenomena occur at three different altitudes and the precipitations occur at two (or three, if we include the odd man out, hail). Of course, they do not share exactly the same differentiae – any more than birds and humans share the same two-footedness (*PA* 1.3.643a3–5) – but they correspond and are the same analogically.

Time or speed of change likewise provides common differentiae. Among the kapnospheric phenomena the factor of time is highly variable, from the extremely rapid shooting stars (with the exception of lightning there is nothing so rapid), through the slower comets to the permanent and unvarying Milky Way. The actual speed of ignition may be similar in each case, but the apparent movement varies enormously.[11] Similarly among the phenomena of condensation, the upper rain and snow are of longer duration, since they are on a seasonal cycle, whereas the dew and the frost are on a short, diurnal cycle.[12] Though time for both great genera is

9 E.g. Hesiod, *Op.* 548–53; Homer, *Il.* 16.384–6; cf. *Od.* 12.403–11.

10 DK 12A11.7. Aët. 3.7.1 (*DG* 374.19–22): "the wind is the flow of the air when the finest and dampest parts in it are moved or melted by the sun." Xenophanes joins wind with the rain: "For when the wet is raised from the sea, the fresh part, separated because of its fineness, forms clouds of mist and causes raindrops to drop through felting and blows out winds" (DK 21A46 [1.125.21–3] = Aët. 3.4.4 [*DG* 371.9–17]). Anaximenes does not bring in the wind except to say that hail has a pneumatic part in it. Antiphon (87B29 [2.344.6–15]) provides an apparently similar theory of interaction of rain and wind.

11 1.7.344a25–9 suggests that ignition speeds may vary.

12 As Olympiodorus 79.28–9 says, the day is a small year just as man is a small cosmos.

long and short, the *measurement* of time is appropriate to each. Since the kapnosphere is affected only or at least primarily by the sphere of the fixed stars, that motion is constant and can mark no real distinctions of time. Granted, there is the sidereal day, but as far as the sphere of the fixed stars is concerned, it provides no variations within its cycle that would affect the process of friction. The concentrated mass of the dry exhalation is the sole determinant of the presence or absence of shooting stars. As a result, time is only longer or shorter. The sun, by contrast, has two specific motions, the diurnal and the annual, and these govern the longer and the shorter time frames of precipitation.

The third common differentia is quantity of material. The stars of the lofty Milky Way, as Aristotle explicitly declares (1.8.346b7–10), attract a vast quantity of material, and we may infer that the lowly shooting stars use very little and the intermediate comets a middling amount. Correspondingly, the lower dew and frost involve small quantities of wet exhalation while the upper snow and rain use large (and paradoxically the hail and the huge rain use very large quantities).[13] Aristotle has clearly used these three differentiae, altitude, time, and quantity (each associated with one of the major causal factors, place, efficient and material cause), to create a corresponding structure between the two great genera. The three differentiae are applied in the atmosphere in a simpler way and in a way that generates the analogy against which Aristotle is able to understand the nature of hail.

Of course, not all differentiae correspond. Differences of heat are specific to the atmosphere. The kapnospheric fires are obviously hot, but the Milky Way does not burn with a hotter heat than the shooting stars; at least the difference between simple friction and the ignition of the dry exhalation has no explicit effect on heat. Heat and cold are, however, differentiae among the precipitations: rain and dew are warm, and snow and frost are cold. Again this is a set of differentiae that arises naturally from the principles in each case. The uniform motion of the sidereal sphere generates a constant friction and a steady heat; the sun with its diurnal and annual changes produces smaller and larger variations in temperature. The distinction between the kapnosphere and atmosphere in this respect is related to their difference in place. The upper region, being the region of fire, is absolutely hot. The common place of water and air naturally will be relatively hot and, influenced by the sun, will alternate between hot and cold. Indeed,

13 347b14–18. Aristotle also bestows difference of degree on the size of the precipitations – ὑετός and ψακάδες – at least in rain (347a11–12).

since both air and water are wet, they can only be distinguished from one another by hot and cold.

Similarly the shape of the exhalation is a differentiating factor only in the kapnosphere. Here the shape of the *sustasis*, its length, breadth, and depth, accounts for the different motions of the fires. Shape is not a difference active among the precipitations, probably because the motion of all forms of precipitation is natural and downwards. (Though perhaps Aristotle missed an opportunity to relate shapes of cloud to precipitation.) The closest thing Aristotle offers is the extent of the earth's surface, whether the exhalation is gathered up from a large area or a small area (1.11.347b19–21).

The systematic correspondences between the kapnosphere and the atmosphere make one omission particularly remarkable. Immediately after the short-term shooting stars, Aristotle discusses the reflective effects of the kapnosphere, the auroral phenomena, but he reserves for the end of the treatise the full discussion of the reflections of vapor, the haloes, rainbows, and mock suns. The reason for this postponement I will postpone until the appropriate time.

Hail

In chapter 1.9 Aristotle constructed an analogy among four forms of condensation – rain : snow :: dew : frost. As in the kapnosphere, these apparently simpler phenomena set the pattern for explanation, and the more intractable one, in this case hail, is treated as an exception or variant. The analogy divides the phenomena into upper and lower, warmer and cooler, and larger and smaller, a scheme into which hail cannot fit, since it has no ground-level correlate corresponding to dew and frost. Aristotle uses the term, ἀνάλογον, only here in the *Meteorologica* (besides 3.2.372a5 in a rather different context). In the biological works, where the term is much more common, it indicates a functional and structural similarity between members of different great genera, for example, between the wings of birds and the fins of fish. I have argued elsewhere that these analogues, in spite of their strikingly different appearances, share at a higher level of generality a similarity of causal structure, both of material and of function.[14] Here in the *Meteorologica* the analogy is drawn *within* a great genus, but all the same Aristotle constructs a systematic set of correspondences between rain and snow, dew and frost, from which hail is excluded on account of its

[14] Wilson 2000: 53–89.

different efficient cause. The analogy, then, becomes the datum against which the problems of hail must be solved.

For this reason alone it is impossible to deploy the principles of "more and less" among the high and the low, the warm and the cold, as they have hitherto been deployed. But hail is also a peculiar and unusual phenomenon. It occurs rarely and hailstones can be unbelievably big (τὸ μέγεθος ἄπιστος 1.12.348a27). It requires, therefore, a rare and unusual cause. Other factors, too, both general and specific, constrain Aristotle's solution to the problem. Since the heat of the sun is the efficient cause of the rain cycle, and heat is affected by the sun's changing position through the course of the seasons, all forms of precipitation, including hail, will be influenced by the seasons. The most surprising feature of hail is that this icy precipitation falls usually in the milder spring and fall rather than in winter (1.12.347b37–348a2), and in warm rather than cold climates (348a3–4). In addition, particles of ice seem to fuse together after they have individually frozen, a very perplexing fact (348a4–13). A successful theory must account for all these features.

Just as he had done for the kapnospheric comets, Aristotle takes these puzzling considerations and confronts them with a combination of his own meteorological scheme and a selective doxography. Anaxagoras, he says, claimed that a hot cloud is thrust up into the cold upper air, where its vapor freezes (348a14–20, b12–14). For this reason hail occurs more in the *summer* and in warmer climates, since the heat raises the clouds further in those conditions.[15] Aristotle rejects this view and argues, again on empirical evidence, that hail is formed at low altitudes and is not found in high places, as snow is (348a20–30). He claims that we know some clouds to be very low and close because of the loud and frightening noises they produce; when they do not make these noises (but are still very low), they often produce prodigious hail.[16] Moreover, bigger hailstones are rough, while the small ones are smooth, indicating that they become worn down on descent. These large hailstones, then, have been formed close to the earth.

From these arguments it is possible to recover something of the method Aristotle employed in joining the doxographical tradition with his analogical scheme. Anaxagoras accounted for the difference between snow and hail by claiming that the intense heat of summer drove the moist clouds

[15] Though Vicomercatus 54H argues that this should make hail no more rare in winter.

[16] Philoponus 124.7–8 draws the connection explicitly. Olympiodorus 92.17–20 says that after the clouds make this sound hail falls.

high into the cold air.[17] But for Aristotle hail does not occur most frequently in the summer; it is a spring and fall phenomenon. Since there is not a lot of heat at those times of year, Aristotle thinks – though he does not explicitly argue – that the hail clouds cannot be driven high up. (He does argue that there is too little moisture in the summer for hail to form [348b26–8].) But what is striking is that Anaxagoras' explanation is not a bad one by Aristotelian standards. It uses principles – vapor, altitude, heat, and season – appropriate to the condensing phenomena of meteorology. Aristotle himself borrows and uses elsewhere Anaxagoras' theory of ἀκτῖνες, the rays that warm the surface of the earth but leave the upper air cold. And the objections Aristotle brings against this account of hail are certainly not beyond his own breathtaking capacity to explain away. His main objection to Anaxagoras is that he had attempted to explain the paradox (παράλογα 347b35–6) of ice in the warm seasons merely by applying the ordinary principles of condensation in a heightened degree. Aristotle, for his part, feels the need to explain the paradox, but he does it by means of a paradoxical cause that cuts against the orderly analogy of the other forms of precipitation with their differentiations of more and less. The fact of the paradox itself persuaded Aristotle that he needed a cause of inversions, *antiperistasis*, specifically the descent, encircling, and compression of the cool, upper, wet air mass by the warm lower air. Indeed he had just used *antiperistasis* for the first time in the treatise to explain the unusual and surprising fact that dew is found around the Black Sea when a north wind is blowing rather than the expected south (1.10.347a35–b7).[18]

At the same time, *antiperistasis*, being the result of an interaction between hot and cold regions, is an appropriate cause for phenomena of condensation. Hail takes up a middle position between the cold and the hot, the upper and lower regions, between the summer and the winter, and the north and south. The process of *antiperistasis* and specifically the sinking of the cold cloud into the warm, lower air, makes hail into an ambiguous phenomenon dualizing between the upper and the lower condensations. Its unique etiology explains why it stands outside the analogy of the other

17 Anaxagoras was at least correct in locating the difference in the efficient cause, as is brought out by a contrast with the theory of Anaximenes (Aët. 3.4.1 [*DG* 370.14–20]), who had given various accounts of cloud, rain, snow, and hail. Hail is different from snow in that some airy stuff (πνευματικόν) is included in the moist (ὑγρόν). This, of course, is a material difference, and Aristotle's system places material differences (largely) at a higher level of generality.

18 Just as he had used the stars in the explanations of both the upper comets and the Milky Way, so here too *antiperistasis* is used twice among forms of condensation in order to increase its utility and its credibility as a generic cause.

four phenomena and provides another reason for rejecting Anaxagoras' claim that hail is formed at very high altitudes.

Finally it is worth noting that although Aristotle develops his account of hail as the odd man out among the forms of condensation, he eventually brings it back into the fold. Not only does it dualize within the analogy but, as it turns out, hail has its own warm analogue. At the end of the chapter Aristotle introduces the heavy rainstorms of warm days, whose very large drops also have their origin in *antiperistasis*. In this case, though, the warm surrounding air does not condense and cool the water so much that it freezes. These storms too are paradoxical in being characteristic of the summer deserts of Arabia and Ethiopia (1.12.349a4–7). In this way hail finds a warm-weather analogue so as to correspond to the other four, and the paradox takes its place within the dominant analogical order.

CHAPTER 8

Fresh waters (1.13–14)

Situated below the kapnosphere and the atmosphere lies a diverse group of phenomena – winds, rivers and the associated climate changes, the shape and zones of the earth, the sea and earthquakes – which seem to have little in common with each other except for their proximity to the surface of the earth. Aristotle's application of his fundamental principles, the dry and wet exhalations, to these phenomena is sporadic and uneven, and his lengthy discussion of the whole group, extending almost to the end of Book 2 (1.13–2.8), is punctuated by several digressions. Our first impression is one of miscellany.

The most important digression results in the winds and rivers being treated out of their natural order.[1] Aristotle begins 1.13 with the winds, because they are higher than the rivers and are made out of the dry exhalation, but he quickly switches to the rivers, since they serve as a heuristic introduction to the winds. Nevertheless, his discussion of the rivers is much longer than is required for that limited purpose, and even when that discussion is completed he does not return to the winds but instead pushes on into a protracted examination of climate change and the nature of the sea. It is only after ten of the standard Bekker pages that we are finally carried back to the winds.

In spite of the digression, though, these chapters reveal on closer examination Aristotle's elaborate attempt to integrate the lower region within the whole sublunary cosmos and to impose upon it an order second only to that of the kapno-atmosphere. His first challenge is to insure that the lower things are of the same basic kind as the upper – homeostatic processes subject to generation and corruption. Aristotle had found this claim denied among his predecessors, and the deniers came in two distinct, but related, groups. The first group explained rivers, for example, as the flow

[1] Olympiodorus says that Aristotle μακρὰν πλάνην πλανᾶται (101.23–4); Solmsen 1960: 400 and Strohm 1984: 155–6 also view this section as a sidetrack and digression.

of water from large underground reservoirs, without saying how the water was generated there in the first place. This basic pattern of "reservoirist" explanation can be found repeated in the theory that the winds are just masses (if not exactly subterranean reservoirs) of air that happen to be in motion. The mistake again, according to Aristotle, is that these winds are only moved, not generated. A second group, the "entropists," thinks that the earth is drying out and will eventually become completely desiccated. Deserts can never become fertile again, and the sea is disappearing and in the process becoming more saline. In Aristotle's mind, there is a close relationship between the reservoirists and the entropists, since for both there is an original supply of material, which is not supplemented by further generation.

Although he treats his predecessors' views with explicit contempt, these views form a leitmotif throughout this section and are the background against which Aristotle pursues his own solution. He claims that the tradition has nothing to offer beyond what anyone off the street (ὁ τυχών) could produce (1.13.349a16). In fact, "the views of οἱ πολλοί [i.e., just anyone] are preferable" (349a26) to those of the sophisticates (τῶν σοφῶς βουλομένων λέγειν 349a20–1).[2] But, all the same, they have created such a tangled confusion that Aristotle has to work out the fundamental problems for himself (διαπορήσαντες πρὸς ἡμᾶς αὐτούς 349a13–14). This prevents a simple exposition of the phenomena and requires Aristotle to pursue a course of argument less demonstrative, more problematic, and more exploratory than his previous.

This is the background against which Aristotle establishes his own generative theory, and its most distinctive feature is the elaborate analogy between the winds and the rivers: both winds and rivers are formed from exhalations that flow in discrete streams generated under the earth. Aristotle's task is to show that the rivers behave in conformity with this analogy and then use the analogy between the rivers and the winds to establish the nature of the winds. But even this task hardly accounts for the lengthy digression. Once the nature of the rivers is established why not return to the winds? The most important reason is that, though the analogy is most marked among the winds and rivers, it is not confined to them. We find additional parallels between the "unnaturalness" of earthquakes and of the sea. Terrestrial waters, fresh and saline, just like the winds and earthquakes, form corresponding systems based on the wet and the dry exhalations. Each

[2] Cf. Herodotus on the theorists of the Nile (2.20) "who wished to be thought wise."

group manifests a basic horizontal orientation and each is best treated as a unit.

Through the structuring analogy between the winds and the rivers, Aristotle recapitulates on the surface of the earth the basic contrast he had established between the kapnosphere and atmosphere. But the fact that the terrestrial phenomena form unified systems introduces an important development. The phenomena of the kapnosphere and the atmosphere were internally distinguished by differences of more and less and were defined by the same basic features (material, height, cold/hot, length of time), and, as such, they form much more unified and independent genera. The horizontal genera, by contrast, are *causally* connected. Rivers are a condition of the sea, and together with the sea are a necessary condition for climate change. Winds serve in turn as causes for earthquakes. By contrast, only in the case of the aurora is one kapnospheric phenomenon dependent on another. The trend is unmistakable. Causal chains are becoming longer the further we descend from the stars. They will become yet longer and cross-generic when we come to the violent phenomena.

The river–wind analogy

Although the strategic importance of the river–wind analogy was well recognized by the ancient commentators, it has not been fully appreciated in the modern literature, perhaps because the analogy itself is insufficiently supported by philosophical argument.[3] Aristotle introduces it as a simple principle in a polemic context at the beginning of 1.13 ("as if one were to suppose that all the rivers were one river" 349a26–7); and even when he comes to discuss the winds in 2.4, he never offers an independent argument to show that they arise from the dry exhalation and relies instead on the assumed analogy with corresponding rivers and their vaporous sources.

Though some bits and pieces of the correspondence can be ascribed to various Presocratics, the full river–wind analogy is probably an innovation of Aristotle himself.[4] Xenophanes includes rivers and winds in the list of

[3] Olympiodorus recognizes the analogy and compares Aristotle's treatment of the winds to Plato's treatment of justice in the *Republic*, where, in trying to determine its nature in the individual, Plato wanders off and discusses the more intelligible justice in the city (100.15–24). For its manifestation in Herodotus, see Romm 2010: 219.

[4] Böker 1958a: 2223 and 2229, though he agrees that the analogy is basically due to Aristotle, points out that at least the water–air analogy is found in Anaximander ῥύσιν ἀέρος and the Hippocratic corpus ἀέρος ῥεῦμα. Theophrastus, *de Ventis* draws out the wind–river analogy on a number of points: winds can, as it were, overflow (34); winds are likened to a rush of water through a gap (29); the dividing of a wind by an obstacle to the dividing of a current of water passing around an obstacle

effects of the sea, but there are no analogies here.[5] Anaxagoras derived the rivers partly from water in the hollows in the earth, but we hear nothing about the relevant winds.[6] Democritus provides the closest thing to an analogy between rivers and winds when he says that the rain, by soaking the earth and filling the hollows beyond their capacity, causes earthquakes – a doctrine that would recommend a reservoir theory of rivers. Correspondingly his winds blow when confined in a narrow space.[7] On Democritus' account both rivers and winds would originate from confined spaces. There is not much here, however, and Aristotle can fairly be credited with the basic structure and elaboration of the analogy.

The river–wind analogy is first introduced and deployed against a medley of previous theories, loosely clustered around the "reservoirist" view, and its basic outlines can best be understood in that context. Some people, Aristotle says, think that the winds are merely air in motion, and that the air condenses to form cloud and rain.[8] Other more sophisticated thinkers add that all winds are in fact a single wind, "on the grounds that the air which moves is in fact one and the same whole."[9] This view might be credible, he continues, if rivers all flowed from a common source, but they do not. If they did, the winds would have one source and flow from it as from a reservoir.[10] Now, Aristotle already has a reason to reject this nexus of theories: he wants the dry exhalation, rather than elemental air, to be the material cause of the winds.[11] His explicit strategy, however, can hardly be called a refutation. He forces his opponents' theories into a closer

(28); the interchange of water and sea breezes compared with the inflow and ebb of water (26); see Coutant and Eichenlaub 1975: xli–xlii. Cf. also *Pr.* 26.36.

5 DK 21A46 (1.125.20–4) = Aët. 3.4.4 (*DG* 371.9–17); and 21B30 (1.136.3–10).

6 DK 59A42.5 (2.16.13–15): τοὺς δὲ ποταμοὺς καὶ ἀπὸ τῶν ὄμβρων λαμβάνειν τὴν ὑπόστασιν καὶ ἐξ ὑδάτων τῶν ἐν τῇ γῇ. εἶναι γὰρ αὐτὴν κοίλην καὶ ἔχειν ὕδωρ ἐν τοῖς κοιλώμασιν. At a more general level Aristotle attacks the schools of Empedocles and Democritus for holding what amounts to a "reservoir" theory for all change (*Cael.* 3.7.305b1–5): "The students of Empedocles and Democritus fail to notice that they make generation of the elements into each other into an apparent generation. For they say that each of the elements is present and extruded – as if generation occurred from a vessel and not from some material – and with the change there is no generation." Similarly at *Cael.* 3.4.303a24–9 Democritus and Leucippus are criticized for explaining the generation of water, air, and earth from one another by the extrusion of the largest atoms. In time the process will exhaust the supply, says Aristotle.

7 *Mete.* 2.7.365b1–6; Seneca, *QN* 5.2 = DK 68A93a (2.106.27–36).

8 Alexander 53.28 and Olympiodorus 97.39–98.2 cite Hippocrates, *Flat.* 3; Lee mentions also Anaximander (DK 12A24 [1.87.35–6] = Aët. 3.7.1 [*DG* 374.19–22]): ἄνεμον εἶναι ῥύσιν ἀέρος, for which see Gilbert 1907: 512; Capelle 1935: 329; and Böker 1958a: 2220. Böker 1958a: 2227–8 also claims the target here is Democritus. As Ideler 1834–6: 1.446 points out, Aristotle says at *Mete.* 4.9.387a29–30: ἔστιν δὲ πνεῦμα ῥύσις συνεχὴς ἀέρος ἐπὶ μῆκος. The same theory is found in a dialectical context, *Top.* 4.5.127a3–12; so also *Pr.* 26.36.

9 The theory has no recorded advocate.

10 It is a possibility mooted at *Pr.* 26.36.

11 Olympiodorus 98.18.

relationship with each other than their conceptual apparatus permits. In his mind, if wind is air in motion, then all winds are one and therefore must flow from a common source. He then supposes that if he can show that the winds do not flow from a common source, the other theories will collapse by *modus tollens*.[12]

In Aristotle's mind, grave consequences follow from accepting the theory that wind is air in motion. For if wind is merely air in motion, the direction in which it blows will be an accident and not an essential difference. There will be, then, no special Boreas or Notos with their special inherent characteristics; there will be no radially ordered windrose to correspond to the rivers that encircle the Mediterranean.[13] And, as chapters 2.5–6 later show, Aristotle thinks that direction is the cause of real and important differences among the winds. This fact in turn highlights a couple of other purposes for the analogy. Since the individuality of the rivers is a manifest and obvious fact, the analogy can help to persuade us that winds must be individual as well. Moreover, if the rivers exist as individuals, they may perish as individuals, and, in consequence, changes in the climate are more likely to be local than pervasive – precisely the result Aristotle is trying to establish.

But Aristotle abuses the analogy to attain his goal against his opponents. He is trading on a straightforward equivocation when he denies that all winds are one in the same sense as all rivers are said to be one (349a20–30); for the claim that all rivers are one obviously means that they flow from a common source or spring, probably the sea, and not, as the analogy with the wind requires, that they form an indistinguishable mass of water or all flow in the same channel. Moreover, Aristotle has set up a straw man. There is no opponent we know of who has claimed that all the winds flow from a single source, with the possible exception of Homer's Aeolus and his bag of winds.[14] Yet, by the time we come to the end of the introduction, Aristotle has taken his stand: the river–wind analogy is a principle, and his purpose is to refute the claim that all winds come from the same vessel-like source (349a33–b2).[15]

12 Alexander 54.16–20 points out that Aristotle's arguments here do not refute the claim that the winds are air in motion.

13 So Olympiodorus 101.7–11.

14 *Od.* 10.19–20 (so Alexander 54.27–30). Placing the winds in a bag is obviously an anomalous situation even in this poetic context. Aristotle may have in mind the same passage as he has for the sources of the sea in 2.1, viz., Hesiod, *Th.* 736–8.

15 Aristotle also considers the possibility that the winds are self-originating (ἐξ αὑτῶν τὴν ἀρχὴν ἀφιέντας [1.13.349b1–2]) as the painters depict them. This is an image that arises in a different context at *MA* 2.698b25–7: "Not even Tityrus or Boreas could move a ship by blowing from within

Theories of river formation

Since no one can see where the wind blows, Aristotle sensibly starts with the more observable sources of the rivers and only much later applies what he learns there to the winds. He argues that the rivers do not flow from water stored in subterranean, cavernous reservoirs but rather result from the continual condensation of wet exhalation beneath the earth. He is certainly right to criticize the reservoir view on the grounds that the size of the reservoirs would have to be enormous (349b15–19), though his claim that the reservoirs would have to be close to the size of the earth is an obvious exaggeration.[16] Whether they would have to be so big or not, it is clear that he and the authors of the theory he is criticizing are thinking of subterranean caverns that would have to be so large as to be liable to collapse.[17]

In favor of his view that rivers arise from the condensation of vapor under the mountains, he draws on a comparison with the clouds (349b19–35), perhaps inspired by clouds' often mountainous shape. Just as the air holds both the vapor and the droplets of the clouds, so the mountain holds the vapor and the droplets from which the springs issue forth.[18] The similarity extends to the mechanics of percolation and condensation. The mountains are like dense sponges (σπόγγος πυκνός 350a7–8), in which little droplets form and coalesce. We should not suppose that the water droplets are suspended in the *air* within the spongy earth. Sponges, after all, do not hold little droplets suspended in the air in their holes; they hold water in liquid form. The water droplets are suspended directly in the earth and consequently the *vapor* is suspended directly in the earth.

Aristotle clearly developed this condensation theory, which relies on a continual process of coming-to-be, in order to refute the "reservoirists."

it, if they happened to blow in the way the painters depict them. For they depict them blowing the wind from themselves."

16 Since the reservoirs have to be above the springs (cf. 2.1.353b26–7), Aristotle perhaps does not literally mean the whole earth down to the center of the universe, but rather just that volume of the earth that is above the springs. Alternatively and more likely, he is thinking of the *Phaedo* where the subterranean caverns are indeed enormous and penetrate the entire earth (111d–113c). Mariotte 1718: 15 points out how much the flow of rivers diminishes in drought. Though Aristotle never claims that the amount of rainfall is insufficient for the volume of rivers, Mariotte (21–4) shows that rainfall vastly exceeds river flow. For the remarkable persistence of the idea that the rivers are independent of rainfall, see Tuan 1968.

17 The rainwater has already separated off from the earth and gathered in a distinct location (ἀποκεκριμένον 1.13.349b26, 29). Such reservoirs would have to be in high ground and have egresses at their lowest reaches to allow the flow until the reservoir is empty. If they are under mountains, they would indeed be unstable.

18 Alexander 58.8–11 brings out this analogy in the context of the reservoir: there can no more be a large reservoir for the rivers underground than there can be for the rain in the clouds.

But, though he thinks that condensation is the most important cause of river formation, he recognizes two other and more traditional causes, which in the end undercut his innovative theory of condensation. Even in his refutation of the reservoir theory, he concedes that rivers often do arise from reservoirs. Indeed at the end of the chapter (350b36–351a18) he admits that there are many subterranean rivers that flow from underground reservoirs. These arise from mountains just as normal rivers do but are caused when water is trapped in the mountain valleys and then forces its way out (ἀποβιαζομένου τοῦ ἄνωθεν ἐπιόντος ὕδατος 351a6–7). In this manner the Caspian Sea, situated below the Caucasus mountain, forces its way out into the Black Sea.

One of Aristotle's arguments against reservoirs reveals yet a third cause for the origin of rivers in addition to condensation and trapped mountain water: of the water that exists in the ground (presumably resulting from previous rainfall) and is not generated (and so has not been condensed), some is observed to ooze forth in a way that is different from the gush we associate with underground reservoirs (349b27–350a13). Aristotle aims this observation against the view that rivers originate in general from cavernous reservoirs, and to that extent the observation is effective. But it causes some collateral damage to Aristotle's theory of condensation, since the percolation of existing water here is essentially the same process as the *last* phase of the normal Aristotelian condensation-caused river. In both cases, the tiny droplets (ῥανίδες) come together into larger drops, which then ooze out of the earth. Aristotle explicitly draws a parallel to the process of condensation into rain (349b30–5). The result is that condensation and percolation are hard, if not impossible, for an observer to distinguish. While the two kinds of river formation have different remote causes, their proximate causes are identical: in both cases, the springs ooze from the earth. And since Aristotle has provided no separate evidence (besides his analogy with the clouds) that condensation proper occurs below the earth, we have no way of knowing which is the ultimate cause of oozing springs and, with it, no assurance that rivers are not just water in motion.

This problem just brings into starker relief the redundancy of Aristotle's theory of condensation. If, as is almost certain, Aristotle believes that the wet exhalation in the earth derives ultimately from the rain ("For [the mountains] receive a great amount of rain that falls" 350a9–10), the condensing of vapor below the earth is a superfluous cause, since the rainwater that soaks into the ground must become vapor before it turns back into water. In this way the cycle is completely pointless.[19] Now, Aristotle might have posited as an alternative that the cold, dry earth transforms into cold, wet water, and thereby supplies the springs. In that case, at least, the cycle of transformation would not have been redundant. Similarly he might have justified the vapor cycle by claiming that mountain springs require vast amounts of vapor, which must rise underground from the surrounding lowlands into the mountains and move up to alpine springs, where they are condensed and flow forth.[20] But there is no evidence that he entertained either view, and his vapor theory is developed only to the extent necessary to refute the "reservoirists." He consistently does his best to avoid the obvious question of where the subterranean vapor comes from.

In sum, the exhalation makes an awkward material cause for rivers. This very fact, however, reveals all the more clearly how important Aristotle thought it was for purposes beyond the narrow explanation of the phenomenon. We have already seen that wet exhalation serves as a more visible counterpart to the dry exhalation that explains the winds. But, given the difficulties we have just seen with Aristotle's account, the exhalation theory of rivers hardly provides a persuasive introduction to the winds. If Aristotle's only intention in taking up the rivers first out of their proper natural order had been to establish the wet exhalation as their cause before turning to the wind, the condensation theory does not give us confidence. Moreover, the condensation theory is not the only alternative to the reservoir. We are accustomed by now to see Aristotle's sublunary world as the realm of Heraclitean flux, and so it is worth considering why Aristotle was not satisfied with the "ever-moving water" of the rain-fed oozing springs. The probable answer is that Aristotle introduced subterranean condensation primarily to provide the rivers with an etiology independent of the rain cycle, and that he wanted to separate rivers from the rain in order to establish a descending hierarchy of phenomena, in which the rivers and the winds have their own corresponding natures, inferior to but independent of those of the kapno-atmosphere. The need to break the rain cycle

19 This did not stop Descartes adapting and clarifying the cycle (1637: VI.255 [Adam and Tannery]): the rainwater flows under the mountain, where it is vaporized by the earth's heat, rises in the mountain, then condenses and flows out of springs. See Martin 2011: 125–47.

20 An intriguing possibility offered by an anonymous reader. Though it would produce a better theory, perhaps the strongest evidence that Aristotle did not have it in mind is 350a4–6, "The biggest springs *neighbor* (γειτνιῶσιν) mountains and high places," that is, they are near, rather than on, the mountains. Moreover, those who make canals (349b35–350a2) are clearly making them at the foot of mountains to which the water is percolating down like sweat. Again, Aristotle's statements that *mountains* receive a great deal of rain and act like sponges suggest that it is the mountains alone that have the capacity for vapor.

pervades these chapters, and, though the wet exhalation may make a poor material cause, the material transformations it requires are an obvious way to accomplish this rupture. Aristotle's rejection of the reservoir theory can be seen as motivated by these same architectonic concerns, inasmuch as the reservoir theory is merely an auxiliary hypothesis to account for why rivers still flow in the summer when there is no rain water to supply them: rain water is stored in the reservoirs during the winter and issues forth during the summer. The only major objection Aristotle has to his predecessors that we cannot trace to the need to break the rain cycle is the claim that all rivers are one. It will receive an adequate consideration in the context of the sea.

Finally, the fact that one cause (oozing of rain water) makes the rivers dependent upon the rain cycle, while another (vapor condensation) isolates them from one another, is intelligible in light of the double perspective we saw in Chapter 5. We can now see that under one cause rain and rivers operate together as parts of a single cosmic whole and under another they can be treated as isolated great genera similar to the great genera of animals.

Aristotle's geography (1.13)

Now that Aristotle has shown how rivers flow from mountains, he proceeds to argue that the largest rivers flow from the largest mountains. In support of this thesis he provides an extended set of examples, far more elaborate than he requires. Springs had long been associated with hills, and, though our evidence is weak, few people apart from Plato (*Phaedo*) seem to have believed that water as such could travel uphill. It should hardly need a lengthy proof to show that rivers start in high places. Nor can one object that Aristotle's claim here is directed against the reservoir theory of the origin of rivers, since conceivably the largest river could flow from the largest caverns underneath the largest mountains. In view of the modest requirements of the immediate thesis, this lengthy *historia* is supererogatory. So why is it here? Aristotle's main purpose is to provide us with a conceptual geography that corresponds to the circle of the winds forthcoming in 2.6.[21] For this reason we must anticipate some of his later remarks, though the full significance of the pairing must wait until that later chapter (see "Disk and sphere" in Chapter 10 below).

Winds, rivers, and their effects together form the middle of the three, paired bands that define the vertical space of the sublunary world. These

[21] Olympiodorus 104.7 calls this a *historia*. For the strongest winds blowing from the largest (wet) sources, see also Hippocrates, *Vict.* 2.38: καὶ τὰ μὲν ἰσχυρότερα τῶν πνευμάτων ἀπὸ μεζόνων καὶ ἰσχυροτέρων, τὰ δὲ ἀσθενέστερα ἀπὸ μειόνων καὶ ἀσθενεστέρων.

phenomena, sandwiched between the kapno-atmosphere above and the minerals and metals below, are distinguished by their horizontal and radial orientation and are located on the threshold of the earth's surface. Aristotle uses his parallel *historiai* of the rivers and the winds to emphasize their similarity to one another. For his immediate purpose he treats the central landmass around the Mediterranean as a flat plane whose compass directions he conceives of as fixed points on its extremities – a very old-fashioned view of the world where Greece and the Mediterranean are at the center.

Aristotle is not being naive or reactionary. He adopts this "flat-earth" view in full awareness of the place of this landmass on the terrestrial sphere (2.5.362a32–b9). In chapter 2.5 he uses the spherical model to argue that his south wind blows from the northern tropic of Cancer rather than from the south pole, and the sphericity of the earth compels him to postulate a southern landmass as a counterpart to his own. The hemispheric symmetry of the globe, then, is a supplement to, rather than a substitute for, the symmetry of the plane, prefigured in the *Phaedo*'s many "ponds."[22] By refusing to set the earth's surface fully in agreement with spherical principles, by allowing his study to proceed on traditional principles, Aristotle makes room for the possibility, unrealized in his own body of work, of a quasi-autonomous science of geography, where the earth can be treated as a flat surface, as it was for the Ionians, without injury to the phenomena. Here Aristotle brings about within the confines of a single science a fundamental change of principles such as would usually require the introduction of a new science. This is remarkable enough. His truly exceptional accomplishment was to transfigure this traditional yet alien structure by integrating it into his theory of exhalation and combining it with his hierarchy of posture.

The traditional geography of the Greeks, at least since the time of Herodotus (4.40), had opposed Europe and Africa.[23] There is some evidence that Ionian maps divided the world of the Mediterranean along the west–east axis running from the Pillars of Heracles (Gibraltar) through the

[22] *Phaedo* 109b, where there are many hollows in the spherical earth, each of which forms its own *oikoumenē*.

[23] Herodotus approves of the twofold division and disapproves of the threefold (4.42). Olympiodorus 104.7–10 notes that Aristotle divides into three rather than two as the παλαιοί do. Aristotle makes the same division at *HA* 7(8).28.7.606b9–20 (noted by Sorof 1886: 24) but divides Europe and Asia (in contrast with Greece at *Pol.* 7.6). Bunbury 1883: 1.395–403 (especially 400) remarks how little Aristotle's geographical knowledge had advanced beyond that of Herodotus. Indeed, if Aristotle still treats the Dead Sea as mythical (2.3.359a17), we should not have much confidence in his knowledge (alternatively, his reference to μυθολογοῦσι may be to Ctesias [Ideler 1834–6: 1.534]).

straits of Messina, Corinth, and Sardis to the Taurus Mountains.[24] Two parallel lines north and south of this axis ran through Europe and Africa respectively and connected the summer solstitial sunrise (northeast) to its sunset (northwest) and the winter solstitial sunrise (southeast) to its sunset (southwest). These solstitial points are more than merely directions relative to the viewer: both for traditional thought and for Aristotle they indicate physical points at or near the extremes of the landmass.[25]

Ephorus in the early fourth century is usually credited with bestowing on the "frame" of the map a rectangular form, where the solstitial points are in the corners, and the wind direction (Apeliotes [east], Notos [south], Zepher [west], Borra [north] and general sun direction (rising, midday, setting and the Bear [of course, the sun is never directly in the northern sky]) identify the sides.[26] The four major perihellenic peoples (Indians, Aethiopians, Celts, and Scythians) are also located along the sides between the solstitial corners. The Indians, for example, live toward Apeliotes (the east wind) from the summer rising to the winter rising. We have in Ephorus, then, a system that combines geography and ethnography in a solstitially oriented, eight-pointed windrose.

This system may have been conventional by the fourth century. If so, Aristotle modifies it by first abandoning the winds as directional points and substituting for them the mountains and rivers, since the mountains, not the winds, are the cause of the rivers.[27] The mountains in turn are located by the solstitial points, which are more appropriate to this office than the winds, being the limits of the sun's movement north and south, and the cause of the rise and fall of the wet exhalation.[28] The tribes are omitted as being likewise irrelevant. Aristotle then transfers the winds to their own diagram (ὑπογραφή 363a25–6), which, like the map of the rivers, has its center in Greece (τὸ ὑφ' ἡμῶν οἰκούμενον 363a29) and is

[24] As Myres 1896: 611–20 has argued.

[25] As Heidel 1937: 18n41, has argued. We see the same division of the landmass in Hippocrates, *Aër.*12–13, where Greece is favored in climate, because it falls between the winter and summer sun. Greece cannot fall between, if the solstitial points are merely directions rather than fixed boundaries.

[26] If we are to trust Cosmas Indicopleustes' (2.78–80) apparent quote of Ephorus. For a photograph of the rectangle in manuscript, see Harley, Woodward, and Aujac 1987: 144. As Kahn 1984: 103n2 points out, the placing of the tropics in a real terrestrial location goes back to Homer (*Od.* 15.404: Ὀρτυγίης κατύπερθεν ὅθι τροπαὶ ἠελίοιο). Herodotus makes use of the same system of indicating direction by winds (e.g., 4.22: πρὸς ἀπηλιώτην ἄνεμον; 4.37: ἐπὶ τὴν νοτίην, πρὸς βορέην ἄνεμον).

[27] Aristotle was not alone in this. Strohm 1984: 159 notes the general tendency of Greek geographers to establish the rivers and mountains as the foundations of their maps.

[28] This should not be pressed too far, of course: the sun's movement is not the cause of the condensation of vapor underground, which evidently occurs in any season, and that is why rivers flow even in the summer.

laid out on celestial bearings that have, as I shall argue, fixed positions on the face of the earth.[29] The map of the winds, however, is circular and more geometrically abstract, locating the points on the circle by letters reminiscent of mathematical diagrams (Apeliotes is Β, etc.).

In his map of the rivers (τὰς τῆς γῆς περιόδους 350a16), Aristotle maintains elements of the traditional west–east axis of symmetry. The Ister (Danube) on this account forms the counterpart to the Nile.[30] Both flow east from sources in their respective western mountains (Ister from the Pyrenees, the Nile from the Silver Mountain), and after flowing the length of their continents both turn toward the central axis line and debouche into the Mediterranean.

But to this binary symmetry he adds another continent, which has the effect of shifting the map toward the circular shape and radial symmetry of the windrose. The landmass now has three divisions, Asia, Europe, and Africa, and each continent is given one particularly large mountain and one particularly large river.[31] The Araxes river of Asia (usually identified with the Syr Darya) flows north into the Caspian but adds some of its waters to the Tanais (Don) River, which then flows into Lake Maeotis (Sea of Azov).[32] This double debouchement parallels the belief that the Ister branches off and sends part of its waters into the Adriatic Sea.[33] Though Aristotle mocked those who represented the landmass as a circle (2.5.362b12–30), he is clearly conceiving of it as an oval.

29 So also the cause for the fact that southern and northern winds are most common clearly indicates absolute direction (2.4.361a4–23).

30 Paired again at 2.2.356a28.

31 For a treatment of Aristotle's geography, especially on the relation of the landmass to the sea, see Sorof 1886. We do not know of anyone before Aristotle who is aware of the eastern rivers flowing from the Paropamisus. Myres 1896 in reconstructing Herodotus' map had continued the axis line of the Mediterranean into the "Persian" map, effectively joining the two maps side to side. Aristotle's arrangement resembles rather a T-O map.

32 There is a general agreement in antiquity that the Oxus (Aristotle's Bactrus, modern Amu Darya) and the Iaxartes (Aristotle's Araxes, modern Syr Darya) flowed into the Caspian Sea. There is only slight evidence that any ancients knew about the Aral Sea (and therefore the question arises whether it existed then or not, see J. K. Neumann 1884). Aristotle (2.1.353b35–354a5) mentions the Hyrcanian and the Caspian Seas as if they were distinct, and Ideler 1834–6: 1.499–500 reads this passage as evidence that Aristotle knew of the Aral Sea. Bolchert 1908: 10 and Bunbury 1883: 401n2 think that Aristotle is confused. Bolchert 1908: 12 adds that Aristotle may have heard of the Caspian Sea from Herodotus and the Hyrcanian Sea from Hecataeus (or a source dependent on him) and thought that they were different seas. Pliny, *HN* 6.18.46 mentions different names for the sea depending on the location. Ptolemy, *Geog.* 6.12–13 mentions that one of the rivers of Sogdiana flows into the Oxus lake, and he locates the middle of it roughly in the position of the modern Aral Sea.

33 Aristotle mentions this in *HA* 7(8).13.598b14–16; it is also in *Mir.* 105.839b9–840a5. Some wonder how Aristotle could have thought that the Ister bypassed the Rhone, which he knew of as being navigable (1.13.351a16–18). He may have thought of it either as a branch of the Ister or as a local south-flowing river like the rivers of Greece.

Aristotle's description of all three continents exhibits the same basic pattern, necessitated partly by his thesis about the largest rivers, but also intended to establish his roughly circular geography. Each of the great rivers is located by its proximity to a great mountain, which in turn is located by a point of the rising or setting sun. The Parnassus (usually identified with the Paropamisus) is located at the winter sunrise;[34] the Caucasus at the summer sunrise. The Pyrenees are set toward the equinoctial sunset. The Chremetes and the Nile are not given sun directions, but they are easy to infer. The fact that the Chremetes drains into the outer sea and that it can probably be identified with Hanno's Chretes River recommend the Senegal River.[35] If the most important source of the Nile originates at the same Silver Mountain as does the Chremetes, then clearly Aristotle has the Nile arising in western Africa, in accordance with the traditional Ionian practice. Clearly without his needing to say so, the Silver Mountain lies toward the winter sunset.

These mountains lie on the periphery of the landmass and are the sources both of rivers that flow toward the Mediterranean (Araxes–Tanais, Nile, and Ister) and of the rivers that flow into the outer sea (Indus, Chremetes, and the Tartessus). The result is more or less a bowl with the Mediterranean in the middle, and it is in this bowl that the winds blow. This view is corroborated by and adapted to Aristotle's description of the depth of the various seas at 2.1.354a11–23, for the earth is tilted from north to south and from Europe into the Mediterranean (354a23–6), and the south no

[34] The identification is problematic in many ways. Bolchert 1908: 42 has argued that it betrays a Ctesian, or at least pre-Alexandrian, geography, and that the name was corrected to Paropamisus by Alexander's expedition. The latter is almost assured: the corruption of Paropamisus into Parnassus predates Aristotle's report, for he calls it the Παρνασσοῦ καλουμένου (1.13.350a19). The only really secure river name in Aristotle's description is the Indus. As Heidel 1937: 42n107 has pointed out, the Paropamisus or Hindu Kush is not placed toward the winter sunrise by any later geographer, but it is placed on an extension of the Mediterranean axis, so that the Paropamisus lies rather at the equinoctial sunrise (due east). It is also peculiar that Aristotle says that the outer sea is visible upon crossing the Parnassus. This would suggest that it is very far south indeed. The best bet for the Choaspes is the Panjkora, a tributary of the Kabul river (accepted by Talbert 2000).

[35] Here I agree with Lee on the most likely identification. Ideler, who suggests the Chretes as our Chremetes, only denies that it is the Niger. As Ideler 1834–6: 1.465n points out, Vitruvius 8.2.6 makes the Nigir river, arising from the Diris in the mountains of Mauretania, the source of the Nile, as does Pliny, *HN* 5.10, who reports Juba saying that the river goes underground for more than twenty days' march and comes out in Ethiopia at the Black Spring. Hanno describes the Chretes as follows: "we settled a colony and called it Cerne. We judge from our voyage that this place lay in a direct line with Carthage. We then came to a lake which we reached by sailing up a large river called Chretes. This lake had three islands larger than Cerne; from which proceeding a day's sail we came to the extremity of the lake, that was overhung by large mountains . . . " Müller 1965: 8 takes the Chretes to be the R. St. Jean (the Tajirit) but this is hardly a great river.

doubt tilts likewise into the sea.[36] It is a scheme that has little place for the inconvenient Euphrates and Tigris, and its general structure is clearest in the case of river-rich Europe.

In addition to the large mountains (two in Asia, one each in Europe and Africa), Aristotle mentions secondary mountains associated with less famous (though not necessarily smaller) rivers. Whereas the Ister flows west to east along the breadth of Europe, the Arkynian and Rhipae mountains to the north form the sources of various rivers that flow north into the outer sea. In addition, Aristotle mentions some of the rivers of Greece, all of which flow south from the Pindus, Scombrus, and Rhodope mountains into the Mediterranean (350b15–18). We see this same pattern in the Araxes–Tanais river complex (the analogue of the Ister), which flows north of and past the Phasis river (the analogue of the local Greek rivers) to drain into Lake Maeotis; and likewise in the Indus, which, like the north-flowing European rivers, flows into the outer sea. It may be, then, that the unidentifiable Aegon and Nyses (350b10–12), which are said to flow from the Aethiopian mountains are north African seasonal rivers or rivers of the area around Carthage (corresponding to the secondary rivers that flow into the Mediterranean).[37]

This extensive *historia*, ostensibly in service of the argument that rivers flow from mountains and that the greatest rivers flow from the greatest mountains, is out of all proportion with its appointed task. Its similarity with the cycle of the winds along with the elaborate analogies between the winds and rivers argue for a deliberate attempt on Aristotle's part to integrate an older Ionian geography based on a disk-like earth into his spherical cosmology. A fuller discussion and defense of this interpretation must await our consideration of the winds.

Climate change and land–sea exchange (1.14)

Now would seem to be the time for Aristotle to return from his "digression," resume his consideration of the winds, and renew his attack on the wind-bag theory that had introduced the horizontal phenomena in 1.13. Instead he ignores this theory (never to be mentioned again) and follows a seemingly meandering course of argument toward the sea and the problem of its salinity. The series of subjects we meet in the next four chapters, however,

[36] There is no need, with Webster (2.2 note 2) and Strohm 1984: 170 to suspect the mention of the shallows outside the Pillars of Heracles (2.1.354a22–3). The passage rather confirms that Aristotle conceives of the Mediterranean as a basin.

[37] Though it is hard to imagine their sources being called the Aethiopian mountains.

are gradually revealed to be guided by Aristotle's concern with his second target, the entropists, who denied the permanence of the terrestrial waters. Only when we realize what is at stake in the homeostatic permanence of the saline sea can we return and fully appreciate his objections to the mountain reservoirs.[38] That is not to say that the sea will provide an explanation or justification for the general application of the doctrine of homeostasis. The homeostatic stability of the sea is in the end a fundamental principle, and Aristotle cannot prove it. But he can and does recommend it by the use of those techniques he had applied to the kapnosphere. For just as in the upper region Aristotle's gradual progress through shooting stars and comets to the Milky Way helped to convince us of the latter's uniformly sublunary and homeostatic nature, so here below, the remarkable causal coherence of the terrestrial waters makes plausible their common homeostatic nature. If Aristotle can show that the earth itself is not simply drying out, then he has eliminated one argument in favor of the gradual disappearance of the sea.

At the same time, the similarity between the kapnosphere and the terrestrial waters goes beyond methodological issues. Aristotle holds the terrestrial waters as a mirror to the heavens: the rivers in their coming-to-be and passing-away track, though more slowly than the comets, the "great year" of the celestial motions; the sea in its isolation, volume, and salinity is as unvarying as the Milky Way. This symmetrical isomorphism between heavens and waters points to an order different from cause and effect and breaks the Ionian illusion that the sea must feed the stars. The entropists are never out of his sight.

The elaborate order this symmetry implies is intended as an alternative to the causal lineage proposed by the Presocratics and should not be mistaken for the effect of the final cause. *Harmonia*, as Aristotle says in the *Politics*, takes the place of soul in inanimate things (1.5.1254a32–3). Since a *harmonia* is something that is dependent on the composition of the parts of the underlying body (*de An.* 1.4.408a5–23), this is tantamount to saying that *harmonia* takes the place of the final cause where there is no final cause.

Still, for all the symmetry and order, the structure of chapter 1.14 is rambling. Aristotle begins by introducing his general theory of homeostatic climate change. The major methodological challenge in establishing such a theory is the long duration of these changes, which makes them difficult for humans to observe. His discussion of this issue is involved and sophisticated. He then turns to those who believe that the sea and therefore the

[38] On the importance of this principle, see Strohm 1984: 157.

world in general are drying up, and he ends with a set of examples that amplify the conclusion of his own theory and illustrate that these changes are localized and not pervasive.

In creating his theory of homeostatic change, Aristotle makes surprisingly little use of the exhalations. He joins together two processes that would seem to be only distantly related, the secular moistening and desiccation of regions of the earth and the formation of land from the sea; and the only reason they are joined here seems to be that they both involve battles between earth and water. On first reflection we might expect a common cause to explain them both, most obviously a battle between the dry and the wet exhalations. But the dry exhalation is nowhere in evidence, and the wet exhalation, though clearly the ultimate cause of moisture in a district, serves mainly as the cause of the rivers and then drops out of consideration. When we remember, however, that Aristotle avoids using the exhalations as opposites to account for opposite effects, we can see that a battle between them was never likely to explain climate change, much less an exchange of land and sea. Besides, since the dry exhalation is needed to explain the ubiquitous winds, it would be ill-suited to explain the localized desert.

The exhalations, then, make unpromising causes for explaining land–sea exchange, and their absence in this chapter indicates once again that Aristotle has some purpose in mind more important than cataloguing the phenomena under their power. Already in the context of subterranean rivers he had shown interest in the interaction between elemental layers and especially in the way in which the irregularities in the boundaries of those layers can cause diverse phenomena. There the mountains, terrestrial intrusions into the atmosphere, are a major causal factor, not only causing rivers, but also being a cause of the division between the kapnosphere and the atmosphere. Here again he makes extensive use of the interaction between elemental water and earth, which explains silting and the formation of lagoons. This is at the same time a recognition of the legitimate role of Presocratic mud in the process of world formation, and a mixed-element correspondent to his own so-called ἀήρ, that combination of airy and fiery elements distinguished in 1.3.[39] For in both cases the nature of the whole depends upon the prevailing part.

It is surprising too that rivers should play such a large role in this chapter. No doubt Aristotle wanted to provide a systematic continuity with

[39] E.g., Xenophanes, DK 21A33.5–6 (1.123.1–9); Empedocles, DK 31A66 (1.295.10–15); Democritus, DK 68B5 (2.135.10–13).

the previous chapter, but rivers also serve a more important function as the causal intermediary between the two species of inter-elemental battle, regional moisture changes and land–sea exchange, and this in turn makes them crucial in Aristotle's dispute with the entropists. According to Aristotle, districts of the earth undergo periodic senescence and rejuvenation according to the amount of wet exhalation present in the ground. A district in its youth is moist and rich in vapors. Toward the end of the chapter he accounts for this initial state by positing a "great winter," a period within a great cycle of time when a vast amount of rain falls and recharges the ground with moisture and rejuvenates it. Since the "great winter" gives rise to vapors that condense to form springs, this cycle is responsible in turn for the death and birth of rivers (1.14.351a36–b8). Abundant rivers, then, are a product of a moist region. The rivers, in turn, flow to the sea where they deposit silt. At first the silt is swampy but, as it accumulates, it dries out, and over a long period of time new dry land is created. The sea, however, is displaced by this silt and necessarily must flood some other low-lying area. But as the moist district grows old and its rivers dry up, the silt will subside and disappear, and the sea will reassert itself in that place and retreat from the other area it had flooded.[40] Rivers, then, are intermediary between vapor and silting: they are the *effect* of the vapor in a rejuvenated district, but they are also the *cause* of silting and alluviation and therefore of the generation of land, and in this way they serve to join causally the two battlefields of earth and water.

The causal chain does not end there. Though Aristotle does not make a sharp distinction between them, the process of lagoon formation is an important variant on and effect of alluviation, both ultimately dependent on the rivers. Around river mouths, sand bars form in shallow areas that eventually enclose lagoons of brackish water, fed partly by the rivers and partly by reflux from the sea. Eventually the rivers find different courses and avoid the lagoons or, in the course of time, dry up as all rivers do. The connection with the sea is cut off by the growing sandbar, and the water in the lagoon permanently evaporates, and thus new land is born. Aristotle sees this process as operative in the Nile Delta itself and as responsible for forming at least part of it. He provides evidence of this in the supposed fact

[40] The outlines and fundamental principles are clear enough in this otherwise very difficult text (1.14.351b7–8). The interpretation of Webster (1.14 note 1) and Lee makes the best sense. They are followed by Strohm 1984: 163 with some reservations. Alexander 60.6–13 thinks that river water physically pushes the sea water away, thus creating land. Webster and Lee save us from this bizarre hydrology. For an account of Mediterranean land–sea exchange from a modern perspective, Horden and Purcell 2000: 304–28.

that the Delta is lower than the Red Sea and that earlier attempts to connect the Mediterranean with the Red Sea were abandoned when it was realized that the Red Sea would flood the Delta (352b22–31). In a more dramatic way, Aristotle sees the area around Ammon in Libya as having been subject to the same process. The dry depression here, below sea level, is simply the result of a large lagoon's having been cut off from the Mediterranean and having dried out (352b31–353a1). This is clearly a process much more powerful than simple silting, but it is not clear what that process is, until Aristotle extends it to Lake Maeotis (Sea of Azov) and even Pontus (Black Sea). He observes that Lake Maeotis is growing increasingly shallow and that it is only boats of shallower draft that can now ply its waters (353a1–7). Olympiodorus may be right in observing that the lake is surrounded by lagoon-like shores, which have the effect of isolating its waters (123.2–7). We may suppose that, as time goes on, the waters of the many rivers that feed it will be diverted more directly into Pontus and that Maeotis will become dry land. In a related way, the Bosporus, which connects Pontus with the Propontis (Sea of Marmara) is becoming increasingly occluded on the Asian side. The process of lagoon formation is making the passage narrower and will result finally in Pontus being completely cut off (120.6–10). It will then suffer the same fate as Lake Maeotis. Once again, then, rivers play a crucial role in the battle between land and sea.

The various species of the genus of moisture need not in their own right have a causal connection among themselves. Rather, the compelling reason for positing a causal connection between vapor, rivers, and the sea is the polemic against the entropists. The apparently paradoxical, but clearly intended, result of Aristotle's scheme is that one district by becoming wet can make a neighboring district dry. In two distinct dynamics the presence of one opposite in one place causes the other opposite to appear elsewhere. The causal system Aristotle has developed makes it impossible for the world simply to dry out. The moistening of a region leads to more rivers flowing, more silt and more dry land appearing: wet causes dry. Alluviation dries out the ground and converts sea into land, and this process in turn inundates other land, including formerly dried out depressions: dry causes wet. Eventually the failure of moisture leads to failure of rivers and the subsidence of the silt into the sea again: dry causes wet. And presumably the sea will then retreat from the other places it has flooded: wet causes dry. In this way Aristotle finds in two different patterns a unity of opposites that preserves the cycle. This result, which could not be achieved without the rivers, is welcome both because it explains why these changes are local, and because it guarantees that they cannot be linear and entropic.

The phases of change, of course, need not be precisely opposed to one another: a currently drying district, such as Egypt (351b28), may still sustain a river that continues to create new alluvial land. Nevertheless, we should expect to find cases where the first moment of the process alone is operative, where the abundance of vapor alone produces a fertile district without rivers producing silt and new land. But this is not the case. The issue of moisture and desiccation is tied invariably to land–sea exchange: "the same places on the earth are not always moist or dry, but they change according to the birth or failure of rivers" (351a19–21). Even in cases where the riverine influence is obscure, Aristotle insists that the same causes are at work. So between the Trojan War and classical times moist Mycenae has become dry and the formerly swampy Argos has become arable (352a9–14). Though Aristotle makes no explicit mention, we must suppose that the local Inachus River is somehow the cause.

We must observe, however, that this process does not affect a single district in a uniform cycle of drying and moistening; rather it requires the interaction of two adjacent districts, the maritime littoral and the inland country. But if rejuvenation depends on flooding low-lying areas, it is hard to see how inland (and therefore higher) districts could ever become moist. For this reason it is surprising that there is no mechanism whereby the presence in one district of wet exhalation, the ultimate source of rivers, directly causes its absence elsewhere. We might imagine some process whereby the desiccation of Egypt has caused the moistening of northern Europe, but instead, we get a universal moistening due to the problematic operation of the "great winter."

Aristotle is aware of this difficulty and tries to overcome it by an opportune turn to methodology (351b8–28). This new approach with its emphasis on human societies and their collective memory justifies Aristotle's leap in evidence back to the barest fragments of our earliest cultural memories and to the sole remembered case of inland moistening, Deucalion's flood.[41] Just as in his discussion of comets he had commented on phenomena spatially remote from sense perception (τῶν ἀφανῶν τῇ αἰσθήσει 1.7.344a5), so here we must contend with enormous spans of time.[42] In service of this point he

[41] As an anonymous reader has pointed out to me, Aristotle owes this approach to Plato's *Timaeus* (22b–e), where the Egyptian priest chides the Greeks for lacking the long cultural memory of the Egyptians. It is worth noting that according to the priest the civilization of the Greeks is regularly interrupted by floods, whereas Egyptians enjoy civilization without cataclysm because the Nile floods in a regular and non-violent way. For Aristotle, by contrast, the whole world is subject to the same processes of flooding and desiccation.

[42] Strohm 1984: 163 mentions the parallel between the "life cycle" of the earth and the life cycle of man.

introduces the curious analogy that likens the cycles of climate change on the earth to the biological process of aging (and by extension, rejuvenation) (351a26–36). This analogy is then joined to a discussion of human society and the destruction of cultural memory.

It is not likely that Aristotle is granting a vitalistic premise to his adversaries merely for dialectical purposes. His language – he calls it the ἀρχή and αἴτιον (351a26) – suggests that he is committed to the principle that the earth is in some sense "alive." Nor is he committing *metabasis* here, since he is careful to attribute the earth's senescence to the proper meteorological principles.[43] In the first instance climate change is caused by the hot and cold in the earth, which are caused in turn by the sun and its revolution (351a30–3). Later we learn that a "great winter" drops sufficient rain to replenish long-term river flows (352a28–33). He even goes so far as to avoid mentioning fossilized animals and plants – already discussed and problematized by his day – as signs that what is now land was once sea.[44] He is clearly eager to avoid real biological principles in his meteorology. But the biological analogy remains and deserves careful attention.

Once again Aristotle's target is the Ionian entropists, who think that the earth came to be at one time and will perish at another as if it were an individual human being. He points out that the life of the individual human is different from that of human society and that of the earth, for all that they may, in virtue of a certain similarity or relation, be said to be alive. Individual humans die, but the species does not; the earth is an individual but is not a member of a species. For this reason the entropists' analogy breaks down. The earth is an eternal individual, and yet it is subject to phases of maturity and senescence. Aristotle, by yoking land–sea exchange strongly to the fates of particular rivers, ensures that the phases of maturity and senescence will be local. The intermediary term that binds together the life of the individual human and the life of the earth, both in likeness and in cause, is the human society. For a society is neither permanent nor, as the entropists maintain, does it simply develop from a primitive form. Rather it follows a Herodotean pattern, where fortunes of peoples rise and fall. This pattern in human society, easily observable, is a sign that the earth goes through similar phases, since a society's phases are caused by

[43] As Strohm 1984: 162 notes, it is "neither meant as a poetic image nor in the 'vitalist' sense of the Stoics."

[44] According to Hippolytus, Xenophanes (DK 21A33.5 [1.123.1–5]) mentions shells found inland and in the mountains; impressions of fish and seaweed in the quarries of Syracuse; and a daphne fossilized in a stone in Paros, a clear indication of a dry period followed by an alluvial inundation, followed by drying. For Egyptian shells, Plutarch, *de Is. et Os.* 367b.

earth's phases (as Egypt dries out, Egyptian society is on the decline). Thus there is life for the individual human, for the society, and for the earth, and life will mean something different in each case. The terms may be focally related; they may even share a family resemblance.[45] Senescence of both individual humans and the earth is caused by drying. But humans do not rejuvenate; and part of being alive for humans is also the capacity for memory (which the earth does not share). Human societies are related to individuals by being aggregates of individuals and sharing a collective memory. When the aggregate breaks up, either because the individuals emigrate or because they become dry and die, the collective memory is lost and the society dies. Societies, of course, are not moist or dry, but their senescence is caused by the moisture or dryness of the earth. Through these considerations Aristotle shows that the relationship between earth, human life, and society is more complex than the entropists had understood, and that granting the earth a certain qualified life does not mean that it has to perish utterly.[46]

While Aristotle may succeed in demonstrating the polysemy of "life," and thereby in showing that the whole earth does not *have to* end up dead, his conception of social memory should not be understood as a positive argument for cyclical climate change, since all the reasons he identifies for the cultural amnesia of climate change are fully consistent with the entropic view. First, whereas sudden changes such as famines, wars, and plagues are immediately felt and easily remembered, the gradualness of climate change makes the process difficult to notice. There is no need to talk of cycles here: the gradual change can be linear and the effect on cultural memory will be the same. Second, the changes involved in wars and plagues are first impressed in memory before they are passed down to the next generation. But with climate no memory of the change can be formed until several generations have passed, the length of time required for people even to become aware of the change. Again, this is true of entropic change as much as of cyclical change. Moreover, when to these two obstacles are added the disintegration and death of the remembering community caused by the change itself, it becomes all but impossible for humans to remember any case in which a moist, fertile land became a desert. All these, however, are challenges to remembering entropic changes, and none of Aristotle's supplementary examples address the full cycle.

[45] For focal relation, see Owen 1960 and Wilson 2000: 116–74.

[46] Strohm 1984: 163 notes some of the elements of this carefully composed section.

The loss of cultural memory by itself merely shows that long-term changes, cyclical or entropic, are difficult to detect, and this inference could only be aimed properly at the null set of those who deny climate change *tout court*. For this reason the loss of cultural memory was clearly not intended to prove that the climate changes on the earth are cyclical. In fact, Aristotle makes perfectly clear that the loss of cultural memory is only partial, and he cites the mytho-historical record to show that humans have moved from one area to another in pursuit of more fertile land. Immigration and emigration of peoples, their coming-to-be and perishing, are facts of record and are caused by climate change.[47] Though the individuals of a population may not remain numerically or culturally identical over time, the populations clearly shift their habitations with the climate rather than merely die out. People shift from place to place, though it sometimes takes an outsider to be aware of it. Clearly, then, there must be a cause for this shift: moisture must likewise shift from place to place. The entropists have failed to realize that the earth, like human society but unlike human individuals, can rejuvenate.

Only after this specific evidence is canvassed may considerations based on more abstract principles be adduced. Recurring to the *topos* of 1.3, Aristotle argues that the earth is small in comparison with the universe (352a25–8). The implication is that without homeostasis the earth would be dried out in no time. Consequently the sea cannot feed the stars. Moreover, even a climate cycle is short in comparison with the eternity of the universe. The earth after being moist for a brief time would implausibly remain dry for the rest of eternity.

While Aristotle attacks the entropists' contention that the world is drying out, he is, as he himself is aware (πλείους 352a22), hard pressed to come up with examples of secular moistening or inundation. But the challenges that face social memory provide him with a justification for lowering his standards of evidence in this case. He cites the very primitive flood of Deucalion, because it draws on the mytho-historical past and marks a break in the cultural memory with the previous period that he had treated as characterized by desertification. The methodological concerns that led to the introduction of social memory allow Aristotle now to find among remnants of tradition his only instance of secular moistening and to provide an account of land–sea exchange that extends beyond the shores of the sea.

47 As Aristotle draws out the relationship between memory and climate change, he is careful to adduce examples that draw on man's cultural relationship to his environment (the subsidiary mouths of the Nile were dug [351b33–5], Memphis was founded [352a1]; Mycenae became infertile and Argos fertile [352a8–14]). Homer is mentioned or alluded to twice in the section (351b35; 352a8–14).

The myth has the added advantage of being related to the Achelous River, whose frequently changing course indicates that it is in the process of silting and in that respect similar to the Nile (352b1).[48] But the flood of Deucalion is less useful in other respects: quite unlike the gradual process of senescence that unfolds over many generations and more akin to famines, plagues, and wars, the flood was a sudden catastrophe, taking place in the single generation of Deucalion. Moreover, the challenge facing any such cause is to explain the capacity of the mountains to store vapor sufficient for the ages. It is strange indeed that a subterranean reservoir, which could never be sufficient to store even a year's volume of river water, is here sufficient to store river vapor for millennia.

Whereas the flood of Deucalion can be dismissed as Aristotle's desperate attempt to find an example of inundation, the great winter, with which it is awkwardly joined, poses problems of a different order. The notion of the great winter, with its analogy to the solar year and the gradual change of seasons, does not suggest a sudden and cataclysmic flood.[49] Moreover, it is a grand celestial cycle and for that reason connects the rivers with the order of the heavens (352a29–31). As such, it is ill adapted to cause the local changes Aristotle seeks to explain and is more suitable for a Stoic cycle of universal cataclysm and *ekpurosis*.[50] The unwieldy causal structure Aristotle brings to bear here, where neither flooding nor the great winter is fully consistent with his aims, is an indication that his concern lay less with the suitability of the cause than with finding a period of moistening sufficient to balance the more obvious cases of drying and so to maintain a homeostasis of local climate changes against his entropist opponents.

[48] For myth as a remnant of past knowledge, *Mete.* 1.3.339b29–30; *Met.* 12.8.1074a38–b14. Ninck 1921: 12 notes the special place of the Achelous River as son of the earth, a prototype of rivers.

[49] As Olympiodorus 119.5–12 has it.

[50] Verlinsky 2006. See Censorinus (*de die natali* 18.11) on Aristotle and the discussion of Van der Waerden 1952 for evidence of Aristotle holding such a Stoicizing view. *APo.* 2.15.98a31–4 mentions storms occuring more at the waning phase of the moon. A great year like that described by Censorinus is not impossible for Aristotle.

CHAPTER 9

The sea (2.1–3)

The ancient biographical tradition tells a story that upon the death of Alexander the Great, Aristotle withdrew from Athens to his estate at Chalcis.[1] There at the Euripus, the narrow strait separating the island of Euboia from mainland Greece, he pondered long on the causes that might explain why the water there flows now in one direction, now in another. Sick to death with anxious thought on the matter and unable to solve the problem, Aristotle took his own life in some unspecified way.

This story is typical of the genre of philosophers' deaths, but it is remarkable in connecting Aristotle's suicide with the sea.[2] The author thereby evokes in the reader the mythical *topoi* of despair at knowledge unattainable and consequent self-sacrifice, of death, and the inevitable journey to the afterlife. These evocations are surprisingly apposite, for, whether through literary conceit or dialectical method, Aristotle fashions his own discussion of the sea as a kind of *katabasis*, a descent to the underworld. The most famous classical example of this motif is, of course, Odysseus' visit to the realm of the dead, which falls midway through his journey and involves a passage to outermost Ocean (*Od.* 11.13–15). In a similar manner Aristotle places his chapters on the sea in the middle of the *Meteorologica*, after his descent from the mountain springs and before he ascends again to the winds, to the violent storms and finally to the high geometrical order of the rainbows and other optical phenomena. And in the very center of this section he describes and critiques Plato's *katabasis* in the *Phaedo*, as Olympiodorus notes. Plato himself, during the dark night of the philosophical soul in the *Phaedo*, in the central section on *misologia*, compares the baffling to and fro of antilogic to the water flowing through

[1] Düring 1957: 345–8, who attributes it tentatively to Hermippus. The story first appears in Justin Martyr (*Cohortatio ad graecos* 34b, Migne, *PG* 6.305); Procopius (8.6.20) provides the greatest detail.

[2] Mulvany 1926 for some of the methods of fabricating Aristotelian biographies; Cf. Chitwood 2004: 48–58, 79–93, 134–40; Riginos 1976: 194–8.

the Euripus (90c4–6). Aristotle could hardly have been unaware of all these associations.[3]

In fact he alerts us at the outset to the dramatic and awesome nature of the katabatic tradition (τραγικώτερον . . . καὶ σεμνότερον 2.1.353b1–2). Awesome is the sea, coeval with the earth and invested with a hoary mythology.[4] Aristotle meets this mythology head on, citing myths and mythographers more times here than in any other section, and treating them with a seriousness that far outstrips their limited scientific value. From the outset, then, Aristotle acknowledges the air of mystery that has surrounded the sea. The tradition is dramatic too, not just in the hands of the poets, but also among the Ionian philosophers who gave the sea a history and a development. Once upon a time, they say, water covered the whole earth, then the sun began to evaporate much of it and fed upon its vapors. Our present-day sea is the remnant of that original water, and it too will eventually be all dried up.

Aristotle exploits the structural features of the traditional *katabasis* in order to mock its pretensions. The ancient theologians (οἱ . . . ἀρχαῖοι . . . διατρίβοντες περὶ τὰς θεολογίας 353a34–5) labored under the mistaken belief that the sea was an important part of the universe, that it and the earth were the most honorable places and that they were the principle and origin of the heavens that surround them. Even those philosophers "wiser in human wisdom" (οἱ δὲ σοφώτεροι τὴν ἀνθρωπίνην σοφίαν 353b5–6) exaggerated the sea's importance by subordinating the heavens to it and feeding the stars on its vapors. While Aristotle characterizes both groups as old-fashioned, he especially ridicules the philosophers for dressing up the old myths in modern disguise, just as Plato in the *Phaedrus* had ridiculed those excessively clever and industrious men (λίαν δὲ δεινοῦ καὶ ἐπιπόνου 229d3–4) who rationalize the myths.[5] For this reason he likens Democritus' theory of the sea to a fable of Aesop (2.3.356b9–17) and chastises Empedocles for introducing a poetic metaphor into a scientific discourse (2.3.357a24–8).[6] Poets and philosophers both treated the sea as a subject of great importance.

[3] Olympiodorus in this context (144.21–145.5) mentions that Plato provides three *katabaseis* or *nekuiai* (*Phaedo*, *Gorgias*, and *Republic*).

[4] For the tradition of the high value of the sea, see Ninck 1921. Olympiodorus 129.24–7 mentions Euripides as tragic in making the sea a major part of the cosmos and making the sun and the heavens come to be from the earth.

[5] Alexander thinks that Aristotle is mocking the theologians (ὑποσκώπτων 66.23–67.1). See also Strohm 1984: 168–9 for the contrast between θεολογία and ἀνθρωπίνη σοφία.

[6] Even the evidence about the density of the Dead Sea may be mythical (μυθολογοῦσι 2.3.359a17). A different version of the Aesop fable is given in Perry 1952: 324–5.

Aristotle must have taken delight, then, in casting these pretensions aside and comparing the sea first to the bitter chyme of the stomach and finally to urine in the bladder. In his hands the traditional order of nature is inverted: the sea is no longer the source and principle of the rivers and other waters; it is no longer food for the sun and the stars, no longer the stuff of the air and winds; it is merely the end of a digestive process, a residue and sterile waste.[7] This *katabasis*, then, helps to make clear why Aristotle bothered in the first place to refute the theologians and their talk of springs (πηγαί) and roots (ῥίζαι) of the sea: he believes that the theologians are the original source of the error, from which the philosophers' entropic interpretation has flowed.[8]

But, more importantly, Aristotle places this mock *katabasis* at the center of his treatise, in order to emphasize the sea's humble place in the grand structural hierarchy of the sublunary world and its low status in a threefold system of water, a system that includes the rain cycle above the earth and the rivers on its surface. In spite of their material similarity, these waters are essentially different from one another and form a hierarchy that Aristotle nowhere explicitly describes but is everywhere intent on supporting. It is his main object to show that, in contrast with the celestially driven, constantly changing and flowing, fresh waters above and on the earth, the sea is a stable, stagnant, and saline residue remote from heaven's ordering influence. Now, it is often more fruitful to watch what Aristotle does than what he says he does, and what he does with the sea – and by extension with all the other phenomena – is to select empirical evidence and endoxic reports in order to further his grand architectonic and hierarchical conception of the sublunary world.

Confinement

Aristotle's first task is to isolate the sea from the rivers and the rain cycle, in order to establish the sea as resting in a determinate place without sources

7 The sea is the ending point for the rivers (τελευτὴ μᾶλλον ὕδατος ἢ ἀρχή 356a35–b1; τελευτῶντες 356a23). Aristotle twice mentions that certain salt water contains no fish (2.3.359a22, 29); the associations with ashes (τέφρα 2.3.359b7) also make clear that the nutrient value of the sea is largely spent.

8 Hesiod, *Th.* 738, 809, 812. For ῥίζαι, Ideler 1834-6: 1.496 points to Xenophanes as a possible source (DK 21A47 [1.125.25–7]: οἱ μὲν γὰρ διὰ ταῦτα ἄπειρον τὸ κάτω τῆς γῆς εἶναι φασιν, ἐπ᾽ ἄπειρον αὐτὴν ἐρριζῶσθαι λέγοντες; cf. ἐκ τοῦ κατωτέρω μέρους εἰς ἄπειρον [μέρος] ἐρριζῶσθαι, ἐξ ἀέρος δὲ καὶ πυρὸς συμπαγῆναι (Aët. 3.9.4 [*DG* 376.14–16]). West 1966: 361 cites Empedocles, DK 31B54 (1.332.22): αἰθὴρ <δ᾽ αὖ> μακρῇσι κατὰ χθόνα δύετο ῥίζαις as also in the tradition of the concept. Cf. Parmenides, ὑδατόριζον DK 28B15a (1.244.6).

or springs. His first endoxic target, the chief of the "ancient theologians" (οἱ ἀρχαῖοι καὶ διατρίβοντες περὶ τὰς θεολογίας 353a34–5), though not explicitly named at the beginning of 2.1, is probably Hesiod.[9] Far below the earth's surface, says Hesiod, the distance a man would fall in a year, a dim region supplies the roots and springs "for the dark-hued earth, and for Tartarus' realm of twilight, *and for the restless sea* and the sky which supports the stars."[10] As West and others have argued, Hesiod probably has in mind here a kind of *apeiron*, "empty, immense and gaping" (*Th.* 740), an aboriginal and indeterminate chaos of stuff, which forms the ground and source for the determinate structure we see on the earth's surface above.[11] Now, we can imagine how the earth might have subterranean roots or sources. By extension too we may conceive, as Aristotle has, of underground rivulets of water. But Hesiod also mentions sources of Tartarus and Heaven. He can hardly have been thinking of determinate root-like structures. If this is so, Aristotle inappropriately projects upon him the kind of determinate springs he himself had just theorized for rivers in the previous book (1.13). Aristotle is, in fact, interpreting Hesiod through the *Phaedo*'s much more determinate description of Tartarus, in which the subterranean channels of water are actual passages rather than inchoate and chaotic sources of moisture. In so interpreting Hesiod, Aristotle should be aware of his anachronism, since he himself deliberately postpones his criticism of the *Phaedo* to the end of chapter 2.2, precisely because Plato's vision is more specific and determinate and subject to more specific criticisms.

Hesiod deserves better than this, but in this first chapter (2.1) Aristotle's purpose is both to introduce the notion that the seawater has settled into a determinate resting place and to distinguish it from the flowing water of the rivers. Whatever Hesiod's sources (πηγαί) might actually have done, Aristotle interprets them as flowing and thus makes them serve as a foil for his own theory. He pointedly confronts the theologians' myths with a philosopher's division (353b18–35), since a definition is the

9 Gilbert 1907: 400n1 and Cherniss 1935: 131n528 mention Thales as a possible target. Antonio Vallisnieri (1661–1730) "rejected the theory that spring water originates from seawater that evaporates and recondenses after being percolated through the earth. He demonstrated with sound arguments that it comes from atmospheric precipitation" (DBS 13.564). In his *Lezione Accademice* he noted that it is contrary to hydrostatic law for water to jump from the sea to the mountains and that water cannot become sweet by passing through the earth, only by distillation (1715: 5). He describes too the work of Perault, Mariotte, Sedilo, and de la Hire in measuring over many years the amount of water in rain, snow, hail falling, and how much water flowed in the rivers per year in order to show that rainfall easily sufficed to supply the rivers (7).

10 Hesiod, *Th.* 736–7; translation in Fränkel 1973: 105.

11 As Solmsen 1950, Fränkel 1973: 106, and West 1966: 361 propose.

first thing he needs to solve this problem (τούτων δ' οὕτω διωρισμένων 353b30–1):[12] all terrestrial waters either flow or stand. If they flow (like rivers) they must come from sources and springs. If they stand, they either have no sources (like lakes and swamps), or if they have sources (like wells), they must be dug artificially. This division confronts Hesiod with an entirely alien conceptual scheme and introduces some of the fundamental concepts Aristotle has developed for confining the sea, in particular the distinction between flowing and standing water. The differentia of the water's movement (flowing–standing) here takes priority over its salinity, as is appropriate for a study of its φύσις (cf. 353a32); but eventually rest and salinity will be so combined that the standing water will be relegated to the salty sea, while the flowing waters, located in the never-ending rain cycle, will be fresh. Standing fresh lakes are ignored hereafter as salinity becomes a more important factor, and the artificiality of wells is forgotten immediately.[13]

Aristotle's anti-mythological division can be interpreted in at least two ways. Neither is without difficulty, and both reveal how much Aristotle is being guided by his architectonic aims. It is clear that the distinction between flowing and standing is important and becomes more important as the chapters proceed. If we suppose that this is the primary distinction and take him at his word that flowing waters arise in every case from sources (τὰ . . . ῥυτὰ πάντα πηγαῖα 353b19–20), how are we to classify rivers that flow from lakes? If we say they arise ultimately from springs, then surely the lakes must arise from springs as well. But Aristotle makes perfectly clear that rivers from sources do not flow from lake-like reservoirs (353b20–3).[14] He clearly does not stipulate a meaning and simply define πηγαῖα by *fiat* ("arising from springs") as rivers that flow, for this would be an intolerable impoverishment of the sense of πηγαῖα.

Alternatively, Aristotle may mean that waters flow (for the most part and excepting rivers that flow from lakes), because they flow *from springs*, that is, they are in immediate causal proximity to subterranean condensation. This characterization is consistent with Aristotle's division and makes not only rivers πηγαῖα, but also artificial wells, whose water likewise is generated

[12] Strohm 1984: 169. It is one of very few explicit divisions in the treatise. At 356b21–5 we also find Democritus faced with the decision whether the vapor falls back as rain or not – again this division is in response to the Charybdis myth.

[13] Indeed Olympiodorus 126.24–7 objects to the theory that salt water is the remnant of evaporation on the grounds that in that case all bodies of water would have to be salty.

[14] So also 2.4.360a29–31: "For we do not suppose that any flow of water at all produces rivers, not even if they have a certain volume; rather the flow must be from a spring."

by condensation.[15] This is an appropriate sense of πηγή, and it is probably what Aristotle intended. But it is still not clear why πηγαῖα must exclude naturally standing water. Aristotle himself recognizes a submarine source for the Pontus at the so-called Depths (1.13.351a7–16). And even if this source is fresh and originates from a lake in the Caucasus – and so is not a source for the Pontus' *saline* water – it still shows that water may well up from below the surface of a standing body of water. Moreover, we only know of the Depths because their waters are fresh; submarine salt springs would escape our notice and so may be, for all we know, a common occurrence.

Aristotle bends other evidence to show that the seas cannot have sources: many seas are isolated or nearly isolated from one another (which seems to assume rather than show the point), and we know the Hyrcanian/Caspian Sea has no sources because the inhabitants who live all around it would have seen them (354a3–5). The implication is that inflowing rivers, of which the Caspian has several, cannot make a standing body of water πηγαῖον. Nevertheless, we might object, the inhabitants would not necessarily have seen *submarine* springs. Moreover, there might be large springs at the very edge of the sea, as we would expect in view of the nature of water tables. Aristotle does not consider this latter a possibility, and yet it reflects a hydrology very similar to that of the artificial wells (and it is hard to see how their artificiality *per se* makes any important difference). Why, then, should seas not have such sources? Is it because these rivers or springs around the sea contain fresh water and so are obviously not sources for the *saline* sea? Such a qualification would successfully deny the sea its littoral sources but would do nothing to prevent fresh-water lakes from having them (and, for the purpose of the division, seas and lakes are the same, differing only by degree, 353b24–5).[16]

In short, Aristotle's division and the evidence adduced for it are unconvincing as an argument that the sea has no sources. He wanted to establish that among natural terrestrial waters, flowing waters are coextensive with waters from sources and that standing waters are coextensive with collected waters. Since the sea is standing, it cannot arise from sources. The argument provides the illusion of philosophical cogency in the face of poetic fancy. Its value lies strictly in its conclusion and has no heuristic value in its own right. It is Aristotle's first step in isolating the sea from other waters and, in spite of its weakness, it forms the basis for further elaboration. As

15 So Alexander 68.29, who thought that the theologians were giving the sea an eternal source (67.24–6).

16 Alexander 68.33–69.6 contends that source-fed bodies of naturally standing water might be possible if they are small, but the sea is not small.

Aristotle moves beyond this general argument and considers the role of the rain cycle and of salinity, the source waters become exclusively associated with fresh water that is always in circulation and always in the process of coming-to-be; and standing waters (now ignoring fresh water lakes entirely) become associated with the standing reservoirs of saline water. Aristotle's main concern, then, is to show that the sea, unlike the rivers, is stationary, and to this he dedicates the second half of chapter 2.1.

Having argued dialectically through the division, Aristotle must now show how cases of apparent sea flow are not flow in the relevant sense, and do not prevent the sea from being a standing reservoir of water. His discussion on this point goes into vastly more detail than the problem requires (compare the vexingly brief account of the sea's salinity in 2.3), but, as becomes clear, this detail helps him to anticipate further developments in the conception of the sea, specifically the notion that the sea is held in a kind of container. He freely grants that sea water moves, especially when it meets narrow straits, but this is just an ebb and flow natural even in a standing body (obviously no need for Aristotle to commit suicide here!). Aristotle thereby adopts and adapts the sloshing motion that Plato introduced – the verb ταλαντεύεσθαι (354a8) picks up the αἰωρεῖται of Plato's *Phaedo* (112b3) – by substituting a horizontal movement for Plato's impossible vertical (ἄνω ποταμῶν 2.2.356a19) slosh.

He also grants that flow is caused by the influx of rivers, most notably the great rivers of eastern Europe, which enter Lake Maeotis (Sea of Azov) and Pontus. This flow continues from shallower seas into ever deeper ones, the Aegean, Sicilian, and finally Sardinian and Tyrrhenian.[17]

The idea that the Mediterranean basin is a kind of container for the sea is emphasized in a digression describing the general slope of the earth from Europe into the Mediterranean.[18] Again the discussion is more detailed than is necessary for proving that the sea does not flow, and it serves a couple of other functions. First, it anticipates the refutation of the *Phaedo*, by showing how the rivers flow into the sea rather than into the ground. It also shows how the Mediterranean forms a basin (for all that the outside sea does too), a proper place (οἰκεῖος τόπος) for the water, and prepares us for the digestive analogies to come.[19] In fact, the observation that a

[17] See Oleson 2006 on possible Peripatetic contributions to knowledge of sea depths.

[18] For land being higher toward the north, Empedocles, DK 31A58 (1.294.18–20) = Aët. 2.8.2 (*DG* 338.5–9).

[19] For the basin, 2.1.354a11–13 and 2.2.355b17–18. For hollows of the sea outside the Pillars of Heracles, *Pr.* 26.5.940b27–9. As Drossaart Lulofs 1930: 10 points out, the sea outside the Mediterranean does not come into account.

sea flows into its deepest part (354a11–13) is selected to anticipate the flow of digestive fluids downward from the stomach. Finally, it fills out the general structure of the sublunary world by placing the Mediterranean at the center of the flowing rivers. Thus, by the end of the first chapter Aristotle has completed the negative and begun the positive steps of showing that the sea is a standing body of contained water not subject to flow.

The second chapter is the formal beginning of the positive moment, the determination of the nature of the sea (φύσις 353a32, stated at the beginning of 2.1); the third chapter is devoted to its most important property, its salinity. According to Aristotle, the problem of the isolated nature of the sea (2.2) arose among his predecessors from the observation – we have seen this active from the very beginning of the treatise (1.3.339b9–13) – that there is no complete layer of water surrounding the earth as there are layers of air and fire (2.2.354b4–17). It was their common opinion that the sea used to cover the earth and is now in the process of drying out.[20] The additional doctrine that the evaporated water feeds the heavenly bodies and causes the tropics of the sun and planets is an early one and fades out quickly after the introduction of theories of elements, depending as it does on the unproblematic transformation of water into fire. Aristotle does not properly distinguish those early thinkers, who posit a naive entropism of the sea and earth, from the later atomists, who place the development of the earth in the context of cosmic cycles and innumerable worlds (and who therefore can hardly be accused of not seeing that the cosmos is vast in comparison with the earth).[21] He exaggerates the degree to which they destroy the rain cycle by the entropic evaporation of the sea, and in service of his architectonic agenda he presents as the only solution to this problem the radical isolation of the sea from the cycle of fresh water (355a21–32).[22] The fresh water comprises the vapor in the air, the precipitation, the rivers, and fresh lakes, which together form a cycle of becoming. The saline sea, by contrast, is an inert and stable body.

20 Anaximander, DK 12A27 (1.88.7–21); cf. Anaximenes, DK 13A7.5 (1.92.13–16); Xenophanes, DK 21A33.5–6 (1.123.1–9); Anaxagoras, DK 59A90 (2.26.42–27.22); Democritus, DK 68A100 (2.108.26–109.17); Diogenes of Apollonia, DK 64A17 (2.54.10–24); Antiphon, DK 87B32 (2.344.20–345.20).

21 The naive entropists: principally Heraclitus, DK 22A11 (1.146.27–36); but also Anaximenes, DK 13A7.5 (1.92.13–16); Anaximander, DK 12A11.7 (1.84.17–21) and Diogenes of Apollonia, DK 64A17 (2.54.10–24). Also see Gilbert 1907: 684–6 and, for further details from Aristotle, 2.2.354b33–355a8.

22 Anaximenes knows that moisture feeds the stars and yet air recondenses (DK 13A7.7 [1.92.21–5]). Xenophanes knows that all things come from earth and water, and yet rain ultimately arises from the sea (DK 21B30 [1.136.8], Anaxagoras recognizes the existence of hail; Democritus too has ὄμβριον ὕδωρ (*Mete*. 2.7.365a1).

There is abundant evidence here for Aristotle's careful distinction. As he repeatedly stresses, the fresh water participates in a daily, that is, short-term, cycle of genesis. The rivers, he says, do not form a single mass (ἀθρόον) nor are they standing (στάσιμον) but are in constant change from day to day (ὡς γιγνόμενον ἀεὶ φαίνεται καθ' ἡμέραν 354b14).[23] Almost everyone, including himself, thinks that the sweetest water is raised *daily* (ἀνάγεται καθ' ἑκάστην ἡμέραν 354b29). The rivers *daily* pour large quantities of water into the sea (καθ' ἑκάστην ἡμέραν 355b22). Aristotle transfers this emphasis on genesis to his food and digestive analogies, and he notes that the sweet parts of the food *quickly* disappear upon ingestion (ταχέως ἀφανίζεται 355b13; cf. ἀναφέρεται ταχύ 355b19); and a few lines later in a bold comparison he likens the evaporation of the river water in the sea to water which, when spread across a table, vanishes as quickly as thought (ἅμα διανοουμένοις ἂν ἀφανισθείη πᾶν 355b29–30).[24] So clearly Aristotle is stressing the ephemeral nature of fresh water, and by contrast, claiming that the sea does not come-to-be on a daily basis. He certainly agrees with his predecessors that some terrestrial waters are involved in a continual process; he simply insists that the sea is not.

The analogy of water evaporating from the table reinforces the isolation of the sea. The fresh water evaporates almost as quickly as it flows into the sea. Indeed we may infer that the rivers actually debouche their waters *onto* the sea as onto a table.[25] This is a very short-term cycle indeed. In fact, Aristotle may intend the surface evaporation of the fresh river water to be so rapid that we never taste the fresh water upon the sea. The salinity of the sea – whose causes must wait until 2.3 – is part of a different, long-term cycle. The two kinds of water never mix.

The separation of fresh and saline is explained by and grounded in the notion of place. In a curious observation Aristotle notes that the sea is not in its own natural place; instead it is in the natural place of fresh water (355a32–4). This is primarily because the light, sweet water is on top and ascends in evaporation, while the heavy sea water remains below.[26] Aristotle's criticism of Democritus provides further clarification: if it was the weight of the sea water that originally made it stay behind when the light water evaporated, that cause should still be operative, with the result

[23] Strohm 1984: 175 notes the emphasis on "daily."

[24] Aristotle likewise foists upon Heraclitus the view that the sun is new every day (νέος ἐφ' ἡμέρῃ 355a14): if the sun shines by the burning of exhalation it cannot remain the same for a moment but is always in change.

[25] Olympiodorus 141.8–20 falsely interprets the table-top metaphor as the whole sea (cup of water) being spread out upon the earth as upon a table, and some evaporates, some soaks into the earth.

[26] Cf. *Pr.* 23.6.932a32–8 where fresh water is on the surface and salt water is below in the lake-like seas.

that the sea will not disappear (2.3.356b17–21). But for Aristotle too the fresh water is always in motion and does not rest, and for this reason it cannot *stay* in its natural place (note that fresh lakes and swamps are now ignored). In this respect Aristotle is once again setting the sea in the broad structure of the treatise. He had earlier used the expression "common affections of water and air" to identify the rapid changes that occur in the rain cycle (1.9.346b17–18). Water does not form a simple unity but has at least two natures, the sea and the fresh water-vapor. Moreover, since the volume of the air is very large in comparison with that of the terrestrial water, he may have thought that much, if not most, of it takes the form of vapor in the air, perhaps enough to cause Deucalionic floods in great winters (1.14.352a31–3), and perhaps enough in theory at least, though never in fact, to inundate the whole surface of the earth with a continuous layer of water.[27] If this is so, Aristotle's theory of the sea and its relation to fresh water is already implied in 1.3, where he mentions the scattered nature of terrestrial waters (339b9–13). In any case, he rejects the view of Empedocles that the sea is the sweat of the earth (2.3.357b18–21) and prefers the view that the sea is a remnant, since we can in this case at least account for where the water has gone.

To illustrate further the relationship of place and weight to the instability of fresh water Aristotle uses a digestive analogy. The food that enters the stomach is sweet, but, almost as soon as it enters, it becomes bitter. This, we know, is the effect of the stomach acid, but for Aristotle the sweet part of the food is quickly separated and distributed to the rest of the body leaving the bitter residue behind.[28] This bitter residue is compared to the sea. But the stomach is not the proper place of the bitter residue; it is the proper place of sweet food (2.2.355b11–15). Likewise the sea occupies the proper place of fresh water. While Aristotle's purpose in using this analogy is to clarify the notion of proper place, it further strengthens the distinction between fresh and saline. As we saw, Aristotle stresses the fact that the light, sweet food evaporates quickly and he implies that the sweet and the bitter are distinct components of the ingested food.[29] Of course, there are many ways in which the analogy is not apt, some of them more significant than others. Aristotle does not mean to imply – at least not yet – that the water

[27] Olympiodorus 126.29–33 goes some distance toward this point by noting that the sea's salinity should vary with the rainfall, being saltier in summer, fresher in winter.

[28] Cf. *Pr.* 2.3.866b19–27 and Theophrastus, *de Sudore* 2 on salt and sweat. Both are separation theories of residue – the sweet in food is used as food and the salty is what is left over and expelled.

[29] Note that the rivers are given a close causal relation to the fertility of the land in 1.14. As Ninck 1921: 27 notes, the spring enlivens because the water itself is living.

that feeds the sea is composed of two different kinds of water, sweet and saline. The rivers are just fresh. Again, whereas the stomach discharges its residue into the intestine, the sea permanently retains the salt water. But in other respects it is surprisingly apt. In his account of sleep in the *de Somno* Aristotle recognizes that the sweet vapors from the ingested food, which rise toward the head, may return back to the lower parts. This ἀναθυμίασις is "driven forward and then turns and changes its course like the tide in a narrow strait (καθάπερ εὔριπον)" (*Somn. Vig.* 3.456b20–1). The passage reinforces the general point that the sea and digestion are closely related in Aristotle's mind and may account for the remarkable interest Aristotle takes in the ebbing and flowing of the sea in our previous chapter. But more specifically it is clear that in sleep the nutriment returns to the center of the body. This is especially noteworthy, because no significant amount of the nutriment is used up in this cycle. As with the rain cycle, all that goes up to the head must come back down.

By understanding the digestive analogy in this way Aristotle can preserve the dignity of his rain cycle – for it is nutriment, not waste – and view the sea as a remnant, though not in the way the Ionians thought. Instead the sea is a fraction of the whole supply of water. Much, perhaps most, of the water is in constant circulation and behaving in a manner quite different from the salty sea. This is why it was so important for Aristotle to establish the rain cycle in Book 1 before tackling the question of the sea, and why he makes use of his own conclusions when refuting the entropists;[30] for without the closed cycle of the rain, we could not understand in what sense the sea is a remnant. In this way too he can give a better explanation for the extent of the sea: it is what is left behind by the circulating fresh water and it preserves the balance between fresh and saline. The Ionian theory by contrast can provide no sufficient reason for why the sea should have its present extent; it is merely at some arbitrary point on the path to complete desiccation. At the same time, this circulating water guarantees the stability of much of Aristotle's sublunary world, for if water covered the entire earth, there could be no dry exhalation (hence no theory of dual exhalations), no comets or Milky Way, and no saline sea. The fresh-water cycle is essential for the sea's isolation and the correct maintenance of its nature. It also prevents fresh water from becoming a mere modification of sea water, as happens on the Ionian remnant theory, for which saline water is the original form and fresh water a derivative (2.2.354b15–16). And

[30] As Solmsen 1960: 422 remarked, Aristotle uses his own theorems on the rain cycle to refute Democritus here (356b28–30): "if one allows the sun to travel in its course, it will continually draw up the fresh water, as it approaches, as we have said, and release it again when it withdraws."

this in turn preserves the natural order of the sublunary world. Moreover, the strict distinction between fresh and salt water also supports Aristotle's refutation of the "sun-feeders" and their entropic error. If only fresh water is evaporating, it is clear that the saline sea cannot wholly evaporate and disappear. And since the fresh rain falls back down, it is clearly not being consumed by the celestial bodies. Such are the effects of Aristotle's strict distinction between fresh and salt water.

But Aristotle also applied the concept of the biological residue to the sea in order to confine it within its present extent in contrast to the constant flow of the fresh waters. Biological residues (in contrast with colliquescences) have places into which they run and where they are collected and stored (*GA* 1.18.725a33–b3), places such as the intestine, bladder, and uterus.[31] "Each of the residues is and becomes that residue only in its proper place. Before that, none is a residue except by great force and contrary to nature" (*GA* 2.4.739a2–4). In a similar manner the seas are confined to restricted places. This conception of a contained proper place – for this is clearly what Aristotle intends – is an inversion of the position Aristotle started with at 2.2.355a33–4, where the sea is in the proper place of fresh water.[32] This inversion must await its justification in 2.3, but for now it helps us understand why the sea does not form a complete layer around the earth and therefore how it is the result of cosmic order rather than an accident of entropy. The sea is contained because of a perpetual balance between fresh and saline driven by the heavenly motion of the sun. Again, the idea that inert residues are contained in their proper places reinforces the division between flowing and stationary waters introduced in 2.1. Indeed Aristotle had already used one of the words for residue, ὑπόστασις (353b23–4), at the beginning of that division, before the question of the sea's nature had been undertaken. It is clear that his arguments against the sources of the sea were intended to prefigure this conception of the sea's proper place at the bottom of the meteorological cosmos.

This balanced and hierarchical order is the result of the movements of the sun, but there is no reason to suppose that the sun moves in such a manner *in order to* produce that order. The sea, its place, and its salinity are merely the spontaneous result of the interaction between sun, water, and dry exhalation. This quasi-biological order that results is not intended to imply the operation of the final cause. Though the sea shares certain structural features with bodily organs, it is not an organ and discharges

[31] Cf. Hippocrates, *Carn.* 13.

[32] For the proper places of digestion, cf. also *Pr.* 3.21.874a22–7. Olympiodorus insists throughout that Aristotle wants to establish the sea as a ὁλότης, which implies continuity and stability (133.8–36).

no function in the cosmic animal, nor did the sun bring it about for the purpose of generating order in the sublunary world. The order is good, but it neither is nor has a final cause.

Salinity

In 2.3 Aristotle moves from nature to property and provides an explanation for the salinity of the sea. This logical order happens also to be strategically convenient, because his particular solution to the problem of salinity poses a considerable threat to his conception of the sea's nature. He prudently establishes its nature before tackling the property of salinity and is thus able to qualify his solution to make it more consistent with the sea's nature. In treating the questions separately, Aristotle follows the path favored by the early Presocratics, who separated the shrinking of the sea from its salination; he therefore feels free to mix and match theories on the two problems, as his predecessors did, who canvassed two explanations for salinity. According to the first explanation, the sea has become salty because some admixture from the earth has dissolved into it. Water has filtered through the earth and picked up salt in the process, as Xenophanes, Anaxagoras, and Metrodorus argued.[33] On this explanation there is no reason to suppose that the sea is shrinking.[34] The second, also that of Anaxagoras as well as of Diogenes of Apollonia, views the sea as a remnant of a larger, slightly saline body, which after the fresh water has evaporated, becomes intensely salty. In this case the shrinking sea and its salinity are joined as cause and effect. While Anaxagoras may have combined both filtration and concentration, Aristotle, threatened by the economy of the joint explanation and its implications, treats them as exclusive solutions. He takes Empedocles to task, because his sweat-of-the-earth metaphor is not decisive between them: Empedocles failed to say whether the salty sweat of the earth arises from

[33] Xenophanes, DK 21A33.4 (1.122.37–8); Anaxagoras, DK 59A90 (2.27.2–7) = Alexander, *in Mete.* 67.17–22: τὸ ὕδωρ τὸ διὰ τῆς γῆς διηθούμενον καὶ διαπλῦνον αὐτὴν ἁλμυρὸν γίνεται τῷ ἔχειν τὴν γῆν τοιούτους χυμοὺς ἐν αὐτῇ. Oder 1898: 273n49 thinks that Alexander's Anaxagoras here (67.21) is a corruption for Archelaos, based on Diogenes Laertius 2.17. In the same passage Alexander attributes the opinion to Metrodorus, DK 70A19 (2.232.40–2); also *DG* 382.4–6.

[34] Anaxagoras, DK 59A90 (1.26.42–27.1) = Aët. 3.16.2 (*DG* 381.20–4): τοῦ κατ᾽ ἀρχὴν λιμνάζοντος ὑγροῦ περικαέντος ὑπὸ τῆς ἡλιακῆς περιφορᾶς καὶ τοῦ λεπτοτάτου ἐξατμισθέντος εἰς ἁλυκίδα καὶ πικρίαν τὸ λοιπὸν ὑποστῆναι. We should note that Alexander (67.12–14 = DK 64A17 [2.54.21–2]), citing Theophrastus, especially comments on Diogenes as holding that the drying out of the sea is the cause of its increasing salinity, and he fails to corroborate Aëtius' claim that Anaxagoras held the opinion as well. Alexander merely contrasts Anaximander, who did not relate the shrinking of the sea (a view Anaximander certainly held) to the saltiness; it changed διὰ τὴν ἔκκαυσιν, a notion that Aristotle also picks up and uses.

the fresh water being taken away or from fresh water being filtered through salty earth (2.3.357a24–32).[35] In fact these solutions for the sea's salinity are exclusive only because Aristotle has supposed for architectonic reasons that fresh and salt water are different species of water and cannot change into one another. This was the force of the analogy in 2.2, where the stomach separates the two liquid components of food, the nutriment and residue.

So long as Aristotle was not focusing on the relationship between the fresh and the saline, the analogies in terms of weight and place were adequate. But now he is compelled to explain why the water is salty, and his solution is under severe constraints of his own making.[36] First the analogies have persuaded him that the sea is salty because of an *addition*, and for this reason the salinity cannot merely be an ἀρχή.[37] Undoubtedly, too, empirical evidence weighed heavily as is indicated by the numerous observations (358b34–359b21). The addition must have been *added* somehow, and an obvious source would be the rivers. But Aristotle, maintaining the hierarchical distinction between the saline and the fresh, insists that the rivers do not add salt to the sea, since, if they did, we would be able to taste their salinity (357a15–21). And the fact that he withholds mention of the many saline springs till the end of the chapter (359a24–b21) is strong evidence that he had structural rather than empirical reasons for denying the rivers' role in the sea's salinity. These saline springs certainly would have been more consistent with the stomach analogy, where the food must contain both sweet and bitter parts. As it is, the stomach is a hard analogy to reconcile with fresh-water rivers. Where did the undigested residue, which is the sea, come from, if the rivers bring only fresh water (357a15–21)? And if the rivers are the "food" of the sea, where does the residue go after the fresh water has evaporated? For Aristotle an acceptable solution must explain the salt without recourse to the rivers. And finally the addition cannot be an entropic one, so if there is an addition of salt, there must be somehow a corresponding subtraction. Aristotle's new "addition" theory must preserve the distinction of weight (and therefore place) between the two kinds of water and must account for these differences by the addition or removal of salt.

At the core of the new solution is a new analogy: instead of a stomach, the sea is now like a bladder. With the focus now on salinity, the strict distinction between fresh and saline, between food and residue, retreats

35 Antiphon's sweat doctrine (DK 87B32 [2.344.20–345.1] = Aët. 3.16.4 [*DG* 381.28–382.3]) is based upon evaporation.

36 For a similar list of these constraints, see Strohm 1984: 176–7.

37 Olympiodorus canvasses this option on Aristotle's behalf (151.19–24 and 154.4–8), citing 357a8 εἴτε ἐνυπῆρχε.

into the background. The new analogy corresponds more closely, because urine is actually salty. The taste of urine arises from the addition of some material to water, because water has been filtered through the body. The analogy also furthers the hierarchical agenda: the salty residue in the bladder is in its proper place now, and there is no question of fresh water being introduced into the bladder in the way the rivers debouche into the sea, since the two-waters view is no longer at issue. However, Aristotle must still solve the most important problems, explain the origin of the salt, and adapt the bladder analogy to the requirements of a cyclical cause.[38] After all, the evaporation and recondensation of liquid are not part of the function of the bladder. Urine flows into the bladder for storage and is discharged again through the urethra. But how does salt water get into the sea if not through the rivers?

The bladder analogy is a set-up for a very brief and tentative solution that makes use of the dry exhalation (358a21–6): the wet exhalations from the earth mix with the dry smoky exhalation to form a brackish mist and then a brackish rain. Probably all rain is slightly brackish, but certain rains, like southern and autumn rains, are especially so. This rain falls into the sea and accounts for its salinity. Since this process would constantly increase the salinity of the sea, if only fresh water evaporated, there is a corresponding release of salty dry exhalation during the evaporation of water from the sea (358b12–16). The amount can hardly be much, since rainwater is quite fresh, and since, as Aristotle points out in detail, the evaporation of saline water leaves salt behind.

Having now introduced this theory, Aristotle might have defended himself against its many troublesome implications (more of which below), but he chose not to, and instead he turns to confirmatory signs (358b16–359b21). In particular he is interested in showing that salinity is due to admixture and that this admixture is the cause of both its taste and its weight. This cause is much more important to him than the details of how the salt becomes mixed into the water. For Aristotle it is a satisfactory response to the σοφώτεροι, especially Empedocles, who could not decide whether salinity was the result of admixture or of residue. By deciding this question and providing arguments, Aristotle *has* provided a cause. Salinity is the result of addition, an addition coeval and continuous with the eternity of the world.

[38] Strohm 1984: 172 notes that the two tendencies in Aristotle's presentation (that salinity is what remains below in the sea and that salinity is what enters the sea from the rain) constitute a dangerous set of compromises. It is also worth citing *Pr.* 13.6.908a28–34, where some of the Heraclitean school are said to make the connection between urine in the body and moisture in the universe, both being produced from the condensation of exhalation.

No doubt there are other more convenient ways of explaining a cycle of salinity. But it is odd to seek a cause for the salinity at all. After all, Aristotle had developed by the end of 2.2 a pretty good modified Ionian explanation for the nature of the sea and its salinity: it is the remnant of a large body of water, the fresh part of which is in constant circulation. On this theory the salt is an aboriginal addition and the salinity is an ἀρχή. The containment of the sea is a τελευτή (2.2.356a35). It is a homeostatic solution inasmuch as it depends on the continuity of the rain cycle. And though the solution may suggest two species of water, the fresh and the saline, the fresh may just happen to be that which cycles on top of the saline. Alternatively, the rivers could pick up some small amount of dry exhalation from their sources, which could then be liberated after the rivers reach the sea in the manner Aristotle describes. This would account for the numerous salt springs he describes and have the added advantage of involving his general material principles.

As it is, Aristotle's solution is redundant, and it is redundant in the same way as his solution for the formation of rivers is. As we saw in 1.13, rain soaks into the earth and there produces wet exhalation, which simply recondenses within the earth to form water and springs (349b19–350a15). There we had a redundant rain cycle within a rain cycle. Here Aristotle is so intent on establishing a cycle of salinity at this point, that on this sole occasion in the treatise he allows the dry exhalation to be recycled. As with the subterranean rain cycle, we have a similarly redundant cycle in which dry exhalation falls together with the rain and reevaporates.[39] It is a cycle for the sake of a cycle. In both cases Aristotle deliberately limits his use of exhalations to a minor cycle in order to maintain the independence of the sea from the rivers and the river springs from the rain. But, if we accept Aristotle's solution, vapor, when it evaporates from both the sea and the land, produces saline rain, which ends up in the sea, whether directly or by first falling on the land. The result is that the sea will receive the combined salinity of both marine and terrestrial dry exhalation emissions but will liberate only its own share.[40] This will simply increase the salinity in the sea, since the sea receives salt from all rain. Moreover since the river water transports the brackish water to the sea, inevitably the rivers too are going to be partly saline.[41]

[39] Alexander 86.10–20 recognizes this problem but says that the salty exhalation that evaporates is not the same as the one that precipitates.

[40] Aristotle might reply, as an anonymous reader suggested, that the earth reabsorbs the salt that falls upon it from the rain. This is a possibility, but that Aristotle did not consider these consequences of his theory suggests that he was interested primarily in the micro-cycle of marine salinity.

[41] So Olympiodorus 157.21–4.

But these redundancies are precisely the key to Aristotle's architectonic motivation. Aristotle wanted to establish cycles for the rivers and for the sea that are at once independent of each other and at least partly independent of the major movements of their respective exhalations. The rivers ultimately depend on the rain, but the wet exhalation engages in its own subsidiary cycle, as we have seen. The dry exhalation must, of course, come originally from the land, and to heaven it must ultimately rise, but first it falls into the sea, causes salinity, and then reevaporates from the sea. Thus while the springs form a cycle within a descending cycle, salination forms a sort of detour loop within the broader ascent of the dry exhalation toward the upper region. And this correspondence and the order to which it contributes in the sublunary world is, I suggest, the explanation for the fact that salinity is the only phenomenon in which the dry exhalation forms a cycle; the release of the dry exhalation from the sea also avoids the problem of the increased salinity that might arise from continual additions. For we may suppose that however much dry exhalation enters the sea from whatever source, it will finally be liberated into the upper region. The cycle within the cycle could not have been attained by the modified Ionian explanation.

The desire for order and hierarchy also plays a role in making the sea and its salinity the effect of a permanent homeostatic process. For in this respect it corresponds to the Milky Way, which is likewise an effect of the dry exhalation and dependent on permanent combustion. The Milky Way is localized in a single, discrete area of the sky; it is the highest, most celestial and perfect of the sublunary phenomena, and associated with the brightest stars (1.8.346a19–23). The combustible exhalation (ὑπέκκαυμα) becomes actualized, that is, it attains its highest form, by being super-heated. The sea is likewise isolated, but it is the lowest of the waters and a residue; it is degenerate and sterile through the addition of the salt, an undigested byproduct like the ash left by a fire (2.3.359b1–7).[42] The sea, then, is the base counterpart of the Milky Way. Its transformation from being the holy principle of all waters and of the heavens themselves into a sterile waste at the bottom of the sublunary world is now complete.

[42] Olympiodorus 157.8–18 describes the difference between this kind of dry exhalation and other kinds, saying also that it is heavier and stays closer to the earth. On the heat of the sea, cf. *Pr.* 23.7.932a39–b7.

CHAPTER 10

Winds (2.4–6)

After the long "digression" Aristotle returns to the winds.[1] His purpose here, as I shall argue, is guided by his systematic analogy between the rivers and the winds, and he spends most of his energy introducing an elaborate apparatus in order to support it. He wants to show that the winds, like the rivers, arise from a source and are not just air in motion. But the analogy draws him quickly into conflict with some of the elements of traditional Greek wind lore that he wants to preserve. The most important of these elements is the remarkable predominance of the Etesians, which blow from the north during the summer months, when, as the Ionian philosophers claimed, the northern snow melts and vaporizes. There is nothing of comparable stature among the rivers, and the Etesians threaten to upset the neat radial symmetry of the river–wind analogy. Aristotle accepts the Ionian explanation (without acknowledgement) but magnifies the less persistent southern Etesians in order to preserve the analogy. This in turn recommends positing a southern hemisphere, a source from which southern winds blow. But if Aristotle accepts a southern hemisphere, he risks producing a single system of winds that embraces the entire globe, again endangering the analogy with his rivers, which are confined to the northern hemisphere. It is a mark of his commitment to the river–wind analogy, I argue, that he retreats from this global system of winds and posits instead two isolated systems of winds, one corresponding to the Mediterranean rivers, the other in the southern hemisphere. These systems behave as disks on the surface of the globe. How Aristotle is able to reconcile the flat disk view of the earth with the spherical is a study in Peripatetic flexibility.

[1] Strohm 1984: 179 maintains that this is in no sense a return to the discussion of 1.13. He particularly finds fault with Aristotle for claiming that the nature of the exhalations used to explain the sea's salinity is so very different from that of the winds: that Aristotle uses the same principle in both cases shows how compelled he was by his first principles.

The vaporous catalyst

But before we turn to our major concern, we must consider the strange material and efficient causes of the wind. Aristotle's investigation begins in 2.4 with two questions. (1) Are the winds just air in motion?[2] (2) Are rain, air, and wind just variations of the same nature (349a19)?[3] He focuses on the second question and answers in the negative that wind and rain are of different natures. His reasons make use of his familiar distinction between the two exhalations and apply his earlier theory – hitherto resting dormant – of the dual nature of the air. Air is represented in one passage as a combination of both exhalations in the manner of complementary tallies (σύμβολα): the hot of the smoky exhalation combines with the moist of the vapor to form hot, moist air (2.4.360a21–7).[4] Elsewhere the exhalations simply coexist: "the moist does not exist without the dry, nor the dry without the moist, but they are all called what they are according to what predominates in them" (359b32–4). In this latter case, air is simply a mixture of the vaporous and the smoky exhalations.

Up to this point in the treatise (with the exception of the sea) the exhalations have operated independently, each producing its own effects. The challenge for our present chapter is to determine how and to what extent they can operate together. Whether air is a tallied combination or a simple mixture, its precise nature cannot be determined from Aristotle's descriptions.[5] It suffices for the refutation of his predecessors that air be composed of two different elements endowed with different capacities, since in that case wind and rain will be distinct and not mutually transformable. When the air is predominantly vapor, there is rain; when it is predominantly smoke, there is wind (360a8–13). This

[2] Cf. *Mu.* 4.394b7–9: "winds arise when cold pushes the dry exhalation so that it flows. For wind is nothing other than a lot of massed air flowing." Cf. also *Pr.* 26.2.940b7 and the opinion of Theophrastus, *Vent.* 29. Olympiodorus 171 understands *Mete.* 2.4 to be entirely written against the ῥεῦμα καὶ χεῦμα doctrine of the Hippocratean *Flat.* 3. Böker 1958a: 2229 agrees.

[3] This is the view of Metrodorus of Chios (DK 70A18 [2.232.36–9]): ὑδατώδους ἀναθυμιάσεως διὰ τὴν ἡλιακὴν ἔκκαυσιν γίνεσθαι ὁρμὴν πνευμάτων θείων.

[4] On the token theory, cf. *GC* 2.4.331a24 and Joachim's note. There the token is the quality that both elements have in common and which therefore facilitates their mutual transformation. In the *Meteorologica*, however, the tokens are the two different generic qualities hot/cold and moist/dry that come together to form the nature of air.

[5] Aristotle's theory of the mixture of the two exhalations is dangerously close in consequence to Anaximander's, according to whom the wind is the rarest vapor separated off from the air, while the rain arises from vapor given off by the earth from the action of the sun (DK 12A11.7 [1.84.17–20]). Böker 1958a: 2243 argues that Aristotle holds the mixture theory of the winds, according to which the air is made up of the wet and dry exhalations, which can separate like chaff from wheat, the heavy rain falling and the light wind blowing.

argument relies on the principle that the dry and the wet are not mutually transformable, and without it Aristotle's claim that winds and rain are of different natures would be nonsense. It is with this modest apparatus that Aristotle answers the second question. Thereafter we would expect him to focus solely on the dry exhalation and its windy effects, and to some degree he obliges us. The most official doctrine, that the wind is made of the dry exhalation and rain comes from the wet, appears in a quasi-definitional formula (360a8–13). One year or one district or one part of a district may be rainier or drier (and therefore windier) than another according to the proportion of wet and dry exhalations in the air (360b3–5). This same doctrine is repeated unambiguously at several points throughout the chapter.[6]

Yet, surprisingly, Aristotle also makes the wet exhalation an important, even essential, component in the production of the wind. This tendency is evident throughout the chapter. From his first introduction of the two components of the air, Aristotle stresses this role (360b26–9): though winds abate while it is raining, they increase again after the rain has fallen. They increase after rain, because the heat in the earth and from the sun causes the rain in the earth to vaporize and this is the stuff of wind (2.4.360b31–2). This is odd. We were led to expect that the winds, being dry exhalations, occur predominantly in hot, dry areas where no rain has fallen, but this is clearly not the case. Indeed, the wind here seems actually to be formed from the wet exhalation. Shortly after, the dry exhalation seems to serve as a carrier for the wet:[7] "the winds stop, because the hot is separated off continually and carried to the upper place. The vapor then cools and becomes water" (360b34–6). Here the dry exhalation blows itself and the moist exhalation around, until it rises and separates off from the moist, leaving the moist to condense and fall. When the moisture falls to the

[6] "The dry exhalation is the principle and nature of all the winds" (2.4.360a13–14). "Because sometimes the vaporous exhalation, sometimes the dry and smoky, is many times the amount of the other, sometimes the years are rainy and moist, others they are windy and parched" (360b2–5). "A wind is a mass of dry exhalation from the earth moving around the earth" (361a30–1). Cf. 3.1.371a5. Though Theophrastus concurs, the dry exhalation is much less important for him (Coutant and Eichenlaub 1975: xliv).

[7] Strohm 1984: 188–9 suggests that the dry exhalation functions in the end as a motor to the often accompanying wet exhalation. He argues (citing 2.8.367a34 in the context of earthquakes) that winds are not usually supposed to be warm because they set the air in motion and the air contains large quantities of cold vapor. He thinks that this is not just an *ad hoc* hypothesis used to explain earthquakes, but a systematically deployed principle. However, he also holds that there is a shift in Aristotle's opinion from the early belief that the winds are essentially made of dry exhalation to the later view, manifest particularly in 2.6, that the ἀτμίς is the material of winds. The Hippocratic *Vict.* 38 affirms that it is the nature of all winds to cool and to moisten.

earth, it cools the dry exhalation in the ground and prevents the generation of winds.

Elsewhere rain plays an important role in the actual generation and nature of the wind: "where the earth receives the greatest rainfall the exhalation must be greatest, like the smoke from green sticks (οἷον ἐκ χλωρῶν ξύλων καπνόν), and this exhalation is wind" (361a17–20). Here moisture is not just carried by dry exhalation; it contributes to the nature of the dry exhalation, since smoke cannot be generated without it. More specifically, it is the melting and vaporization of frozen water that generates the smoke, which then blows as Etesian winds from the damp and rainy north in the summer (2.5.362a5–7). Moisture, then, is essential in generating the winds. And it is consistent with this odd form of reasoning that intense heat will wither (μαραίνειν) the winds, since "it dries up the earth too quickly before a mass of exhalation can be produced" (2.5.361b14–20). The earth's rapid desiccation could not prevent the production of winds, unless moisture were a necessary condition.[8]

Chapter 2.5 provides clearer indications of a catalytic role for moisture in the production of the dry exhalation: "that which is frozen or contains no dryness in it does not produce fumes; but when the dry thing has moisture, this does produce fumes when heated" (362a9–11). A passage from *Meteorologica* 4.9 confirms and clarifies this notion. Here θυμιατά (commonly and etymologically translated as "fume producing bodies") are described in such a way as to recommend their association with our dry exhalation: the θυμίασις of a woody body is smoke (387a32–b1) and causes things to be colored (χρωματίζει, cf. *Mete.* 1.5.342b4 in the context of the aurora). Now, θυμιατά also include such liquids as fat, oil, and wine (387b6–9), but what is important is that fumes in general, and not just smoky fumes, contain some moisture, since fuming "is the release of the dry and the moist together due to burning heat" (387a30–1). That is, moisture is essential for the production of fumes. We might be tempted therefore to infer that water vapor is an essential component of winds. In this same passage, however, Aristotle sharply and repeatedly distinguishes the moisture of fumes from water vapor (e.g., ὕδωρ οὐ θυμιατὸν ἀλλ' ἀτμιστόν 387b8–9). Whereas when water evaporates it leaves the bodies that contained it behind, in the process of fuming, the bodies that produce fumes also disappear along with the fumes (387a24–5). Clearly, fumes are not moist in the same way that vapor is. But is the moisture in wood not

[8] Cf. 2.5.362a2–3: "when the sun is closer [to the north] it dries the earth before the exhalation can form."

water?[9] It is, but the similarity with the passage at 4.9 leads us to conclude that water in the wood (and so in the earth) contributes to the smokiness of the dry exhalation independently of its capacity also to produce vapor. The moisture in the earth catalyzes, as it were, the production of the dry exhalation. This is corroborated by the distinction Aristotle draws between the action of the sun in evaporating water on the surface of the earth and drying out the earth itself by heating (2.4.360a6–8). It is the latter action that properly produces the fumes.

The balance of evidence suggests, then, that moisture is not the material of the winds. It merely allows the earth to fume, and as such, is a catalyst and carrier. This theory, which seems to have its origin in the study of fuming bodies, has the added advantage of explaining why the winds are not flammable. The moisture that carries them prevents combustion until the fumes rise into the kapnosphere and leave the wet exhalation behind.

Finally, we should note that the interaction between the wet and dry exhalations here is prefigured in the salty sea. Both winds and sea are mixtures, whose components can separate: the sea is a mixture of water and dry exhalation, just as is the wind. The light fresh water separates and leaves the salt behind, though some of it disperses into the air. In a similar manner the dry component of the wind separates and enters the kapnosphere. The wind's interaction with the sea becomes even more involved when we turn to earthquakes.

The efficient cause

Among the many curiosities of Aristotle's account of the winds is the fact that the earth itself serves as a source of generative heat (πολὺ πῦρ καὶ πολλὴ θερμότης 360a6) in addition to the more usual sun. We hear about this source of heat for the first time in these chapters and we hear about it on several occasions.[10] One obvious reason for its introduction here is that

[9] Fuming is the separation (ἔκκρισις) of dry and moist together, and yet it does not become wind (4.9.387a28–9). For green wood, 2.4.361a19; cf. 3.4.374a5. Webster, followed by Düring 1944: 95, found 387a24–6 ("For vapor is a moist excretion into air and wind" [ἔστιν γὰρ ἀτμίς ἡ . . . εἰς ἀέρα καὶ πνεῦμα ἔκκρισις]) difficult, because they did not appreciate the important and ambiguous role moisture plays in the generation of winds. We might be tempted to infer, as Webster does, that these fumes are different from both the moist and the dry ἀναθυμιάσεις. But among the fumes of oily liquids and fats Aristotle also includes the fumes of wood, which figure prominently in his discussion of winds.

[10] The same fact is mentioned again 2.4.360b30–2: "after it rains, the earth is dried out by its own internal heat and by the heat from above and produces exhalation, and this is the stuff of the winds." For heat in the earth and from the sun, 2.5.362a6. Again in the context of earthquakes, 2.8.365b25–7; 367a9–11.

winds need to be generated in all seasons, even during the winter. If they were generated solely by the heat of the sun, winds would be absent during the cold season, or at least there would be wider seasonal variations in their production than we see. But a second reason derives from the fact that the winds serve as a counterpart to the rivers. We have seen how the rivers have their origin in springs, formed from the vapor condensing in the *cool* of the earth. In an analogous manner the winds arise from the vaporization of fumes in the *heat* of the earth. Both have, that is, an underground source of vaporization or condensation. In both cases, too, the constancy of the subterranean temperature accounts for the continuity of the flow, in spite of the seasonal variations caused by the sun's movement.

More curious than this is the fact that winds and rivers both move in a horizontal plane. The transverse motion of the rivers, for their part, was so easy to understand that Aristotle had provided no explanation: their horizontal movement was simply the result of the descent of water being resisted by the sloping earth. The nature of the dry, windy exhalation, however, is to rise vertically, and there is nothing obvious that could divert its motion sideways in a way analogous to the rivers. How then to account for the winds' motion? According to Aristotle, the bulk of the air follows the "movement" (πᾶς ὁ κύκλῳ ἀὴρ συνέπεται τῇ φορᾷ 361a24–5), and we may presume that by this he means the movement of the kapnosphere and ultimately of a celestial sphere. A sign of this fact is that when there are clouds in the sky, we can see the movement in the upper air, even if the wind has not yet been felt here below (361a28–30). Even on a calm day the clouds are always moving. The motion of the air above pulls the winds along. As a result, the characteristic flow of the winds will be, as with the rivers, a combination of factors. But while the rivers combine their natural downward motion with a passive constraint, the natural upward motion of the winds is subjected to an active lateral force.[11]

On the standard interpretation, the upper movement that carries the winds along is the diurnal motion of the heavens. But on this account, as has frequently been pointed out, there could only be east winds and they would have to blow continually at the angular speed of the sun.[12] This choice of mover, coming shortly after a discussion of the prevalence of the

[11] Alexander 93.26–35 notes as an objection to Aristotle's theory that the wind does not move sideways by a natural motion (see Chapter 6 above on ὑπὲρ φύσιν motion).

[12] As early as Alexander 93.26–35. Philoponus 37.30. Moreover, while Aristotle says that the motions of heaven cause the upper region to rotate, thus dispelling the clouds, he expressly says that the motion does not extend below the level of the mountain peaks (340b33–341a1). This is in explicit contradiction to his etiology of the winds at 2.4.361a22–5. It is worth noting the curious parallel between the remote effect here of the celestial sphere on the winds and the remote effect of the stars

north and *south* winds, is surprising. No doubt Aristotle's ingenuity could invent some subsidiary causes – eddies for example – to account for the other directionalities, but none is forthcoming.[13]

This interpretation, if we accept it, shows Aristotle's problem-solving at its most compartmentalized. We have to suppose that he could explain why the winds move horizontally without considering the direction in which they blow. It also puts him in danger of making the winds into moving air, a view he had criticized in his predecessors. After all, if "the whole bulk of the air" (πᾶς ὁ κύκλῳ ἀήρ 361a24) is moved by the heavenly motion, and the windy exhalations are merely carried along with that bulk, then the exhalations will be just as much in motion as the air. Wind will then be indistinguishable from the movement of air, and this whole objection to his predecessors' doctrine will be pointless. Finally, the standard interpretation throws the river–wind analogy into confusion: winds, like the rivers, are meant to be distinct from the environment around them and are meant to flow *through* the atmosphere as the rivers flow through the land. On this interpretation the motion of the wind is indistinguishable from that of the surrounding air.

Aristotle could have explained the winds in the same way as he explained the oblique motion of some low shooting stars: the heavier vapor above may push the dry exhalation down as it tries to rise, resulting in a sideways motion. Similarly, as Olympiodorus reports, he might have followed Empedocles in claiming that the winds move slantwise (λοξή) "because the earthy and the fiery have opposite movements." Theophrastus also explained the directionality of the wind by a similar struggle of warm and cold.[14] But Aristotle does not have two such conflicting motions available, since at this altitude both exhalations rise, and ejection is reserved for unusual and violent motions. Finally, he might have achieved maximum correspondence with the rivers by claiming that the winds arise from the earth but are impeded from their ascent by some relatively heavy air mass

on the formation of the comets and Milky Way. The repetition is a sign that Aristotle recognized such a remote effect.

[13] Olympiodorus 175.14–30, following Ammonius, obliges, saying that the interface between the vapor and the smoky exhalation at the level of the mountaintops is hard for the wind to penetrate, so it is forced back in different directions and causes the winds.

[14] Stüve suggests with justification that Olympiodorus 102.2 (DK 31A64 [1.295.4–6]) miswrote Empedocles in the place of Theophrastus. See also Strohm 1937: 260–5. According to Olympiodorus 97.5–10, Theophrastus in his *Meteora* said that winds are composed of fiery and earthy parts that struggle with one another and produce a sideways motion. Theophrastus, *Vent.* 22: "if air were self-moving, being cold and vaporous by nature, it would move downwards; if it were moved by heat, it would move upwards. For the motion of fire is naturally upwards. In fact, the motion is in a sense a compound of both because neither prevails" (Coutant and Eichenlaub trans.).

above the earth, just as flames are sometimes diverted from vertical ascent by an overhead obstruction. But, as he observed, the clouds are often seen moving even before the air on the earth's surface, so the upper air is clearly not impeding the flow of the winds. These considerations probably led Aristotle instead to choose a celestial cause and take the opportunity to rebind this phenomenon to the science's higher principles. But is there a better celestial cause than the disastrous diurnal motion, which condemns him to unvarying east winds?

First, in view of the fact that the dry component of the wind eventually leaves the wet and departs upwards (2.4.360b34–6), we may conclude that the winds rise in spite of their horizontal movement, and for this reason Aristotle calls their path "oblique" (λοξή 361a23).[15] It is the same expression as is used to describe the ejected shooting stars, which are a combination of natural and forced motions (1.4.342a24–7). The motion of the winds, then, will more closely mirror the motion of rivers, which slowly descend in their horizontal flow.

Second, it is worth pointing out that Aristotle nowhere explicitly says that the heavens move the winds; he says only that the source of motion comes from above (ἄνωθεν 361a27, 29, 32). This lack of specificity on his part may be intended to conceal his uncertainty, but the fact remains that the upper lateral motion must come from somewhere, and the only motion Aristotle discusses is that of the celestial spheres.

An alternative to the diurnal motion suggests itself from an incidentally related use of λοξός to describe the ecliptic circle and the associated cycle of the seasons. If, as seems probable, the motions of the heavens do ultimately drive the winds, Aristotle still does not tell us which motion it is (πᾶς ὁ κύκλῳ ἀὴρ συνέπεται τῇ φορᾷ 361a24). In view of the fact that this passage is sandwiched between discussions in which the north wind predominates and which argue that the winds blow from the north and south, it may be preferable to suppose that the motion in question is the annual movement of the sun north and south along the ecliptic. Indeed the phrase, φερομένου τοῦ ἡλίου κύκλῳ (359b34), clearly refers to the sun's annual course through the zodiacal circle, not its diurnal motion.[16] Such an interpretation would bring Aristotle into line with a common Greek doctrine, which was likewise

[15] As Webster; not Lee's "horizontal." So Olympiodorus 178.20–8 for the combination of motions. λοξός is distinguished from πλάγιος in [Aristotle] *Mech.* 25.856b9–10 as "diagonally" to "side to side." It is used to describe the path of the sun, λοξὸς κύκλος, at *Met.* 12.5.1071a16 and *GC* 2.10.336a32. Other relevant texts include *Pr.* 25.14.939a39–b4 and 26.48.945b30–4.

[16] "When, therefore, the sun in its circular course approaches the earth, it draws up the moisture with its heat; when it withdraws, the vapor thus drawn up is condensed again into water" (359b34–360a2).

aimed at explaining the *seasonal* character of the Etesian winds.[17] It would make especial sense of Aristotle's claim that the winds are weak when they first start out from the north or south (2.4.361b2–5) since the sun moves more slowly near the tropics than it does at the equator. Objections, of course, remain: the seasonal movement of the sun north and south is rather too slow to account for the rapid speed of the winds. And if Aristotle did intend the annual course of the sun, he should have mentioned it alongside the sun's heat as the cause of hindering and fostering the winds in 2.5. Most importantly, the proposed solution does nothing to prevent winds from becoming mere movements of air. But even this is not beyond remedy, if we consider how Aristotle emphasizes the smoky nature of these exhalations. He was certainly impressed by the fact that plumes of smoke often seem to move independently of the ambient air and seem to have an internal integrity and tension. If these plumes stretch up to the kapnosphere, they may be drawn along by it as if they were strings of taffy pulled up from the earth.[18] In this way, their motion might be considered independent of the ambient air.

On this view, the winds combine their efficient causes in a remarkable way. We have already seen how the *motion* of the celestial sphere

[17] Böker 1958a: 2239–40 brilliantly interprets as the action of the Etesian winds Anaximenes' attribution of the turnings of the planets (ἄστρα) to their being pushed by the dense and striking air. Although Anaximenes himself does not make the attribution explicit, it is made explicit by Metrodorus of Chios (DK 70A18 [2.232.36–9]): "the Etesians blow when the air which had been condensed at the north flows toward the summer tropic together with the sun as it withdraws." Diogenes of Apollonia (DK 64A17 [2.54.10–20]) notes the connection of the winds and the turning though he does not make the relationship clear: "the moisture is vaporized by the sun and from it arise the winds and the turning of the sun and the moon, on the grounds that on account of these vapors and the exhalations these bodies made their turnings where there is an abundance of it for them." Arguably this is better suited to the feeding interpretation of the turnings than to the blowing. But it is not clear that anyone here thinks that the winds blow the sun. Metrodorus seems to think that the sun releases the winds to blow. Theophrastus, *Vent.* 2 gives the sun an important role in the north and south winds: "the north wind and the south wind are strong winds and blow the longest time, because most air is forced to the north and to the south, since both districts are athwart the motion of the sun from east to west. The air there is dislodged outwards by the effect of the sun. This accounts for the extreme density and cloudiness of the air. It is collected in quantity on both sides, so that the flow is greater, more continuous, and more frequent. This is the cause of the force, the continuity, the amount and other such features of the winds" (Coutant and Eichenlaub trans.). For Anaxagoras (DK 59A42.11 [2.17.1–3]): ἀνέμους δὲ γίνεσθαι λεπτυνομένου τοῦ ἀέρος ὑπὸ τοῦ ἡλίου καὶ τῶν ἐκκαιομένων πρὸς τὸν πόλον ὑποχωρούντων καὶ <ἀντ>αποφερομένων. Gilbert 1907: 519n1 takes ἀέρος as the same as τῶν ἐκκαιομένων. For modern descriptions of Etesians, see Mariolopoulos 1961: 43–4 (on daily variations in speed) and Carapiperis 1951 and 1962b; see also his discussion of the ancient evidence, 1962a.

[18] The taffy-like nature of earthquake winds is alluded to at 2.8.366a6–8 where the exhalation is said to be continuous (συνεχής), and therefore the whole wind tends to follow its beginning impulse. This may be reminiscent of Empedocles (DK 31B54 [1.332.22]: αἰθὴρ <δ' αὖ> μακρῇσι χθόνα δύετο ῥίζαις).

furnishes the efficient cause of the kapnospheric phenomena (shooting stars, etc.), and that the *heat* of the sun, its advance and withdrawal, is the efficient cause of the rain cycle. Now, wind is explicitly and repeatedly said to require both smoke and vapor, so it is appropriate that it should have causes most closely associated with both the kapnosphere and the atmosphere and should have them in their accustomed order of excellence. For this reason the seasonal *locomotion* of the sun moves the wind, while the *heat* of the sun acting in due season on the earth (together with the heat in the earth) works in a more humble way to produce their material.

This interpretation does not make Aristotle's theory good, but it does make it better, and it has the added advantage of supporting the architectonic analogy he has developed between the winds and the rivers, since it provides the winds with a more river-like flow.[19] And even where the exact correspondences break down, similar considerations are in play. So Aristotle was especially intrigued by the temporal properties and found it remarkable that rivers are of long duration while winds are intermittent both daily and seasonally. This fact required explanation, because winds are permanent and constant features, just as rivers are. Aristotle needs, therefore, to explain why there are interruptions in their flow: why the rivers stop flowing through the periodicity of the great year; why the winds stop blowing through the periodicity of the seasons. Rivers are stopped by the secular heating and drying of a district. Winds are stopped by the seasonal cooling and freezing of a district. But some correspondences go unexplained. While it is reasonably clear why the springs should gather together to form rivers, it is not at all obvious why the sources of the winds, whose natural motion is upward, should likewise gather together in the absence of channels in which to flow (2.4.361b1–8).

In order to explain the winds' intermittency, Aristotle needs an efficient cause that will stop the exhalation, on both a diurnal and a seasonal periodicity. The sun, generator of both cycles, is the obvious candidate. In a none too systematic discussion at the beginning of 2.5 Aristotle identifies a number of ways in which the sun influences the production of exhalations. First, its absence produces cold, which simply extinguishes the exhalation.

[19] Aristotle returns to the analogy twice in this chapter: "for we do not suppose that any flow of water at all produces rivers, not even if they have certain volumes; rather the flow must be from a spring. The same is the case with winds. For a great mass of air could be moved by some great falling object, but have no beginning or source" (2.4.360a29–33). Again, "the winds are generated from many exhalations coming together little by little, just as rivers gather their waters together from many sources. They are both small at their actual sources but gather strength as they go" (2.4.361b1–3).

Conversely, its presence can produce extreme heat, which in turn gives rise to two distinct but related effects. The first is withering (μάρανσις): when the exhalations from the ground are relatively weak, their lesser heat may be overcome and dispersed by the greater power of the sun.[20] In this case the exhalation is produced but is then superheated and dispersed so as to be ineffectual as wind, perhaps because it rises too quickly. In the second, the intense sun dries up the ground itself before any exhalation can be produced. This second cause confirms, as we saw, the role of the wet exhalation in the production of wind, for the intense heat of the sun drives off the terrestrial moisture necessary to produce the smoky exhalation, "just as a small amount of fuel thrown into a large fire is often first burnt up before making any smoke" (2.5.361b17–20). The heat evaporates the moisture before it can have a catalytic effect.

The heat of the sun depends on the season of the year and the time of the day, and these allow Aristotle to account for the winds' periodicities.[21] Because the sun is particularly hot at the summer tropic around the time of the heliacal rising of Orion, the wind is calm. Inter-seasonal periods of calm may also occur, because in the spring the exhalations have not yet been generated (because it is still too cold) and in the fall they have been exhausted and have not yet been replenished in the earth (2.5.361b27–30). Oddly, during these times, too, weather is unsettled, because there is a change of the season and the sun is reversing direction.[22] After the summer solstice the sun turns south a bit, and the heat abates. The exhalation is no longer dispersed, and so a wind can form. During the night when the heat of the sun abates further, the water in the ground freezes again, so preventing further exhalations until the following day. The winds that blow from the north are called Etesians, and Aristotle is not entirely consistent about their cause. In one passage they blow most after the solstice, because that is when the sun has had time to make its heat felt (362a20–3). Yet in an earlier passage (362a2–7) the intense solstitial heat was said to dry the earth too quickly for the exhalation to form, and only when the sun withdraws can the melted water produce fumes.

These Etesians are the foundation of Aristotle's doctrine of the winds. He extends the doctrine by drawing an analogy between these northern

[20] On the withering or consuming effect on some winds, Theophrastus, *Vent.* 15–16.

[21] Strohm 1984: 184 notes the parallels between the rain and the winds relative to seasons. Since both are governed by the sun and the seasons, we have a parallel between the two exhalations (dry winds and wet rain) governed by the rhythms of the year.

[22] Hippocrates, *Aër.* 1 for seasons being very varied at their changes; also *Pr.* 26.13. For the period between the rising of Orion and the rising of the Dog as being treacherous for sailing, Polybius 1.37.4–5; Lehoux 2007: 7.

and the less apparent southern Etesian winds, and this leads in turn to a discussion of the southern hemisphere.[23] Aristotle begins with a problem: why do the northern Etesian winds blow continuously after the summer solstice but there are no corresponding southern winds after the winter solstice (2.5.362a11–13)? In fact, he replies, there are southern winds, but they are weak and therefore less noticeable. They start to blow seventy days after the winter solstice for reasons that correspond to those in the north. They do not blow as continuously, because seventy days after the heat of the solstice evaporation is confined to the surface moisture, which is easily evaporated, and what is frozen under the earth would require greater heat to melt. This cause, of course, leaves open the question of why they do not blow more strongly at or near the time of the solstice, but it indicates at least the strength of the correspondence between north and south and the similarity of cause in both cases.[24]

Disk and sphere

The correspondence between the north and the south Etesian winds and especially their common origin in the melting ice (2.5.362a26–8) leads Aristotle naturally to the preliminary view that the southern winds are generated in the Antarctic region just as the northern winds are generated in the Arctic (361a4–22). For this view there is considerable evidence in 2.4. Since the sun cannot reach the areas beyond the tropics, an abundance of moisture is preserved there and is available for the generation of the winds (a17–18). The south wind then must blow from a place far south of the winter tropic (the Tropic of Cancer).

This preliminary theory makes it easy to see why north winds blow in the summer and south winds in the winter, depending on where the sun is melting the ice. But at the same time it forces the introduction of the southern hemisphere, about which Aristotle has hitherto been silent. Before this point he had no reason to invoke the hemispheres even in explaining the season-dependent rain cycle, for when the sun advances to the north, it heats and evaporates the moisture; when it retreats from the north, the vapor cools, condenses, and falls as rain. The hemispheres are irrelevant in the explanation, because what happens in each hemisphere stays there and does not affect the other. This is not so with Aristotle's Etesian winds, since they blow from one polar

[23] Cf. *Pr.* 26.10.

[24] Alexander 100–1 argues that since the sun in the winter goes far south of the area from which the southern Etesians blow, they are slow to start and are intermittent.

region to the other. For this reason Aristotle must invoke the southern hemisphere. For this reason too, the antipodes, the corresponding southern landmass, has to be invented, since wind cannot be generated without land.[25]

The progress of Aristotle's thought is curious but clear: in theorizing about the generation of winds in general, he hits upon the clearest example, the northern Etesians blowing in the summer. As mentioned above, these had been the subject of discussion by Ionian philosophers. Democritus, for example, had explained the summer-time flood of the Nile as due to the contribution of the moisture-laden northern winds.[26] From these Aristotle turns to the problem of the southern Etesians and invokes the corresponding Antarctic. But the evidence for the south wind among the Ionians is much weaker. Diogenes of Apollonia mentions the sun's turnings in general without reference to north or south.[27] Anaxagoras mentions only one pole from which the wind advances and though Aëtius credits him with a theory that the Nile's flood is due to the summer snowmelt from the mountains of Aethiopia, this still says nothing about the winds.[28] And since Aethiopia's seasons are the same as ours, this melting snow cannot account for the winter Etesians. It is only with Hippocrates of Chios and his spherical earth that we find a reliable source of moisture beyond the southern tropic.[29]

Aristotle accepts the Ionian account of the generation of the northern Etesians, but in extending it to the south he was obviously motivated by his belief in the spherical shape of the earth and the possibility of a corresponding Antarctic zone. Indeed the southern Etesians are only really consistent with a spherical earth, in which our south wind blows to us from below the equator. Now, Aristotle might have rested content with a single global system of winds, but the consequences of this view are in

25 Aristotle does not explicitly make the point, but Mariotte 1718: 26 points out that there can be no winds from the sea on Aristotle's explanation.

26 "Democritus believes that the clouds are heavy with watery vapor produced by the process of the melting and scattering of snows at the time of the summer tropic. When these clouds impelled by the Etesian winds reach the southern area and the country of Egypt, they pour down heavy rains which provide for the lakes and the Nile" (DK 68A99 [2.107.33–108.2] = Aët. 4.1.4 [*DG* 385.8–17]).

27 DK 64A17 (2.54.13–17).

28 DK 59A42.11 (2.17.1–3); 59A91 (2.27.23–4) = Aët. 4.1.3 (*DG* 385.5–7). Hippolytus, as printed by DK at 59A42.5 (2.16.15–16) reads τὸν δὲ Νεῖλον αὔξεσθαι κατὰ τὸ θέρος καταφερομένων εἰς αὐτὸν ὑδάτων ἀπὸ τῶν ἐν τοῖς ἀνταρκτικοῖς χιόνων. But ἀνταρκτικοῖς is an emendation for the MSS ἀρκτοῖς, preserving which would bring the theory into line with Democritus'. Seneca (*QN* 4.2.17) reports the same theory as Aëtius. In any case, Aethiopian mountains are very different from the Antarctic. An account of Anaximander, Anaximenes, and Anaxagoras on the wind is to be found in Böker 1958a: 2219–22.

29 *Mete.* 1.6.342b36–343a20.

tension with his more elaborated theory that the winds, like the rivers, are centered on the Mediterranean basin.[30] He therefore revises his theory by placing two disks (the inhabited world in the north and the antipodes in the south), each with a complete and independent system of winds, on the surface of the spherical earth (2.5.362b30–363a1).[31]

"How, when and where did world, earth and globe unite?"[32] Arguably it was in *Meteorologica* 2.5–6, where Aristotle combines two models of the earth – the sphere and the disk. Each model has an equator and poles, each has summer and winter tropics; and each has the ever-visible circle of the stars. In appearance they closely correspond, but their functions are quite different. The sphere in 2.5 represents the earth in its entirety; the lines fixed on it are projections of celestial lines. Aristotle introduces this sphere apparently to show that our south winds cannot in fact blow from the southern hemisphere as he had just maintained. The projected lines divide the spherical earth into five zones: the torrid zone between the tropics, two temperate, inhabited zones between the tropics and the ever-visible or never-visible circle, and two frigid zones between the ever-visible/never-visible circles and the poles.[33] Our south winds, he says, come from our tropic and not from the southern hemisphere. But the apparatus he introduces here is far in excess of what is necessary to prove his point. His only real argument is that even the north winds do not blow all the way across our zone, and we feel them as strongly as we do, only because we live toward the north (362b35–363a4). *A fortiori* the south wind could not blow all the way from the southern hemisphere. In short, his most important purpose in introducing the zones is to develop a common vocabulary between spherical and flat views of the earth and promote a reconciliation between them.

In 2.6 the disk Aristotle introduces has a function apparently independent from that of the sphere (see Figure 10.1: Disk and sphere). It is a windrose, the circle around which are ranged the winds in their system of directions. This circle is clearly not a model of the spherical earth; it is the plane of the horizon with the observer at the center. And Aristotle is explicit that the scheme holds for the southern as well as for the northern

30 The contradiction between the two views about the origin of the south wind is noted by Alexander 93.11–19 and 100.3–16.

31 For the north–south division, Hippocrates, *Morb. Sacr.* 13(16), where opposition of winds and their qualities is a major theme.

32 The question is asked in Raaflaub and Talbert 2010: 2.

33 Aristotle uses the term τμῆμα for zone. Autolycus is the first to use the term ζωνή (*de ortibus et occasionibus* 2.5).

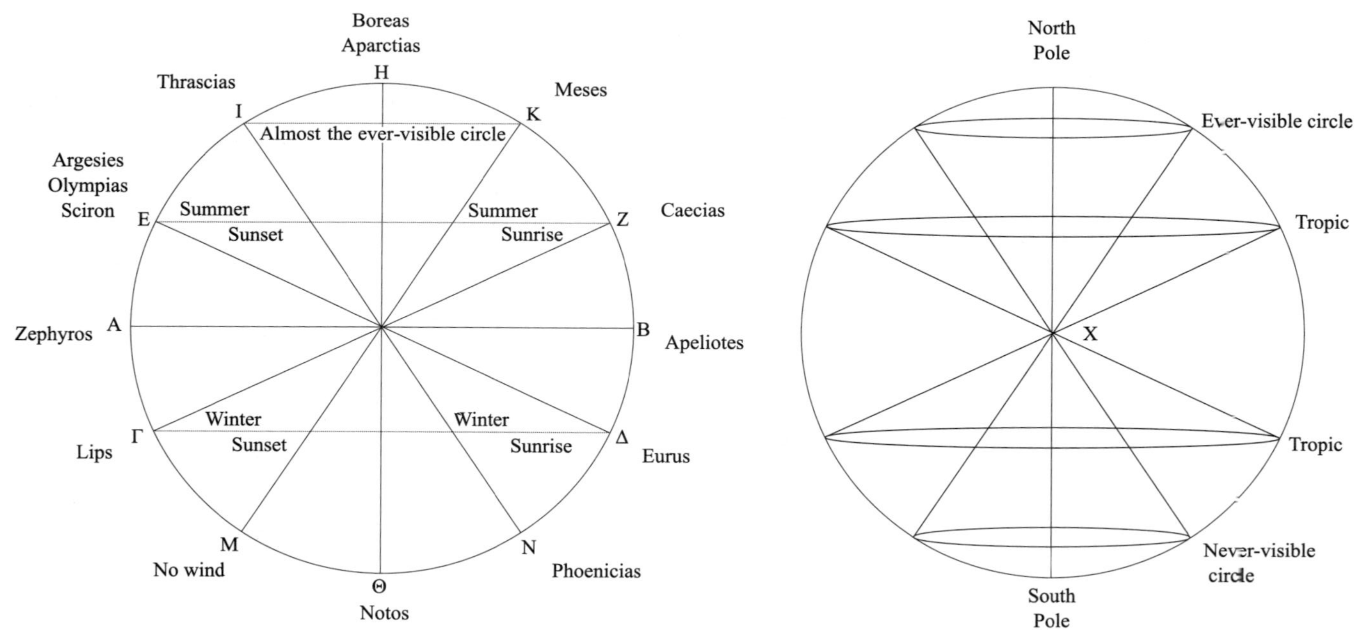

Figure 10.1 Disk and sphere

hemisphere (2.6.363a29).[34] And yet the disk has all the same divisions as the sphere, not into zones certainly, but into points of direction. Aristotle was careful to choose points and lines that could transfer from one model to the other. So, for example, he defines the habitable zones as that part of the surface of the spherical earth in between the bases of two cones, the common apex of which is at the center of the sphere (2.5.362a32–b5). The use of cones here seems unnecessarily involved, given the modesty of his purpose, but the reason for the apparatus becomes clear in 2.6: the cones are the three-dimensional correlates of the lines in the windrose which join each wind to its opposite (see figure). As a result, the cross-section of the spherical earth will be divided into the same twelve sections as the windrose.

Again Aristotle's windrose begins with the traditional eight Ionian points. Six are given by the sunrise and sunset at the equinox, the summer solstice, and winter solstice; the intersection of the meridian and the horizon provide the last two, the north and the south. These points, which for the Ionians had been relative to the observer – though they did not see it that way – become absolute when transferred onto the sphere. We saw how the Ionians' map frame used sunrise and sunset points on the observer's horizon to designate the tropics and the equinox.[35] As a result the northern tropical line, which connects summer sunrise to the summer sunset, is north of the observer. When projected onto the sphere, however, it lies to the south, and the winter tropic (the Tropic of Capricorn) is relegated to a hitherto unimagined southern hemisphere. Now, Aristotle insists that his diagram of the winds (ὑπογραφή) takes a circular form merely for the sake of clarity (2.6.363a26–8). This is usually interpreted as a defense against his own attack on those who depict the inhabited world as round.[36] And yet he also claims that it represents (even if by convenience and some distortion) his inhabited world, for the southern landmass will correspond to this one (363a29–30). This circle is not arbitrary then but corresponds, with appropriate qualifications, to the roughly circular Mediterranean *oikoumenē*.

Onto both the disk and the sphere Aristotle places the ever-visible circle, a feature that has generated confusion in both contexts, because its function

34 Olympiodorus 183.21–185.13, whether carelessly or on purpose, understands Aristotle's diagram of the zones of the earth as if it were a horizon (ὁρίζων). This is an indication of how similarly Aristotle constructed them.

35 Attested to in Hippocrates, *Aër.* 3.

36 E.g., Alexander 107.18–25. It may also be a gesture mindful of Herodotus' mockery of those who draw the earth as a perfect circle (4.36).

has not been correctly understood.[37] The ever-visible circle is the circular portion of the northern sky that revolves around the north pole and never sets below the horizon. Unlike the Arctic circle, which has a fixed and absolute location, the ever-visible circle increases in size as the observer approaches the pole and it has the same angular radius as the observer's latitude. It is dependent on the observer as the tropics and poles are not, and differs from them also in that for the northern observer there is no corresponding southern ever-visible circle, only an invisible circle. As such, it is the "odd line out" on the sphere. Yet, though its extent is relative to the observer, it still remains a celestial circle that can be projected onto the spherical earth and can be given (for any given observer) a determinate position. To this extent it is fixed.

Aristotle may have chosen the ever-visible circle instead of the Arctic circle because the Arctic circle was too far north to mark the limit of the habitable zone. But if this choice succeeded in making his temperate zone more habitable, Aristotle generated further problems for the windrose; for the addition of the ever-visible circle destroys the elegant symmetry of the traditional eight-point windrose, since there are no fully developed southern winds (except for the local and doubtful Phoenicias, 2.6.364a4) at the points of the never-visible circle. And though the ever-visible circle is certainly relative to the horizon of the observer, as is appropriate to the disk, it cannot provide easily observable points on the horizon as the sunrises and sunsets do: the ever-visible circle, as it revolves around the north pole, touches the horizon only at the due-north point, and in order to get the points Aristotle wants we have to drop the tangents to the circle perpendicularly onto the horizon.[38]

Again Aristotle may have chosen the ever-visible circle for his windrose, because he wanted a more equal division of the northern quadrants than the Arctic circle with its 23.5° from north would have afforded;[39] for the

[37] Aristotle says that the north winds almost correspond to the ever-visible circle: "the chord IK nearly corresponds to the ever-visible circle but is not exact" (363b32–3). The power of the model is all the clearer where it has to be forced onto the phenomena. Gilbert 1907: 544 did not see the conflict between the solstitial-horizontal and the projectional systems, as Rehm 1916: 37 points out. Rehm himself (43–4) calls it a rather illogical combination of the sphere, the Ionian drum, and the circular horizon.

[38] As Olympiodorus 187.15–25 and Rehm 1916: 36 note.

[39] As Thompson 1918 has rightly argued, the solstitial points are relative to an observer at approximately the latitude of Athens, which has the effect of breaking up the quadrant into three almost exactly equal parts (Thompson does not, however, address the issue that the ever-visible circle never intersects the horizon). Rehm 1916: 36–41 thought rightly that Aristotle was working with two different systems. They are different, to be sure, but they are commensurable by dropping the tangents of the ever-visible circle to the horizon.

ever-visible circle allows the secondary northern winds (Meses and Thrascias) to be 37° away from north along the horizon (for the observer at Athens), and this would help in discriminating among the northern winds.[40] No doubt Aristotle had these considerations in mind when making his choice, but a more satisfactory reason is to be sought in the reconciliation of the disk and the sphere.

Aristotle is looking for a set of terms that will designate both absolute locations on the real surface of the spherical earth and points on a flat-earth horizon relative to the Greek observer notionally at the center of the Mediterranean *oikoumenē*.[41] He deliberately seeks a vocabulary to capture both worldviews, just as he does, for example, in his discussion of the multiple senses of substance in the *Metaphysics* or of change in the *Physics*. It is a well-known feature of his method that he begins with a term in a more familiar significance and moves by abstraction and addition of conceptual material to a new significance. Famously, for example, substance is a combination of form and matter and signifies individual, usually living, things. But on further investigation through analysis and abstraction, substance comes to signify form primarily and then activity. Through this transformation of meaning, substance has changed its significance radically, and yet it still maintains close causal and terminological connections with the original starting point. In a similar way, I suggest, Aristotle is fitting the flat-earth view, which is more immediately familiar to our sense perception, into a more comprehensive and causally prior spherical view of the earth.

Aristotle is aware of the tension his horizontal phenomena – the rivers and the winds – have created for the spherical worldview of the heavens and the upper *meteora*. He is striving here for a reconciliation and compromise between the Ionian flat earth and the Pythagorean sphere. The ever-visible circle is superior to the Arctic circle in facilitating this reconciliation, because only the ever-visible circle *both* exists relative to the flat-earth observer *and* has a determinate location on the sphere; the Arctic circle by contrast, has no significance at all for the flat-earth horizon, and cannot be transferred from a sphere.

This double worldview is corroborated in Aristotle's description of the shape of the habitable zone cut off by the tropic and the ever-visible circle, which he describes as a τύμπανον (2.5.362a35). Here the tambourine shape designates the section of a sphere. But Aristotle also uses a related term

[40] Neither Rehm 1916 nor Thompson 1918 notices this fact. The points are not meant to be precise anyway. Aristotle says that the winds blow from around these points (363b31–2), a fact that indicates that the winds are being made to fit the schemes.

[41] For the notion of Greece at the center, see *Pol.* 7.6.1327b18–33; Cole 2010: 207.

(τυμπανοειδής) to describe the drum-shaped earth of the Ionians.[42] The shapes are distinct but they are closely related and, by using a related term, Aristotle may want to indicate that the Ionian view, though erroneous on a cosmic level, is nevertheless useful in the science of winds and rivers so long as it is subordinated to the truer and more encompassing reality of the sphere.

His immediate reasons for restricting the winds to each hemisphere and for the consequent fusion of flat and spherical worldviews are to be sought in his systematic analogy with the rivers; for the rivers that Aristotle mentions form a system around the Mediterranean confined to the northern hemisphere, and southern rivers were unnecessary for that inquiry. In order to preserve the analogy, our winds must likewise be confined to the northern hemisphere and a different set of winds must be posited for the south. This analogy was one of Aristotle's main reasons for disrupting the traditional eight-wind symmetry with the ever-visible circle: within our hemisphere both rivers and winds are more prevalent in the higher than in the lower latitudes, and northern winds predominate in his windrose just as the rivers of Europe are more numerous than the rivers of Africa.[43] Finally, as is usual with Aristotelian analogies, there is a common cause for this correspondence – the greater moisture in the north supplies the material for both winds and rivers – a common cause that further unifies the two genera.[44]

By the end of 2.5 the isolation of the winds in each of the hemispheres from one another is complete. Aristotle concludes with a refashioned explanation of our south wind: it blows from the torrid zone and so cannot depend on melting snows as the northern Etesians do.

> That region, because of its proximity to the sun does not have water or pasture land [or more probably "snow" cf. 362a18 and 364a8–10], which produce Etesians on account of the thawing [or "freezing"]. But because that region is much larger and more open, the south wind is greater and more dominant than the north wind and reaches further here than the north wind there. (2.5.363a13–18)

The cause of the southern Etesians, which do not arise from the south polar region, is appropriate to the sphere and, since they are different

[42] *Cael.* 2.13.293b34; cf. Leucippus, DK 67A26 (2.78.11) = Aët. 3.10.4 (*DG* 377.2).

[43] Aristotle may also be thinking (though he gives no indication of this) that since the *oikoumenē* is in 3:5 proportion of north–south breadth to east–west length (2.5.362b23), the distance along the north is much longer than the east and west and therefore has room for more winds to blow from.

[44] As I argue in Wilson 2000, Chapter 2 analogues share causes by analogy, but these causes are also common at a higher level of abstraction.

from the northern Etesians, they modify the internal north–south symmetry of the disk, with which Aristotle started his investigation: the basic geography of the disk remains subject to the corrections introduced by the sphere but now the predominance of the north and south winds in each of the landmasses prompts the question of where the east and west winds are strongest. Aristotle places them between the northern and the southern landmasses (perhaps, as Lee suggests, in the Indian Ocean), where they blow in alternating fashion, "just as here the north and south winds blow" (2.5.363a4–8). Aristotle's silence on the causes and attendant problems in this conception (why should they alternate? from what landmass do these winds arise?) show that these winds were of only secondary interest.

Finally, as with the rivers, part of the individual nature of the winds consists in their origins and directions, and, as a result, neither the winds nor the rivers are indiscriminately one mass.[45] Aristotle follows folk tradition in this regard and accepts as a necessary consequence the windrose and the disk conception of the landmasses. To this he adds and stresses the winds' diametrical oppositions, an element likewise of the folk conception. Diametrically opposed winds are specific opposites (ἐνάντια κατ' εἶδος 2.6.363a30–1) indicating that their form is dependent on their position.[46] Though this opposition is also found among the rivers, especially in the Nile–Ister correspondence, it is much weaker there. Opposite winds interact with one another as the rivers do not. Opposite winds cannot blow at the same time; rather, since the movement of the sun is their cause, they usually (though not universally and exactly) blow at opposite seasons (364a31–b3). Underlying this interaction is the principle of blocking, according to which one wind can stop another. The stronger wind (frequently the north winds) will block and prevail.[47] But winds that blow more or less from the same direction can blow together, for example Eurus and Caecias (364a30). As part of the same pattern Aristotle simplifies the winds into a basic north–south symmetry. The winds are, for all their peculiar characteristics, basically north or south and differentiated as hot

45 Olympiodorus 99.2–3 says that the places from which the winds blow are the specific differentiae. So also *Met.* 8.2.1042b21. That each wind has its proper place by nature (and that this determines its other properties) is stated explicitly as a general principle by Theophrastus (*Vent.* 2); but for the most part the position of winds was a minor issue for Theophrastus and for Seneca (Strohm 1984: 188).

46 Strohm 1984: 186 citing Rehm 1916: 41n2; Böker 1958b: 2342. Cf. Aeolus' paired sons and daughters (*Od.* 10.5–7).

47 Cf. *Pr.* 26.4 "why do the alternating winds blow? Is it for the same reason as causes the currents in straits (εὔριποι)?"

and cold (364a13–22).[48] The west wind, being colder, is a north wind. East winds are lumped in with the warmer south winds, because the eastern region gets more heat than the west (364a24–7).[49] In this respect again Aristotle is preserving the Ionian tradition. The difference in nature of the winds according to their direction is attested in the Hippocratic Corpus, where, for example, hot winds are those between the winter rising and the winter setting of the sun, and the opposite ones come from the opposite direction.[50] In his pursuit of universal types of winds based on their direction, Aristotle downplays the rich tradition (cf. fr. 250 Rose) of local winds. Indeed the only indication that he is interested in local winds comes early in 2.4 in his discussion of winds in localities and districts (2.4.360b5–7). As we shall see, local winds are being saved for an important role in earthquakes. These facts – the pairing of the winds, their assimilation to north and to south, and the absence of local winds – all provide further corroboration of Aristotle's commitment to the windrose and the disk as basic structuring devices of the surface of the earth.

So ends Aristotle's lengthy account of radially oriented, horizontal phenomena, in which his main purpose has been to organize a diverse group of *meteora* into a structured opposition with the upper phenomena of the kapno-atmosphere. Until this point the ordering influence of the sun has been strong – less so for the radial than for the upper phenomena – but hereafter the principle of order weakens and the forces of disorder and violence gain the upper hand.

48 Böker 1958a: 2231–2 discusses tendencies in other authors to balance north–south and east–west, including the Hypernotians and the eastern version of the Blessed Isles.

49 Aristotle is much abused by Webster for the statement that east winds are warmer than west winds, but *Pr.* 26.1 may provide a clue to his thought: "the parts towards the east are higher than those towards the west, as is shown by the size and depth of the sea towards the west." Cf. Hippocrates, *Aër.* 6 for the notion that places facing west have more extreme weather. The argument bothered neither Alexander nor Olympiodorus. For some of the later fortunes of the idea, see Shcheglov 2006.

50 E.g., *Vict.* 38 and *Aër.* 3–4.

CHAPTER II

Earthquakes and stormy phenomena (2.7–3.1)

Earthquakes, on Aristotle's conception, take up an awkward position between the winds just discussed, and stormy phenomena, such as thunder and hurricanes, that are to follow. They are said to share a common nature (τὴν αὐτὴν φύσιν) with both groups, since the same natural substance, the dry exhalation, "is wind on the earth, earthquake inside the earth, and thunder in the clouds" (2.9.370a25–7). Earthquakes are affections of the winds, which move inside the earth and shake it.[1] As such, they naturally follow the winds in the list of subjects to be discussed. But if we expect earthquakes to be produced by the kinds of winds we have just learned about, we will be disappointed, for in fact the winds that cause earthquakes are of a quite different sort. As we compare the properties of winds, earthquakes, and stormy phenomena, it becomes clear that they are not just modifications of one another differing merely in the places they are found. The dry exhalation itself displays a different nature in each case. Indeed the passage just quoted immediately goes on to say that if the dry exhalation "flows in one way it is wind, [if] in another it causes earthquakes; and when in the clouds it changes and separates while the clouds come together and condense into water, it causes thunder and lightning, and also the other phenomena of the same nature as these" (2.9.370a27–32). The behavior of the dry exhalation that produces earthquakes, then, is not the same as that which produces winds or violent phenomena. And though Aristotle does not emphasize these differences in the exhalations as such, they offer clues to the architectonic order that he seeks to impose. In fact, it is the position of earthquakes within the order that best explains why they are assigned to the dry rather than to the wet exhalation, for they belong with the winds in the analogy with the rivers and the sea, but they are associated

[1] Affection (πάθος): 2.7.365a15.

with volcanoes and share in the violence of such phenomena as lightning, which clearly depend on the flammable dry exhalation.[2]

After discussing what Aristotle owes to his predecessors, I want to argue that he conceives earthquake winds as chaotic and disordered in contrast to what I shall call "normal" winds, and that he explains their violence by means of the theory of the "strait" which has the effect of bringing them into correspondence with the sea at the bottom of the world.

Predecessors

The violence that characterizes earthquakes receives Aristotle's attention at the beginning of chapter 2.8, where he argues dialectically in favor of the dry exhalation as their cause and against his predecessors' alternatives (365b28–366a5). He objects that only the dry exhalation in the form of wind – not water or earth – has the necessary violence and power of impulse to cause an earthquake, and he proves this by the process of elimination. The argument establishes that dry exhalation is the fastest, most violent, most penetrating and fine of the materials and so is best able to produce the earthquake effects. Democritus, by contrast, thought earthquakes were caused by excess water sloshing around in the cavities of the earth, or by dry cavities that attract water from wet areas (2.7.365b1–6).[3] Anaximenes held that successive moistening and drying caused the earthen hills in subterranean cavities to collapse and shake the ground (2.7.365b6–12). Aristotle interprets both of these theories in such a way that Democritus becomes the champion of water, and Anaximenes of earth, for the claim of efficient cause, though there is some evidence that Aristotle fashioned his reports to achieve this result. At least, according to Seneca (*QN* 6.10), Anaximenes gave air, water, and fire some role in producing earthquakes in addition to the collapsing of earth. It is possible that Aristotle simplified both Democritus' and Anaximenes' accounts in order to create a single advocate for each of the candidate elements.[4]

[2] Volcanoes get short shrift in the *Meteorologica* (2.8.367a9–11), being a more Italian phenomenon (so Sudhaus 1898: 53, who notes also *Mu.* 6.400a31–2: αἱ δὲ [φλόγες] πρὸς ἑσπέραν ἐκ γῆς ἀναβλύσασαι).

[3] For two interpretations on where the superfluous water goes (out of the earth or into other parts of the earth), see Strohm 1984: 193. Cf. Seneca, *QN* 6.19 on Metrodorus for another theory based on resonance of sound; similarly *Pr.* 26.2.

[4] A tradition dependent on Theophrastus provides explanations from all four elements (see the translation from the Arabic in Bailey 1947: 1747–8); so also Seneca, *QN* 6.5; see Williams 2006: 138 for Seneca's report of Democritus (6.20.1), which makes water, air, and both water and air the cause.

However that may be, Aristotle's argument that wind alone has the vehemence necessary to cause earthquakes is a poor one, since it ignores the destructive force of tsunamis and landslides and seems solely aimed at establishing the claim of the dry exhalation without much thought to the further consequences. After all, a fine element such as air is more likely than water or earth to pass through the earth without shaking it violently.[5] Moreover, the fact that earthquakes are common where the ground is porous and cavernous (366a23–5) clearly shows that the earth has different densities in different places, and while this fact explains how the wind can travel through the porous earth until it strikes the denser parts, such porousness is likely to allow water to produce earthquakes just as easily as it allows the wind. Again, the sea is said to have passages under the earth (διαυλίζεται 366a27), which it shares with the wind. So, if wind flows through passages just as sea water does, its fine, penetrating capacity seems hardly necessary to account for the shaking. Most importantly, though, this whole question of penetration seems unnecessary in view of the theory of "straits" with the larger subterranean passages it seems to imply.

The argument from penetrability and violence is intended to eliminate water and earth as efficient causes, but it cannot touch the first theory in the doxography of 2.7, namely that of Anaxagoras, which in fact most closely resembles Aristotle's own (2.7.365a19–35); for Anaxagoras' aether, like Aristotle's dry exhalation, is a penetrating element. Its natural motion is up and, according to Anaxagoras, it penetrates the hollows of the drum-like earth from below. Rain falling from above clogs the earth's pores, and the rising aether thus trapped shakes the earth.[6] It is possible that Anaxagoras intended this aether to account for volcanoes as well – Aristotle draws a similar corollary (2.8.367a9–12). But Aristotle is silent about what he owes to this theory and attacks Anaxagoras, first incidentally for his drum-like earth, then more pertinently on the grounds that Anaxagoras does not explain any of the times or places where earthquakes typically occur, for according to Anaxagoras they could only occur during rainy seasons,

[5] This is something that Aristotle himself admits at 2.8.368a19–25, where very fine subterranean winds cause noises without earthquakes.

[6] A variation on this account has *air* entering the earth from above. Diogenes Laertius 2.9 (DK 59A1.9 [2.5.34]): σεισμὸν ὑπονόστησιν ἀέρος εἰς γῆν. See Capelle 1912a: 429. Hippolytus reports (DK 59A42.11 [2.17.4]): σεισμοὺς δὲ γίνεσθαι τοῦ ἄνωθεν ἀέρος εἰς τὸν ὑπὸ γῆν ἐμπίπτοντος. τούτου γὰρ κινουμένου καὶ τὴν ὀχουμένην γῆν ὑπ' αὐτοῦ σαλεύεσθαι. Again the similarity to Archelaus' theory is striking. Seneca (*QN* 6.9.1) reports that fire (probably aether) is involved in the shaking of the earth. For a discussion, see Gilbert 1907: 298–9 and n2; Guthrie 1962–81: II.310.

when the entire earth, rather than just a restricted locality, is shaken. It is remarkable that Aristotle never mentions the theory of Anaxagoras' student, Archelaus, which is still more similar to his own: according to Archelaus, earthquakes are caused by the wind returning to the earth, and unlike Anaxagoras' aether, which moves up by natural motion, Archelaus' wind is carried down.[7]

Such is the background against which Aristotle establishes the claims of the dry exhalation. It was a common feature of all ancient theories to posit subterranean passages and caverns. The dispute concerned only the source of the impulse within the passages, and how the passages affected that impulse. None of the theories mentioned by Aristotle seems to involve forces beyond the natural motions of the materials involved: Anaximenes' earth falls and shakes, Democritus' water falls and shakes, and Anaxagoras' aether rises and shakes. Aristotle, perhaps following Archelaus, is unusual in bestowing on the wind a special impulse to move down into the earth.

Aristotle transforms this tradition by adapting many of the causes he finds there. He found fault with Anaxagoras, as we just saw, for not specifying the places and seasons in which earthquakes were likely to happen (cf. his criticism of Anaxagoras' theory of hail), and consequently he follows Anaximenes, who had suggested droughts and rainy periods as especially prone to earthquakes; for according to Aristotle's description in this chapter, winds are most prevalent during droughts and rains (2.8.366b2–3). But at the same time, Aristotle finds fault with Anaximenes for restricting earthquakes to places of drought and rain, on the grounds that some droughty and rainy places are more prone to earthquakes than others (2.7.365b13–16). He avoids this criticism for himself, as we shall see, by positing a multiplicity of causes. From Democritus, for example, he accepts a subsidiary role for subterranean water. Water may build up in the subterranean cavities and increase the pressure on the wind (2.8.366b9–15), which in turn may cause water to break forth from the earth (368a26–34). And of course, wet seasons produce more wind. Aristotle even extends the role of water to the sea, which had no role in the other accounts.

[7] Seneca, *QN* 6.12: "Archelaus . . . says as follows: winds are carried down into cavities of the earth; then, when all the spaces are filled and the air is thickened as much as it can be, the moving air which comes in on top of it compresses the air that was there first and pushes it and with frequent blows first packs it together then forces it out. Then in seeking room the air unblocks all the narrow passageways and tries to break out of its enclosure. In this way it comes about that the earth is moved, when the moving air is struggling and searching for a way out" (Corcoran trans.). Pliny, *HN* 2.81.192 and Ammianus 17.7.11 also think earthquakes occur at times of windlessness. See Capelle 1908: 615.

Normal winds and earthquake winds

But these adaptations play only a small role in Aristotle's overall conception of earthquakes, and the best way to approach his theory is to return to our opening observations and consider how the winds that cause earthquakes (earthquake winds) are different from the winds of the previous chapters (normal winds). Normal winds were global, or at least hemispheric, and formed an orderly windrose embracing the inhabited world. The Etesians and their seasons, which dominated that discussion, now disappear in the context of earthquakes. Earthquake winds are local, since earthquakes are local, and that fact becomes an important explanandum. While Anaxagoras may have made the aether shake the whole disk of the earth, for Democritus and Anaximenes, as well as for Aristotle, earthquakes are localized.[8] They originate in the earth,

> and so all the exhalations have an impulse toward one place. The sun does not have the same kind of influence over them that it has over winds above the earth. The result is that they flow in one direction whenever they get their start from the motion of the sun. (2.8.368b19–22)

The distinction between the local and the global is probably the main reason why Aristotle ignores local winds in the earlier chapters, reserving them for this occasion, and, as we shall see, this fact is closely related to the sun's ordering influence on the normal winds.

Again, Aristotle's account of which seasons are more prone to earthquakes is different from what we find in his chapters on the normal winds. As we learned, the normal winds, especially the northern Etesians, are most frequent in summer, after the rising of the Dog Star (around July 17). Earthquakes, however, occur most frequently in spring and autumn and during rain and drought, since these produce the most wind (366b2–4).[9] These are also the seasons, that is, not the winter, prone to violent phenomena (3.1.371a3–8), so it is reasonable to suppose that the winds associated with hurricanes are similar to those associated with earthquakes.

[8] So on Anaxagoras, Sudhaus 1898: 54. For this reason perhaps Aristotle notes the absence of places and seasons from his account.

[9] According to 2.6, hurricanes (ἐκνεφίαι) occur most often in autumn and secondarily in spring, and the north winds, Aparctias, Thrascias, and Argestes, cause them by falling upon one another (365a1–5). Similarly the time of earthquakes is noted using seasons (μετόπωρον, ἔαρ 366b2); the Etesians are timed by the more celestial solstices and heliacal risings (361b35–6). Aristotle does not correlate the seasons with the heliacal risings and makes no effort to bring the seasons and times of winds into synchrony with the earthquakes. Again, it is odd that the summer brings calm weather (366b5–6), since this is prime Etesian season.

But the earthquake winds are most distinctive in having a self-impulsive motion lacking in the normal winds. The earthquake wind is continuous and "follows for the most part its initial impulse so that its impulse is entirely inward or entirely outward" (2.8.366a7–8). This fact has influence on many of the earthquakes' other features. Normal winds can be stopped, prevented, or thwarted, but they cannot reenter the earth as earthquake winds do. Again, earthquakes are common at noon, because the sun locks down the exhalation (κατακλείει τὴν ἀναθυμίασιν 366a16), a process quite different from what we find with normal winds where the heat of the sun consumes the exhalation.[10] Clearly the process of locking down the earthquake winds is a sign of their impulsive power. Again, Aristotle describes as an indication of earthquakes, a long, straight cloud appearing in the sky. Such a cloud appears when the wind has gone into the ground, leaving the upper air tranquil (367b8–12). But this is a direct contradiction of the mechanism of normal wind movement, in which it is precisely the upper air that moves the winds, rather than the other way around. It is abundantly clear, then, that earthquakes are not caused by the motion of the upper air or heavens.

In retrospect, it may seem odd that Aristotle did not bestow this power of impulse on his normal winds, since it would allow him to jettison their implausible efficient cause.[11] But we can now see that the choice was strategic, for it shows that earthquakes are disorderly phenomena and do not partake in the circular geometry of the rivers and the normal winds.[12] When the wind is above the ground, it is subject to the motive power of the sun (2.4.361a31–2), which makes it travel in various orderly directions. Underground, however, it acts by its own disorderly motions. For this reason the directionality of the winds is of no account in earthquakes, though the dry exhalation maintains its taffy-like ductility. Earthquake winds, then, are less formal, more material, than normal winds; they are, as it were, failed winds – an expression one must use with caution lest it be thought that earthquake winds were striving for normalcy as their final cause. In this sense the study of earthquakes is a descent into the

10 Alexander 118.20–1 confuses the two.

11 Strohm 1984: 197 incorrectly claims that Aristotle uses the term πνεῦμα only when the exhalation has impulsive force. In fact, Aristotle uses ἄνεμος and πνεῦμα pretty indifferently: ὁρῶμεν γὰρ ἐνίοτε ἅμα πλείους πνέοντας ἀνέμους, ὧν ὅταν εἰς τὴν γῆν ὁρμήσῃ θάτερον, ἔσται πνεύματος ὄντος ὁ σεισμός (2.8.366a9–11); ἰσχυρῶς κινεῖ ῥέων ὁ ἄνεμος καὶ προσπίπτων (366b14); δῆλον ὅτι οὐκ εὐθὺς ἅπασαν ἀνήλωσεν τὴν ὕλην, ἐξ ἧς ἐποίησε τὸν ἄνεμον, ὃν καλοῦμεν σεισμόν (368a9–11).

12 Olympiodorus 1.21–3: the dry exhalations by nature follow the upper movement, but when they are produced contrary to nature and remain in the earth, they make earthquakes (cf. 3.12–13).

material realm and might be compared to *de Generatione Animalium* 5.1, where Aristotle considers the affections of living matter in the absence of the operation of any higher, ordering, cause. It is fitting that earthquakes should be likened to pathological pulses and tremors in the body (τέτανοι, σπασμοί 2.8.366b26), the digestive gases and the frisson attendant upon urination (366b20–1).[13] For similar reasons, too, earthquakes are given irregular causes, the ebb and flow of the sea (366a18–23) and the eclipse of the moon (367b19–30).[14] It is the disorder of earthquakes that explains in retrospect Aristotle's concern for the geometry of the winds, for he was eager to impose an order on the winds and rivers in contrast to the earthquakes. Earthquakes, of course, maintain their material continuity with the winds but break with the sphere and the disk, which will only be recovered in the dénouement of the rainbows and the haloes. In this sense earthquakes belong in a group with the violent phenomena.

The theory of the strait

Aristotle uses the concepts of impulse and localization and adds to them analogies from digestion and the sea in order to develop his theory of the "strait," the alternating movement of a fluid through a narrow passage. This theory had played no major role in his account of winds but is absolutely essential for an understanding of earthquakes.

There are a couple of ways relevant to our present consideration in which the winds can produce earthquakes. First, the wind may actively push back into the earth:

> the greatest number and largest earthquakes occur when there is no wind. For the exhalation is continuous and follows for the most part its initial impulse, so that its entire impulse is toward the inside or the outside. (2.8.366a5–8)

Second, Aristotle adopts a rudimentary theory of pressure according to which the wind may be sucked back into the earth. In this case it does not need its own impulse. During the night,

[13] For digestive gases, Hippocrates, *Flat.* 7. According to Alexander 118.34–6, tremors on urination are caused when the "πνεῦμα, reciprocally changing places in a mass from outside to inside on account of the fact that the πνεῦμα which was pushed out by the urine, is confined in narrow spaces." For why we shiver after urination: *Pr.* 8.8; 8.13; cf. *Pr.* 33.16.

[14] For wind at eclipses, *Pr.* 26.18. Alexander 120.27–30 notes that it is strange how the moon, being unaffected, can receive heat from the sun (or how the aether in between can). At 121.4–28 he notes the contradiction between the two theories for earthquakes associated with eclipses (as does Strohm 1984: 198) viz., windlessness and wind. Alexander solves the contradiction by invoking different times before the eclipse. Three hours before the eclipse the moderate heat releases the winds; very near full eclipse, the cold confines the winds.

> the flow turns back inward, like an ebb in the opposite direction of the flow outward, especially towards dawn. For it is then that winds normally begin to blow. If, then, the impulse of their motion happens to change direction like the Euripus and turns inwards, it makes the earthquake more powerful because of its mass. (2.8.366a17–23)

This passage reintroduces the Euripus, the narrow strait that separates Euboia from mainland Greece, through which the direction of water flow frequently changes. We must assume that the wind ebbs or is sucked back into the earth by the lower pressure there. Narrow passages connect the atmosphere with subterranean spaces, and the wind, by passing back and forth, causes turbulence and earthquakes. It is clear that the strait is an important component of the explanation. The wind cannot produce an earthquake simply by hitting the surface of the ground, since in this case any strong wind would jostle the earth. The strait serves to intensify and amplify the flow (366a23–b2).[15] It redirects the wind and is the balance point or a neck through which the fluid flows back and forth. Whether the movement of wind is a flowing or an ebbing, one of the spaces on either side of the strait must clearly be confined in order to sustain the positive or negative pressure. In the case of the flow from the earth it is easy to see that the spaces in the earth provide the necessary confinement. In the case of the flow from the atmosphere, Aristotle, as we have seen, is willing to bestow on the *pneuma* a certain impulsive force of its own, though he also provides a mechanism for local atmospheric confinement.

In his explanation of why earthquakes are local (2.8.368b12–20), Aristotle refers back to his discussion of the localization of drought, wind, and rain in 2.4, where he described the air as a combination of dry and wet exhalation. These exhalations may separate spontaneously, or the wet may be blown out by the dry or vice versa by *antiperistasis*. He stresses that they come together (συνέλθωσιν εἰς ἕν 368b16) and start together (ἐφ᾽ ἓν ὁρμᾶν b19–20). Thus there can be a local separation of the wet and the dry exhalation, and this accounts for the localization of the weather and, by extension, the confinement necessary to account for earthquakes.[16] We are to suppose that these air masses are mutually confining and that in a similar way the earthquake winds have a tendency to collect together and increase their

[15] Cf. Theophrastus, *Vent.* 29: "for a wind passing through a gap is always more forceful and vigorous, like a current of water. Concentrated, it has more thrust. That is why, when there is a calm elsewhere, wind is always present in narrow passages" (Coutant and Eichenlaub trans.). So according to Olympiodorus 204.17–18, the winds become more intense by being gathered together into one.

[16] Alexander 124.15–28 notes that the reason for the localization of earthquakes is that the exhalation from the neighboring districts migrates to the one where the earthquake occurs.

strength above the earth, that is, the atmosphere is capable of producing confined spaces similar to those within the earth. In order to show that he has confinement in mind, Aristotle adds a digestive analogy: the upper gut may be dry while the lower gut is wet, or vice versa.[17] The ebb and flow occurs because of the narrowing in the digestive passage, and in these and the various hollows (κοιλία) local disturbances can occur.

The passage from the *de Somno* that we have already discussed in the context of the sea (*Somn. Vig.* 3.456b20–4) provides further evidence that both confinement and a narrowing are necessary for the effect to occur:

> for it is necessary that the exhalation be pushed forward some distance, then turn around and change course, like a strait. For the heat of each animal naturally goes up and when it gets to the upper places, it turns back again in a mass and descends (ἀθρόον πάλιν ἀντιστρέφει καὶ καταφέρεται).

It is important to note that what is being described here is not evaporation and condensation, since there is no change of state. The same material goes forth and returns and the movement takes place through a narrow strait between two confined spaces. The return of the exhalation is explicitly likened to the movement of water through a strait, but it is not clear whether the reversal of flow is caused by the strait itself or by the confinement of the exhalation within the body, or by both. The most probable answer is both. In the sea the strait unquestionably plays a role, and though it is certainly harder to argue in this context that the action of a strait depends on confinement (the sea is huge and open), Aristotle admits that there are slight flows in the open sea (caused by the movement of the water into the deepest area of the sea), and these slight flows may become amplified in the straits. The very fact that the sea is confined as a standing body of water contributes to its slight, chaotic movement and this in turn sets up the flows through the straits (2.1.354a5–11).[18] In sum, the interplay of analogies suggests that earthquake winds return to the earth both by their own impulse and because of the pressure of a local confinement.

The relation between the sea and earthquake winds is not restricted to analogical similarity. Aristotle clearly intended a causal interaction as well. The ebb and flow of the sea itself, as its water presses back upon the subterranean winds, contribute to the force of earthquakes (2.8.368a34–b12). The association of tsunamis and earthquakes, too, is a sign that

[17] 2.4.360b22–6. Aristotle goes on to invoke *antiperistasis*: οὕτω καὶ περὶ τοὺς τόπους ἀντιπερiίστασθαι καὶ μεταβάλλειν τὰς ἀναθυμιάσεις.

[18] It is worth noting that the association of gases with tremors in the body is well attested, but it is typically the entering cold wind that brings on the shivering (*Pr.* 8.8, 13; 33.16). Since the earth cannot shiver in this sense, Aristotle must have recourse to a different mechanism, that of the strait.

earthquakes are caused by the wind – it is the wind acting upon water that causes tsunamis.[19] And wind is sometimes the cause of water bursting forth from the earth after some earthquakes, just as on the sea the wind is the cause of waves. Finally, calmness in the air and on the sea are indicated by similar signs: the thin straight, regular cloud and the straight, regular waves (367b14–19). In these ways Aristotle is developing an association between sea and earthquakes that goes back to the Earthshaker Poseidon.

But the analogy is hardly perfect. Aristotle, as we have seen, describes the earthquake wind as having an impulse of its own, in virtue of which it returns into the earth. By contrast, the sea, while it is described as "flowing back and forth frequently" (ταλαντεύεσθαι δεῦρο κἀκεῖσε πολλάκις 2.1.354a8), lacks this active impulse: it is merely the sloshing of water back and forth consequent upon the first impulse.[20] Indeed the sea's inertia was, in Aristotle's mind, a good reason to deny this kind of impulse, and the relative inertness of water in comparison with air is established at the outset of both arguments – the suspect eliminations that argue in the one case that the sea is a standing body of water with no springs (2.1.353b19–35), and in the other that the wind is the most intense and delicate of the elements (2.8.365b28–366a5). Earthquakes are less honorable because they are disorderly; the sea is less honorable, because it is a residue.

It is hard to see why the theory of the strait was necessary to Aristotle's explanation of earthquakes *per se*. Earthquakes might simply have been caused by dry exhalation trapped in the earth by some kind of indigestion, and no exhalation need ever have *returned* to the earth's surface. And though, according to Aristotle, a windless calm frequently precedes and accompanies earthquakes, there is no need for the theory of the strait, since the sun could trap the winds within the earth. After all, some earthquakes, such as those that occur around noon, are caused by the shutting up of the winds. These do not depend on the strait, but rather on the blocking of the passage. Alternatively in time of rain when the hollows of the earth fill with water, the dry exhalation may be caught in a restricted space and forcibly compressed (366b9–14). Now, if the theory of the strait was unnecessary to explain earthquakes, it may have been introduced for architectonic reasons. Alone among the phenomena of the treatise, the sea and earthquakes are compared to digestive processes and, as we saw in the case of the sea, the comparison is part of a normative evaluation. Alone, too, among the phenomena, the sea and earthquakes involve movement through straits.

19 On the famous destruction of Boura and Helike, see Higgins and Higgins 1996: 69–70.

20 Nor does Plato describe the source of the impulse in the *Phaedo* (112b).

This is unlikely to have been a coincidence. Their similarities point to the conclusion that both are negative correlates to the higher, more orderly river–wind pair. The sea is an inert and sterile residue contained in a place not wholly natural to it, conceptually and physically distinct from the living waters of the rivers that ring it round in a circle. The earthquake winds are unnaturally trapped below the earth, distinct in their dynamic qualities from the winds of the circular windrose. Both sea and earthquake winds can work together to unleash great destructive forces. Certainly Aristotle did not lavish on this correspondence the sustained effort he made in order to assimilate winds and rivers, but the similarities are frequent and striking enough to convince us that he had it in mind.

Stormy phenomena

Aristotle makes clear at the end of chapter 2.9 that winds, earthquakes, and violent phenomena are all made from the same dry exhalation (370a25–32), and, though he summarizes their differences, his description hardly does justice to the great variety of ways in which this dry exhalation is characterized. With normal winds, as we saw, the exhalation is drawn along by the celestial movement. Earthquake winds have their own impulse, which is amplified by the effect of the strait. With stormy winds the dry exhalation is in itself relatively inert and exhibits only its natural tendency to rise. What gives it its violence and unnatural motion is the wet exhalation, which surrounds and ejects it. As such, the violent phenomena arise from an interaction between the two exhalations where, in a certain sense contrary to the natural order, the wet exhalation rather than the dry serves as the efficient and moving cause. The dry exhalation is now a passive recipient of external force.

Aristotle lays out his theory of the violent phenomena, specifically thunder and lightning, at the beginning of chapter 2.9, and only after its lineaments are drawn does he consider the contributions of his predecessors. His own theory, though, anticipates the problems generated by the doxography, and as we proceed it also becomes clear that he chose to treat thunder and lightning first because the traditional accounts of them were the greatest challenge to the primacy of the unignited over the ignited dry exhalation, and consequently to the set of differentiae that govern Aristotle's entire genus of violent phenomena.

Thunder and lightning have their origin in ejection, and though the process of ejection is often referred to as ἔκκρισις (369a27, 36; alongside the more forceful ἔκθλιψις 369b5), it is clearly not a slow seepage of dry exhalation from a sponge-like wet exhalation. Aristotle invokes again, as he had for the lower shooting stars, the analogy of the pip pressed out from between the fingers in order to convey a more dramatic separation, one requiring a sharp distinction between the wet and the dry (2.9.369a21–4). In general the wet exhalation, as it rises in the atmosphere, releases its heat and forms a condensed layer, through which dry exhalation ascending from below cannot penetrate. This layer, as it condenses, draws itself together like a sack around a body of dry exhalation, which is then pushed downwards (2.9.369a12–29). All the violent phenomena begin in this way, but thunder and lightning are the simplest and most common form. I shall take them up in turn to illustrate the priority of the unignited over the ignited exhalation and their role in structuring the entire genus.

Thunder is caused when this dry exhalation is pushed around and strikes the condensed cloud. Whether the impact is contained wholly within a sack-like cloud or whether the mass exits the sack to strike a neighboring cloud is not clear from Aristotle's description. Thunder and lightning are certainly contained in the surrounding clouds (τοῖς περιεχομένοις νέφεσι 369a28–9) but, given the amorphous shape of clouds, it is probably sufficient that the dry mass be ejected and strike some cloud, whether another part of the same cloud or a different cloud. The important thing is that the thunder and lightning stay up among the clouds. Aristotle's description of lightning and the distinction he wants to draw between lightning (ἀστραπή) and the thunderbolt (κεραυνός) point in this same direction. He describes a delicate and weak fire, which is often produced when the exhalation strikes the side of a cloud (369b3–7), and this is the lightning we see flashing inside a cloud. It differs from the thunderbolt, which is ejected and strikes the ground (3.1.371a17–b14). It is likely, therefore, that the thunder-generating masses that produce lightning also remain within or among the clouds.

But how do these local intra- and inter-nebular ejections operate? Is there a medium in the sack *through* which the *sustasis* of dry exhalation is pushed? And if there is, is it made of dry exhalation or something else? And if it is made of dry exhalation, does the *ejected* dry exhalation push through it to the other condensed cloud or does it merely push its neighboring dry exhalation into the condensed cloud, so that the neighboring mass causes the thunder? Aristotle favors the former view and makes clear as he proceeds through his descriptions of hurricanes and whirlwinds that the ejected mass flows forth continuously through a surrounding medium, whatever that medium may be. As such, these ejections share a taffy-like ductility with the normal and earthquake winds and with the *sustaseis* of the kapnosphere.

Basic to Aristotle's explanation of violent phenomena is the priority of the unignited ejections over the ignited.[21] His reasons for insisting on this priority, though never explicitly articulated, are several, the first having to do with economy of explanation (others will appear in due course). The genus comprises six phenomena (thunder, lightning, hurricane, thunderbolt, whirlwind, and firewind), but an account that makes the fiery form prior must explain all other forms as modifications of that fire. Anaxagoras, who explains lightning, thunder, thunderbolt, whirlwind, and firewind, treats them all as modifications of aether, and all of them are fires.[22] Whirlwind must be treated improbably as a fire, and hurricane is not discussed. In Aristotle's mind, more phenomena can be explained more economically by supposing that all six are winds that can sometimes be ignited. But the challenge to this way of proceeding is the powerful tradition that treats thunder as an effect of the quenched fire of lightning, and for this reason Aristotle tackles this challenge first.

We see the lightning flash and wait for the clap of thunder to follow. The appearances suggest that lightning precedes the thunder, and that therefore lightning somehow causes thunder. But is that the true causal order? According to both Empedocles and Anaxagoras lightning is a pre-existing fire, caused by the direct descent of the rays of the sun or the aether from above.[23] Thunder is the noise caused by the impact or extinction of that fire.[24] Aristotle's critique focuses more on the supposed priority of lightning than on the generation of thunder from it. Anaxagoras in particular renders the descent of aether from above the clouds an unexplained and unnatural motion and is in no position to say why the clouds are important.[25]

[21] This is contrary to his example in *APo.* 2.8.93b7–12, where thunder is the extinction of fire in the cloud. There is no reason to suppose that Aristotle was ever committed to this theory.

[22] DK 59A84 (2.25.25–9) = Aët. 3.3.4 (*DG* 368.10–18). Aristotle owes some of the features of his phenomena to Anaxagoras: the thickness of the whirlwind (πολυσωματωτέρῳ πυρί) and the cloud being mixed in the firewind (νεφελομιγεῖ).

[23] Aetius (3.3) lists eleven opinions Aristotle could have known (Cleidemus, whom Aristotle does mention, is not on Aëtius' list; Anaximenes is credited with a theory similar to Cleidemus': when the sea is cut with oars, it glistens (DK 13A17 [1.94.5–7] = Aët. 3.3.2 [*DG* 368.1–3]). Certainly Aristotle's targets were among the most recent. In the rest of the doxographical tradition there is a strong identification of lightning and thunder with the wind and clouds.

[24] For Anaxagoras, the immediate cause is impact (DK 59A84 [2.25.25–9] = Aët. 3.3.4 [*DG* 368.10–18]); for Empedocles, either impact or extinction (DK 31A63 [1.295.1–3] = Aët. 3.3.7 [*DG* 368.28–33]).

[25] As Curd 2007: 224 has noted, "it is reasonable to think that this motion is ultimately related to the original revolution of the whirl that produces the cosmos in the first place." We would not know from Aristotle's report that Anaxagoras had discussed and accounted for five of the violent phenomena that Aristotle treats. It is Aëtius who provides these details: the hot falls into the cold (that is, the aetherial falls into the airy) and causes thunder by the noise, lightning by the color against the black of the cloud, the thunderbolt by the frequency and magnitude of the light, the whirlwind (τυφών) by the greater body of fire, and the firewind by the mixture of cloud (3.3.4 [*DG* 368.10–18]).

Aristotle admits that the movement of lightning is unnatural, but by making the material cause arise from below in the form of the dry exhalation he can more readily account for why the clouds push it back down. For this reason, too, Aristotle can show why clouds are associated with thunder and lightning. This is the second reason that leads him in turn to insist that thunder precedes lightning and that the unignited exhalation in general is prior to the ignited. (A third has to do with the analogy with earthquakes, as we shall see.) Moreover, dry exhalation has less of a tendency to rise than pure fire, and for this reason it is easier to force downwards. Its ignition can then be a secondary explanandum.

In service of the same basic point, Aristotle next attacks Empedocles' theory that lightning is caused by the sun's rays falling into the clouds. Here he must tread carefully, since the rays of his own sun (ἀκτῖνες) also descend to the earth by some unexplained mechanism. He objects instead that,

> the account is too careless. For the cause of thunder and lightning and other such things must always be separate and distinct, and their coming-to-be must also be. But [Empedocles' idea] is very different: for it is as if someone should think that rain and snow and hail first preexisted and then were separated [from the cloud], rather than being generated – as if the condensation made each of these to be ready at hand. (2.9.369b27–34)

This description suggests that Empedocles' cloud is a sack from which all manner of violent phenomena simply emerge at random. Aristotle's objection on this interpretation is that Empedocles fails to give a distinct set of conditions for the generation of each phenomenon and merely claims that it rains because rain comes out of the cloud, that there is lightning because lightning comes out of the cloud. This objection may have some merit, since Empedocles' cause for lightning, the extrusion of light rays from the cooling and condensing clouds, seems adequate to produce many other phenomena. It is equally likely, though, that the randomness of production is not as objectionable to Aristotle as the simple fact that the phenomena preexist in a bag or reservoir, and that it is no explanation simply to pull them out of the bag without an account of their generation. This criticism seems apt, since the sun's rays are stored in the cloud, until it condenses and extrudes them.[26] Indeed, it is hard to imagine how lightning itself could be stored (so again 2.9.370a24–6). At a general level,

[26] Alexander 130.7–10 takes the κεχωρισμένον and ἀποκεκριμένον to refer to Empedocles' incorrect understanding of the phenomena in the clouds, that is, according to Empedocles, the lightning, thunder, etc., will all have to be separated and preexisting in the clouds.

an elementalist like Empedocles will necessarily explain such changes as extrusions rather than as generations. But Aristotle's objection does not seem to be aimed at this general level; it seems to aim, rather, at the specific meteorological impossibility of storing rays in a cloud. Though Aristotle's lightning, like Empedocles', is generated by extrusion, he makes sure to generate the fire from the dry exhalation rather than produce the fire ready-made from the cloud.

The focus of Aristotle's second objection is more obviously physical but likewise focused on the generation of fire (2.9.370a1–5). Parity of reasoning, he argues, leads us to conclude that denser media, such as water, should be able to capture the sun's rays just as well as cloud does, and these rays should likewise be released as lightning when the media cool. But since freezing water is never seen to emit lightning, Aristotle contends that rays cannot be released from clouds either. There is then no storage of rays, and lightning must instead be generated on each occasion. Here the objection is strictly the issue of storage – being versus becoming – and is not about the failure to give distinct conditions for each of the phenomena.

Aristotle's own theory of thunder and lightning, then, though it is largely of a traditional form and follows the Presocratics closely, avoids some obvious difficulties and defends the priority of thunder over lightning essential to the overall structure of the genus. This is what chapter 2.9 was supposed to do, but it leads us to expect that the manner and extent of the exhalation's ejection from the cloud and its non-ignition and ignition are the primary differentiae of the genus. At a phenomenal level, at least, this is accurate. Thus, while the masses that produce thunder and lightning remain locked among the clouds that push them, the more dramatic phenomena of 3.1 achieve either complete ejection (thunderbolt, hurricane) or incomplete ejection (whirlwind and firewind) from the cloud. Cutting completely across this threefold distinction is the binary division of ignition and non-ignition. Inasmuch as the masses are made of dry exhalation, they may either burn (lightning, thunderbolt, and firewind) or not (thunder, hurricane, and whirlwind). The application of this second differentia to the phenomena is, of course, essential to the success of Aristotle's project, since it formalizes the priority of the windy, dry exhalation in its unignited form over the flame that sometimes results from it and clarifies the difference between his own theory based on exhalations and those of his predecessors, which had made the flaming body primary.

But though the use of these differentiae produces an accurate phenomenal account of the genus, a more careful consideration of the other violent phenomena in 3.1 suggests that there are more fundamental causes at work.

Aristotle begins by contrasting the dispersed and delicate puffs of thunder and lightning (σποράδην . . . λεπτομερέστερον 3.1.370b6–7; cf. λεπτῇ καὶ ἀσθενεῖ 2.9.369b5) with the massed and massive extrusion of the hurricane (ἀθρόον καὶ πυκνότερον 3.1.370b7–8). In respect of mass, the other three phenomena of the chapter, the thunderbolt (πολύ 3.1.371a18), whirlwind (συνεχές 3.1.370b30–1), and the firewind (given its affinity to whirlwind), belong with the hurricane. It is the quantity of the exhalation and its massing or dispersal that determine whether or not it remains contained in the cloud: the larger the quantity of exhalation, the less contained the effect will be. Quantity, then, is prior to and the cause of containment and ejection.

Similarly, both thunder and lightning are caused by the same delicate wind, whereas the unignited winds are thick – at least hurricane is explicitly so (πυκνότερον 370b8), and whirlwind, dependent on it (3.1.370b29–31; 371a1–3), is probably so as well. In all likelihood Aristotle intended delicacy to be the cause of flammability. Certainly differences in delicacy of texture account for different forms of thunderbolt: if the wind is very fine (πάνυ λεπτόν 371a19) the thunderbolt is "gleaming," and if it is less fine, it is "smoky." The exhalation that makes up the firewind is also more delicate (λεπτότερον 371a16). The texture of the wind, then, is prior to and the cause of its inflammability. The remarkable result is that the causes all turn out to be affections of the dry exhalation as such, and this confirms Aristotle's contention that ignition is a secondary modification.

The classification of violent phenomena, then, is organized by their own causal nature, but at the same time they give rise to a careful set of correspondences with earthquakes and the kapnospheric phenomena. The violent movement of the mass of dry exhalation within the hardened cloud is analogous to the movement of the dry exhalation within the earth that causes the earthquakes. This analogy in turn brings several other correspondences to light. Thunder in the clouds is an obvious explanandum among stormy phenomena, and its presence there prompts Aristotle to find an analogous explanandum in the earth, namely the bellowing associated with earthquakes (2.8.368a14–17). It is noteworthy that, just as thunder precedes the lightning, the noises associated with earthquakes occur often before the earthquake itself and sometimes even without the earthquake, though, of course, the causes of the precedence of the noise in both cases are different – noise precedes earthquakes, because the sound penetrates the earth before the impact, and the dry exhalation is often too fine to cause more than a noise. Nevertheless, the phenomenal correspondence remains. We do not observe a cloudquake corresponding to the earthquake, though no doubt they occur.[27] There is, however, a direct correlate of lightning and thunderbolts found under the ground. Aristotle mentions (2.8.366b31–367a11) that earthquakes sometimes do not cease until "the moving wind has broken out of the earth like a hurricane." These eruptions often involve fire, and he cites as an example the volcanic islands of Lipari. "We must suppose that this is the cause of the fire that is generated in the earth: when the air is broken up into little bits (εἰς μικρά) and is struck, it catches fire" (2.8.367a9–11). This is clearly the terrestrial analogue of lightning, which is likewise extruded in small portions (κατὰ μικρά 3.1.370b5) and ignited on percussion. The correspondences, then, between the earthquakes and other violent phenomena are extensive and convincing – terrestrial bellowing : thunder :: subterranean fire : lightning :: volcanoes : thunderbolts :: wind outbreaks : hurricanes – and they provide the third reason why Aristotle insisted on the priority of the unignited exhalation: earthquakes are not caused by fire and are accompanied by fire only under exceptional circumstances.

It is possible, too, that Aristotle views violent phenomena, just like earthquakes, as less orderly in their seasons. At least, he describes conditions rather than times when they do *not* occur: the whirlwind does not arise when there is a north wind (3.1.371a3), nor does the hurricane when there is snow (371a4). The reason is that these phenomena are hot and dry exhalations, whereas north winds and snows are cold circumstances. No doubt, the violent phenomena depend on the annual cycle of rain and drought, though we would never know it from Aristotle's description. To that extent, the violent phenomena are not regularized into the cycles of the seasons. At 2.6.365a3–5, by contrast, where normal winds are used to explain hurricanes, it is when the north winds (Aparctias, Thrascias, and Argestes) are blowing, during the autumn and spring, that hurricanes most frequently occur. In this context hurricanes are caused by conflicting winds (τῶν ἄλλων πνεόντων ἐμπίπτωσιν ἕτεροι 365a4), described in a way similar to the whirlwind, though there is no explicit mention of circular motion; rather, this conflict of winds flows directly out of the doctrine of the opposing winds, and it is a generically different etiology from the disorderly form of hurricane in 3.1, which is a forced outburst from a cloud. Hurricanes that arise from the conflict of normal winds will be more orderly in their times and seasons.

Aristotle is also contrasting the violent phenomena with the placid and more distantly related kapnospheric phenomena. Just as shooting stars are

[27] For the explicit analogy, Pliny, *HN* 2.82.192.

described as scattered in small quantities and in many directions (κατὰ μικρά τε καὶ πολλαχῇ διεσπαρμένα 1.4.341b33), so the exhalation of thunder and lightning is extruded in small portions and is scattered and occurs all over (πνεῦμα . . . ἐκκρινόμενον κατὰ μικρὰ μὲν καὶ σποράδην διαχεόμενον καὶ πολλάκις γιγνόμενον 370b4–6). So also the ejected shooting star, explicitly likened to lightning (1.4.342a13–14), forms an intermediary in the *scala naturae* that is intended to emphasize the dignity and worth of the kapnospheric phenomena. Nevertheless, with the exception of the ejected shooting star, the mode of ignition, having a celestial origin, is importantly different. The movement of the kapnospheric fires is only apparent because the flame moves through the combustible material. Among the violent phenomena the dry exhalation is itself moved and moved contrary to nature and is ignited because of that movement.

At an even more general level Aristotle integrates the violent phenomena into the scheme of the whole treatise by making room for a circle, an imitation of a more perfect order and, as is appropriate, one created by forced motion.[28] The whirlwind (τυφών) turns because of the deflection of the wind within the surrounding cloud (3.1.370b17–31). Aristotle compares this to the swirling of wind in alleys and portals, where, he says, the first part of the wind hits a narrow entrance (or a contrary wind, cf. 2.6) and is turned aside. The following parts do likewise until a circle is formed. The restriction of the space where the wind blows is reminiscent of the action of the strait and its intensifying influence on earthquakes. Whirlwinds in the clouds are produced just as they are in alleys but are combined with a downward motion. The text here is not particularly clear:

> On the ground whirlwinds occur for these reasons, and in the clouds the process is similar at the beginning, except that, just as, when there is a hurricane (ἐκνεφίας), a wind separates from the cloud and forms a continuous stream, so here the cloud follows the continuous wind constantly (οὕτως ἐνταῦθα ἀεὶ τὸ <νέφος> συνεχὲς ἀκολουθεῖ [τοῦ νέφους] Thurot's emendation), but because of its thickness (πυκνότητα) the wind cannot separate from the cloud and is turned at first in a circle for the stated reason, and then is forced down because of the continuing process of condensation (πυκνοῦσθαι) in the cloud in that place where the heat departs. (3.1.370b27–371a1)

It is entirely the nature of the material exhalations that accounts for the characteristic motion. The density or thickness of the hurricane wind, perhaps thickened by the wet exhalation prevents it from easily disengaging

[28] Aristotle might have been content to call it a δίνη (as he does 370b22), but he repeatedly calls it a κύκλος (b22, 26, 32).

from the condensing cloud.[29] In this respect the whirlwind is similar to the hurricane wind in being massed and dense (ἀθρόον καὶ πυκνότερον 370b7–8), though perhaps the whirlwind is even thicker. The hurricane (ἐκνεφίας), at least, breaks forth cleanly from the cloud, though, like the whirlwind, we must suppose that it slips out of the lower, uncondensed part of the cloud. The whirlwind itself is caused on account of the ἀντίκρουσιν τῆς δίνης (371a11). Again the expression is obscure, but perhaps the stream of ejected exhalation, as it strikes the surrounding air, is deflected in its descent, all the while pulling the clouds down behind it. The firewind is treated as a simple modification of the whirlwind, by the addition of ignition on account of its delicate texture (371a15–17). These, having no direct celestial causation, are the most degraded forms of circles among the *meteora*.

With the stormy atmospheric phenomena Aristotle brings to an end the second great cycle of meteorological phenomena that had begun with the rivers and the winds. Its structure and symmetries, its correspondences and analogies, are in general less straightforward and more confused than the ones we found in the kapno-atmosphere. Although the pairing of rivers and winds is comparable to the order in the upper region, the role of the sea is more difficult to make out. Earthquakes partly correspond to the sea but partly share common traits with the stormy phenomena. One need only compare the clear correspondences among the blooded genera with the polyvalent associations among the bloodless testacea, crustacea, and insects in Aristotle's biological works to be assured that, far from being an objection to the architectonic interpretation of the treatise, such a tendency is fully consistent with Aristotle's conception of the *scala naturae*, in which the lower members become less distinct both in themselves and from each other.

[29] πυκνότητα and πυκνοῦσθαι in this passage clearly refer to different processes, since the condensation of the cloud (πυκνοῦσθαι) accounts for the ejection of the dry exhalation. There is no reason, then, to suppose as Lee does that πυκνότητα refers to the clouds' thickness. Indeed, in the absence of the definite article it is more likely to attach to the subject, πνεῦμα.

CHAPTER 12

Reflections (3.2–6)

The atmospheric reflections – the haloes and rainbows, rods and mock suns – constitute the last major genus in the treatise, and on first impression they seem wholly out of place. They are not the effects of lightning or hurricanes; nor are they properly situated after the winds, since they require a calm and windless environment (3.3.373a24–5). Rather they are dependent on the air-borne wet exhalation, and their proper etiological place is earlier, immediately after the rain cycle. In fact, if Aristotle had followed the order of topics he used in the kapnosphere, where the aurora, a reflection of a torch, immediately follows and depends upon the shooting stars and their kin, he would have placed the reflections immediately after the rain cycle. His order here undercuts this simple symmetry and is unusual among the small sample of meteorological writings that survive to us. The doxographies preserved in pseudo-Plutarch and Stobaeus, in spite of the confusions in their order, both maintain a distinction between atmospheric (rain, etc.) and terrestrial (e.g., earthquakes, sea) phenomena, and reflections are placed immediately after the rain cycle.[1] Seneca, whatever the broader order of his subjects may actually have been, places reflections among the lights of the heavens, just as the *de Mundo* does.

But if we find fault with Aristotle's placement of the reflections, we must admit that the previous discussion of lightning, thunder, firewinds, and so on, likewise atmospheric phenomena, was also out of place. If

[1] Ps.-Plutarch (*DG*) gives Milky Way, comets, violent phenomena, rain cycle, rainbow, rods, winds, winter and summer, earth, shape of the earth, placement of the earth, inclination of the earth, motion of the earth, differences of the earth, earthquakes, sea, ebbs and tides, halo (*sic*: for which displacement see Diels, *DG*: 60–1). Stobaeus (*DG*) gives Milky Way, comets, violent phenomena, rainbow, rods, haloes, parhelia, rain cycle, winds, earth, shape of the earth, whether the earth moves or not, earthquakes, sea, ebbs and tides, waters. In both cases there is a basic division into three: the upper fires, the atmospheric phenomena, then the terrestrial earthquakes and sea. *De Mundo* places reflections after the stormy phenomena: precipitations, winds, violent winds, lightning, reflections, upper fiery phenomena, volcanic phenomena, earthquakes. It is impossible to tell what Posidonius' order was from the list provided by Diogenes Laertius 7.151–3, since the list is drawn from several sources. But see Theiler 1982: 2.216–17; Kidd 1992: 305–6.

the architecture of the treatise follows a downward tendency, Aristotle should have placed these phenomena after the precipitations and followed earthquakes with minerals and metals. It would not have disturbed the basic order greatly to have reflections and the stormy phenomena coming between precipitations and rivers. But the actual order of earthquakes and other violent phenomena is guaranteed by the introduction (*Mete.* 1.1), where they are placed last and, in spite of their atmospheric location, the stormy phenomena are made to follow naturally upon the earthquakes in virtue of their violence. Though there is no such natural sequence that leads from violent phenomena to reflections, the separation of stormy phenomena from their atmospheric kin suggests that other hierarchies of order are at work.

Aristotle almost certainly postpones the reflections for architectonic and etiological reasons.[2] In composing his treatises, it is his practice to delay a subject that is both highest in value and in some sense beyond the proper limits and methods of the science. In the *Nicomachean Ethics*, the paean to the contemplative life is saved for the last book; in the *Physics* it is the first cause of motion. Now, the reflective phenomena cannot be accorded so elevated a status, but there are reasons to consider them special. In the *Meteorologica* in general, and particularly in the space between the kapnosphere and the earth, circular phenomena form a leitmotif, as we have seen, but up until this point Aristotle has treated them merely as physical facts. He continues this theme with the reflective phenomena but treats them for the first time as geometrical facts and chooses only those geometrical theorems that concern their circular nature. We learn why haloes appear as full circles around the heavenly bodies, and why the rainbows never appear as more than a semicircle. At the same time, as I shall explain in further detail below, the geometrical treatment here introduces new, more exact, principles and a less physical method, which mark reflections out as special.[3] This new method also helps Aristotle to show that he was *au courant* with the latest developments in optics, for which deficiency he had criticized Cleidemus and others.[4]

In other respects, too, the reflections take us beyond the standard meteorological principles. These phenomena reintroduce in a novel way the celestial bodies, which have been retreating in importance since the rain cycle. The sun, moon, and stars return, no longer as moving causes, but

[2] Strohm 1984: 206 comments on its surprising inclusion in the science.

[3] *Met.* 13.3.1078a36–b2: "the greatest form of the beautiful are order, symmetry and the determinate, which the mathematical sciences chiefly display."

[4] Cleidemus: 2.9.370a10–19; others, perhaps Hippocrates of Chios: 1.8.345b10–22.

distantly, as images in a mirror. By contrast, even the aurora, likewise formed by reflection, was the image of a torch, itself a kapnospheric phenomenon, rather than a reflection of the sun. In that case, both the reflective surface and the object of vision were products of the dry exhalation. But here, the celestial bodies, *bona fide* principles when they serve as efficient causes, seem only incidentally meteorological inasmuch as they are merely objects of reflection, a point further emphasized by Aristotle's observation that rainbows can form around lamps (3.4.374a20–9). In a manner directed against Plato, these facts allow him to conceive of reflective phenomena as πάθη of a πάθος, the shape and size of reflections from a state of vapor, so that one of the most ontologically tenuous objects in the treatise is responsive to the most precise and abstract methods.[5] Finally, Aristotle's major concern in these chapters, apart from the geometrical properties, is color perceived through the human visual ray, properly a psychological, not merely physical, subject, and he explicitly directs us to his discussion of sensation for further elaboration of the principles involved.[6] Reflections, then, are furthest removed from meteorological principles and are the most accurate and beautiful of the science. They merit a place apart.

And yet problems are sufficiently numerous to generate suspicions about the section as a whole. Reflections are nowhere mentioned in the "table of contents" in *Meteorologica* 1.1, not even in their physical aspects, and this has been seen as grounds to doubt the authenticity of the whole section.[7] The significance of this fact, however, can be easily exaggerated. It is clear that the introductory chapter is just as much a description of the topics of traditional meteorology as it is an enumeration of the parts of Aristotle's subject. Moreover, rivers merit no mention in the list, though they are a major component of Aristotle's sublunary system; and it is by no means clear what "the parts and the forms of the earth" (γῆς ὅσα μέρη καὶ εἴδη 338b25) correspond to in Aristotle's account. So 1.1 is only in the roughest sense a table of contents. Again, chapter 1.7 (344b17–18) promises an explanation of haloes, and though the promise is in the character of a footnote and not deeply embedded in the text, there is no reason to doubt its authenticity. If it is genuine, it clearly points forward to the discussion

5 Cf. *Pr.* 12.3.906b5–6: "... especially if the rainbow does not have a nature, but is rather the affection of a reflected visual ray."

6 "Let these things be studied from the properties of perception. For the account of them is peculiar to perception" (3.4.374b15–17).

7 A consideration raised by Knorr 1986: 108: "If these considerations do not quite rule out the authenticity of *Meteorologica* Book III, they surely deserve more serious attention than Aristotle scholars tend to afford them. At the least, we recognize that the geometric locus problem in the passage on the rainbow is an importation from geometric practice, not Aristotle's own discovery."

of Book 3. For these reasons, the arguments for doubting the whole section on reflections are weak.

Admittedly, the physical explanations Aristotle offers for reflection and color formation display a bewildering variety, even incoherence, but there is nothing unprecedented in the rest of the treatise. The physical causes adduced here are no more contradictory than those we found among the kapnospheric phenomena or in the origin of rivers, though the problems of Aristotle's geometry merit a separate discussion, which will come to a different conclusion. First, however, it is necessary to take account of the physical causes.

The physical causes of reflection

Aristotle explains the physical properties of reflections before he explains the geometrical, and his focus here is on the details, the differences among the reflections and the causes that give rise to them. He begins (3.2) by supplying a list of properties that need explaining (colors, shapes, times, places), just as he had done for hail.[8] But, unlike in that previous discussion, he is not intending to attack a predecessor's theory. The basic explanation is not in dispute, since everyone agrees that sun and vapor are necessary elements of reflection. Although Anaximenes in the sixth century was still innocent of the nature of reflection – for him the rainbow is caused when the rays of the sun are unable to penetrate the thick dark cloud – by the mid fourth century the basic causes had at least been mooted. Anaxagoras had identified the phenomenon as a reflection of the sun; Metrodorus, though he did not treat it as a reflection, was the first to give an account of its colors.[9] What Aristotle found among his predecessors were not so much errors and confusions as unsophisticated theories that did not take adequate account of the differences among the various kinds of reflections. Aristotle's method is to catalogue these differences and then provide explanations adequate to account for them.

Before turning to Aristotle's explanations for these differences, a brief comment is warranted concerning the history of the visual ray,

8 As Olympiodorus notes: 217.23–7; 223.23–5.

9 Gilbert 1907: 605–6. Anaximenes: DK 13A7.8 (1.92.26–7); 13A18 (1.94.11–18). Xenophanes: DK 21B32 (1.136.14–15): ἥν τ᾽ Ἶριν καλέουσι νέφος καὶ τοῦτο πέφυκε, πορφύρεον καὶ φοινίκεον καὶ χλωρὸν ἰδέσθαι. Anaxagoras says that the rainbow is the reflection of the solar radiance from a dense cloud and that it stands constantly opposite the celestial body that it mirrors (DK 59A86 [2.26.3–6] = Aët. 3.5.11 [*DG* 373.28–374.5]). For possible contributions of Philip of Opus, see Boyer 1956. According to Metrodorus, when the sun shines through the clouds, the cloud appears dark and the beam of light appears red (DK 70A17 [2.232.29–35]).

which Aristotle takes entirely for granted. According to Presocratic theories of vision, the eye actively emits a visual ray toward the object viewed; similarly, there is a theory of light propagation, according to which light proceeds from a bright source and illuminates other objects. Rays were a convenient way of explaining both phenomena and were thence extended to reflection.[10] At least as early as (or as late as) the Pythagoreans, the *visual* ray was thought to emanate from the eye, reflect off a smooth surface and proceed to the object of sight.[11] Empedocles held that the *light* ray becomes weaker when it is reflected, and for this reason the sun's light reflecting off the moon loses its heat and brilliance.[12] Anaxagoras likewise held that the rays of the sun reflect off the earth but peter out in the upper air, accounting for the cold at high altitudes.[13] The weakening of the reflected *light* ray seems to have been discovered first, and it is not until Hippocrates of Chios and his theory of comets, that we have evidence for the failure of the *visual* ray to reach its object after reflection.[14] Aristotle, in turn, explicitly mentions the weakness of the visual ray (3.2.372b8; 3.4.374a23; 3.6.377a34–b2): the visual ray when strained to a distance becomes weaker, and when it is weakened, whether simply because of distance or because of reflection, colors are produced.[15] This theory is troubling, because elsewhere Aristotle explicitly rejects it. At *Sense and Sensibilia* 2.438a25–32 Aristotle objects that the visual ray's coalescence with the object does not explain sight (for

10 Aristotle notes that Empedocles uses both ray theories (*Sens.* 2.438a4–5), viz., the ray coming forth from the eye and the ray emanating from the object of sight. Plato (*Ti.* 45c, 46b) provides the first evidence for the term ὄψις in the sense of visual ray (Mugler 1964: 293–5 [ὄψις 4]).

11 There is little evidence for theories of reflection in the fifth century. The Pythagoreans say that mirroring occurs through the reflecting back of the visual ray: "For the visual ray is stretched out and moves as against a bronze object, and meeting and striking against a dense and smooth [surface] it turns back upon itself doing something like the stretching out of the hand and the turning back to the shoulder" (Aët. 4.14.3 [*DG* 405.15–22]).

12 Plutarch, *de facie* 929e = DK 31A42 (1.329.16–330.13): although the citation is not a direct quotation of Empedocles, a direct quotation is embedded in the passage. The passage is clearly doxographical: the weakening of the ray is the only way in which Empedocles' position differs from that of others.

13 On the authority of Alexander 49.13–15. Democritus, using an effluence theory (ἀπόρροια), held that if there were pure vacuum, and not air, around us, the emanations or images sent from the visible objects would reach the eye unblurred (*de An.* 2.7.419a15–17 = DK 68A122 [2.112.22–3]). *Sens.* 2.438a5–9 mentions Democritus' theory that vision is a reflection (ἔμφασις).

14 Aristotle is our only source for Hippocrates' theory, but there is no reason to doubt, in spite of the similarity between Hippocrates' theory and his own, that Aristotle is accurately reporting his source. Four times he uses the expression, "the *human* visual ray" or "*our* visual ray" (ἡ ὄψις τῶν ἀνθρώπων or ἡ ἡμετέρα ὄψις), both when discussing Hippocrates' theory and when describing the probably Hippocratean theory of the Milky Way (343a3, 13, 19–20; 345b11). On the other thirty-seven occasions ὄψις is used in the treatise, the bare term ὄψις is used without any reference to humans. The reference to humans is clearly peculiar to Hippocrates.

15 On the weakness of the visual ray in Aristotle, see Beare 1906: 66–7, who notes that the phenomenon is not mentioned by any of his predecessors.

how can something that happens outside the eye *be* perception? How can light coalescing with light *be* perception?).[16] Again at *de Anima* 2.7 we find the very different theory according to which color is the actuality of the potentially transparent medium.[17] Important, too, is the fact that Aristotle does not invoke the visual ray in his explanation of the aurora, where instead light (φῶς) is reflected through the smoky exhalation (1.5.342b6). All these facts present serious challenges to the credibility of the visual ray in the chapters on reflection and have even led some to dismiss the section as spurious.[18] However, the visual ray theory is used widely in the discussion of the physical aspects of reflection (in contrast to the geometrical), even in the introductory chapter (3.2.372b8), which must be genuine, if any passage is. Moreover, the theory is used elsewhere in the *de Caelo* without any caveats or qualifications.[19] But decisive is Aristotle's own explicit statement that he is accepting this new theory, because it is useful: "we must believe (δεῖ λαμβάνειν τὴν πίστην) from what has been proved about the visual ray that the visual ray is reflected from air just as from water and from everything else that has a smooth surface" (3.2.372a29–32). The theory, then, is a convenient way to explain how these colors are caused by the very act of reflection and are not images of something colored in itself. The idea may have originated in the observations that light is weaker after being reflected and that ricocheting rocks have a weaker impulse.[20] As with the modern wave and particle theories of light, Aristotle found that both theories could be useful in different contexts.

In illustration of the weakness of the visual ray, Aristotle adduces the curious example of a man whose weak eyesight causes him to see his own image in front of himself (3.4.373b2–10). The man's visual ray emanates from his eye but is too weak to pierce the air and is reflected back to him. In view of the fact that Aristotle is introducing rainbows here, we should

[16] Aristotle's account of color in the *Sense and Sensibilia* aims at explaining why we see the colored surfaces of things but makes no attempt to adapt this theory to the facts of reflection.

[17] This color affects the eye, which is potentially the color of the visible object. Nothing comes forth from the eye. Again, at *de An.* 3.12.435a5–10, Aristotle rejects the visual ray theory saying, "concerning reflection, rather than supposing that the visual ray goes forth [from the eye] and is reflected, it is better to suppose that the air, as long as it forms a unity, is affected by the shape and color. Now on a smooth surface the air acts as a unity. For this reason it in turn moves the faculty of sight, just as if the impression in the wax penetrated right to the other side."

[18] So Olympiodorus argues against those who deny the authenticity of the *Meteorologica* on this basis, 4.27–5.10. See the excellent discussion in Merker 2002: 195–201.

[19] *Cael.* 2.8.290a17–18: "the visual ray when stretched to a great distance wavers because of its weakness." See Elders 1965: 221. On the differing power of sight among animals, see *PA* 2.13.657b24.

[20] Cf. *Pr.* 16.13. Aristotle does not consider such subtleties as the fact that the ray is weaker after a head-on reflection than after a glancing blow. The angle of incidence has no explicit influence on the strength of the reflected ray. The ray simply cannot travel very far after it has been reflected.

rather expect to hear that this man sees the world as if through rose-colored glasses, his visual ray becoming red through weakness. Instead he simply sees an image of himself. Nevertheless, the immediate point of the example is to show that the visual ray can be reflected by the air or water back in the direction from which it came, and that reflection is related to weakening. This charming illustration, however, suggests that Aristotle is trying to have it both ways: the invalid's visual ray is weak and is therefore reflected easily; but the act of reflection itself weakens the visual ray.

Be that as it may, the general theory of the visual ray is not the focus of his attention. It is instead the differences among the reflective phenomena that dominate his discussion. Most troubling is the apparent isolation of one property from another and the series of seemingly *ad hoc* explanations Aristotle provides for them. The explanation for why rainbows are colored rather than white (3.4.373b34–374a10), for example, seems surprisingly independent from the explanation for why they have the particular colors they do (red, green, and blue, 3.4.374b30–3), and this in turn seems to be a separate issue from why the secondary rainbow is less bright and has its colors in a reverse order (3.4.375a30–b15). One wonders why one common cause applied in various ways does not explain all three.[21] Again, when explaining why mock suns and rods appear beside the sun and never above or below it, Aristotle employs a series of strange and casual ancillary causes.

Although these criticisms are justified to a certain degree, they fail to take into account the deeper unity and integrity of Aristotle's method. Just as he explained the manifold appearances among the kapnospheric fires and among the stormy phenomena through a small number of causal differentiae, so here he first describes the colors, shapes, positions, times, and other properties of the reflective phenomena and then subjects them to explanation through two closely related major differentiae and three ancillary factors.[22] Throughout his discussion, it is the differences and natural divisions among the reflections that govern his explanations.

Ultimately the visual ray is responsible for the appearance of all forms of reflections, but (I) the visual ray is affected primarily by the reflecting or veiling material through which it passes. (II) Ancillary conditions also explain the location of the reflecting surface and therefore explain the position of the reflection. (III) Second in importance only to the reflecting

[21] So Merker 2002: 194 notes, Aristotle nowhere indicates how he will combine or reconcile these principles.

[22] Aristotle mentions three (3.2.372a19–21): "they differ in the manner of the reflection, and in surface, and how the reflection makes its way to the sun or other bright object."

or veiling material are the differences in the distance the visual ray has to travel after reflection from, or penetration of, the vapor; (IV) of less importance are the goals of the visual ray, whether it is to the sun, moon, or some other bright object; and (V) the juxtaposition of one color next to another, which may change the appearance of both colors. Let us examine each of these in turn.

I. Vastly the most important factor is the surface from which the visual ray is reflected, and the surfaces are best classified by the effects they have on the visual ray. (a) Generally, the very fact of reflection, regardless of surface, can cause a darkening or coloring by the weakening of the visual ray (3.2.372b6–9). The same effect is produced when the visual ray penetrates through a vapor, and Aristotle says both causes have equivalent effects (3.4. 374a4–8). (b) But some surfaces darken or color the visual ray more. Thus the larger drops of water associated with the formation of rainbows cause the light to be colored (3.4.373b13–17; 374a1–3, 9–10).[23] The larger drops also account for the darkened appearance of haloes when their mist is turning to rain (3.3.372b22–5). It likewise accounts for one of the kinds of rods (3.4.374a16–20).[24] (c) Double reflection naturally weakens the visual ray more than a single reflection. When a white cloud, through which a visual ray has penetrated or reflected, is viewed in a body of water, it takes on rainbow colors. This effect provides a second account for the formation of rods: the rods may appear when a bright, white cloud near the sun is reflected in a patch of moisture (3.6.377a31–b4; cf. 3.4.374b24–8). (d) In another cause for rods, obscurely related to the previous, rainbow colors may appear, when the visual ray is reflected from a surface that is unevenly dense or unevenly moist (3.6.377b4–13).[25] The unevenness of the surface apparently disperses and weakens the visual rays.[26]

Conversely, a surface may mitigate the darkening and coloring effects of reflection. (a) Thus a reflection from or through air or delicate mist

[23] The dark colors of the rainbow are produced by the large drops of water (ψακάδες or ῥανίδες), which are in the process of forming rain and serve as better mirrors to the visual ray (3.4.373b15–17). Unlike the mist, they are genuinely water (albeit small particles of water), and since they are water, they are dark and difficult for the visual ray to penetrate. The fact that the darkness of a lamp's soot causes a *purple* color to appear implies that the *red* of the rainbow is the result of the less dark raindrop. There are intimations of this theory at *GA* 5.1.779b28–33 and 780b8–12 where water gets darker the deeper it is (see Sorabji 1972: 293). This is also a theory mooted by Seneca, *QN* 1.3.3.

[24] Aristotle does not expressly distinguish subspecies of rods, but he describes three distinct causes.

[25] Stothers 2009: 36 suggests that the rod is a colored mock sun; so Olympiodorus 210.5–6.

[26] Ideler 1834-6: II.276 sees the same principle operative in the haloes, which would produce colors if their vapors were not even.

produces the white appearance of the halo (3.3.372b15–17; 3.4.374a1–3).[27] (b) If this mist is dense but does not yet have large drops and is very even and smooth, it can reflect a group of visual rays traveling *en masse* (ἀθρόας 3.6.377b18–23), which may then fortify one another and avoid being weakened.[28] This cause accounts for the bright appearance of the mock sun (3.6.377b22–3). (c) Similarly, when the visual rays are reflected from a large and extensive surface such as the outer band of the first rainbow, the effect can be the bright color, red, rather than the purple that might be expected (3.4.375a1–4).[29]

II. Occasionally Aristotle draws on the physics of vapor to explain the location of certain kinds of reflecting surfaces, and consequently of certain kinds of reflective phenomena. (a) Thus the halo cannot fully become a rainbow, because the sun's heat does not permit a water drop of a size sufficient to produce color to persist long in the halo (3.4.374a11–15). (b) The halo, too, appears close to the earth, because the wind is calm there and allows the vapor to remain in place (3.3.373a23–5). (c) The mock sun and rod cannot appear too close to the sun, because the sun would dispel the mist (3.6.377b30–2). (d) Likewise the mock sun and rods cannot appear under the sun near the earth, since the heat of the sun would dispel the vapor there (3.6.378a6–8).

III. Of great importance in the classification of reflective phenomena is the distance that the visual ray travels after reflection. With two significant exceptions, the distance of the reflected visual ray explains the location of phenomena, and like the physical condition of the vapor just discussed in section (I), distance acts mainly to indicate impossible locations. (a) So Aristotle admits that mists occur opposite the sun, but he argues that haloes cannot appear there, as rainbows do, because the visual ray cannot effectively reflect from small mirrors to reach the sun (3.4.374a1–3). (b) Similarly the mock sun and the rod cannot appear very far away from the sun because the visual ray cannot reflect so far (3.6.377b32–4). (c) Again,

[27] The whiteness of the halo is primarily caused by the ease with which the visual ray penetrates (3.3.374b15–17) the smooth (ὁμαλής), small particles of vapor (ἀήρ and ἀτμίς), which are condensing (συνισταμένου 372b16–17) but have not yet condensed and achieved their nature in rain (τὴν φύσιν ἀπολαμβάνειν τὴν αὑτῆς b21–2). The vapor must be fine, for otherwise the light would be colored or, where the visual ray is reflected to a star or other less bright celestial body, entirely invisible.

[28] For the concept cf. *Pr.* 15.6.911b20–5.

[29] As Webster notes at 375a3: "There is no sense in this, but the nonsense is probably Aristotelian." It begs the question, of course, why the red stripe is bigger, the green smaller, for the fact of the circumference is not a sufficient reason.

the mock sun and the rod do not appear beside the midday sun, because the visual ray simply becomes dispersed there (3.6.378a8–11). This may be the opposite of the "*en masse*" principle that operates near the earth to keep the visual rays concentrated and the mock sun bright.[30]

In two closely related and exceptional cases, the length of the reflected visual ray plays a role in the formation of color, more commonly associated with the reflecting surface. (d) Since the visual ray has to travel further to the sun from the second, outer rainbow than from the first, inner rainbow, the second rainbow is duller (3.4.375a32–b3). (e) Analogously each of the color bands within the second rainbow – red, green, and blue in order – is progressively further from the sun and consequently appears darker (3.4.375b3–9). In addition, we find in the geometrical proof in 3.5 (and spurious for that reason, as I shall argue) the implied principle that variations in the ratio between the length of the ray of incidence and ray of reflection cause various colors to appear.[31] This principle is absent from the more physical contexts.

IV. The goal of the visual ray plays an important, but straightforward, role in distinguishing the phenomena. (a) So haloes can appear around the sun, but they appear more frequently around the moon and bright bodies (3.3.373a27–32). Sun haloes, however, do not differ in any significant respect from moon haloes. (b) The rainbow can appear opposite the sun and, occasionally, the moon (3.2.372a21–9). Moon rainbows are usually white (3.4.375a17–20). (c) Mock suns and rods are exclusively associated with the sun.

V. Finally there is the principle of contrast: when one color is seen against the background of a darker color, it appears brighter.[32] It was used to explain chasm auroras (1.5) and is employed again here (a) to explain why the red color in rainbows, when seen against or next to green, appears yellow. Aristotle uses the principle to defend his claim that the weakening of the visual ray can produce only three colors, despite the fact that more colors are visible (3.4.374b33–375a28). (b) In addition, moon rainbows are usually white, because of the contrast with the night sky (3.4.375a17–20).

[30] Also very obscurely, rods and mock suns cannot appear obliquely to the midday sun, because the visual ray does not go under the earth (ὑπὸ τὴν γῆν Fobes; πρὸς τῇ γῇ Alexander paraphrase [3.6.378a9–11]).

[31] If Aristotle was aware of the law of reflection, according to which the angle of incidence equals the angle of reflection (cf. *Pr.* 16.13), he makes no use of it here. See Boyer 1959: 40–5.

[32] First attributed to Metrodorus: DK 70A17 (2.232.29–35).

In this way the bewildering variety of reflections can be accounted for through five causal factors. It is clear from this survey that the nature of the reflective surface is the most important of these factors and we can now see why. Aristotle chose it over the length of the reflected visual ray, because only in this way could he explain the distribution of white and color among the phenomena. The fact that the white halo is closer to the sun than the colored rainbow would initially have tempted Aristotle to account for their color difference chiefly by distance, but since the colored rod is found in the same place relative to the sun as the white halo and the white mock sun, clearly the length of the reflected visual ray cannot be the primary cause of color. Accordingly, Aristotle chose the reflective surface and argued that, while mist and raindrops could be found in either location, raindrops were rarer in the immediate proximity of the sun, and that mist, when located opposite the sun, was inadequate to reflect the visual ray. The reflective surface, then, is the main determinant of color, and the role of distance in the production of color is restricted to the very narrow problem of the second rainbow. As with the kapnospheric phenomena, it is the differences among the phenomena that are driving the search for causes, and by examining those differences we can detect a more orderly pattern of explanation.

The geometry of haloes and rainbows

In two chapters devoted to the shape of the halo and the rainbow (3.3 and 3.5), we find a new method quite alien from the rest of the treatise. Here alone in the *Meteorologica*, indeed alone in the Aristotelian corpus as it survives to us, we are offered extended proofs with an apparatus of geometrical construction. Our major concern here is with the authenticity of the text. Our conclusion will confirm an expurgated text unified in geometrical interest and style.

The first construction (3.3) aims to prove that the halo forms a circle around the sun or other luminous body, and the second (3.5) shows that the rainbow may never appear as more than a semicircle. The proofs are peculiar in two ways. First, they are both more complicated than is necessary for their purpose and could largely have been substituted to greater effect with non-technical arguments. Secondly, the proofs are mediocre or very poor in their form, their explanatory content, and their consistency with other claims about the phenomena.

In view of these shortcomings, it is no wonder that these chapters have been suspected of being interpolations. Knorr entertained the claim that the whole section on reflection was not Aristotelian but authored perhaps by someone in his school.[33] This is clearly going too far, as I have already argued.

It is also difficult simply to eliminate the geometrical chapters from the section on reflections as a whole. In the list of the various properties associated with reflections found in 3.2, the geometrical properties are deeply embedded and prominent, and as a result we expect to learn why the halo is a circle (371b22–3), and why the rainbow is never more than a semicircle (371b26–7). And if geometrical properties have a place among the core *explananda*, we must accept the basic technical geometrical explanations of 3.3 and 5, especially since Aristotle does not offer any non-technical alternatives.

Moreover, Aristotle grants geometry a place in physics and makes clear elsewhere that the physicist is required from time to time and under limited circumstances to explain the mathematical properties of physical objects. The role of mathematics in Aristotelian physics has generated a vast literature, which it would not be profitable at this point to consider in any detail.[34] Some brief comments, however, are needed in order to shed some light on Aristotle's proofs. According to the *Physics*, it is the task of the physicist to study certain geometrical properties (τῶν συμβεβηκότων καθ' αὑτά) of the heavenly bodies, such as the sphericity of the earth (1.2.193b25–30), and Aristotle echoes this language at the beginning of our section (3.2.371b21–2) where he lists geometrical shape among the properties of reflections (τὰ πάθη καὶ τὰ συμβαίνοντα). In fact, in the *Posterior Analytics* Aristotle cites our specific case:

> as optics is related to geometry, so another science is related to optics, the science concerning rainbows. For it is the job of the physicist to know the fact and the student of optics, either of optics itself or a specific branch, to know the cause.[35] (*APo.* 1.13.79a10–13)

The more physical science is subordinate to or under the more mathematical science as, in this passage, optics is subordinate to geometry, and the study of rainbows (iridology) in turn is subordinate to optics. The reason

33 Knorr 1986: 108.

34 For such matters see McKirahan 1978 and 1992; Lennox 1986; Wilson 2000; Hankinson 2005; Cantù 2010.

35 Barnes' ROT translation, "for it is for the natural scientist to know the fact, and for the student of optics – either simpliciter or mathematical – to know the reason why" ("either of optics simpliciter or of mathematical optics" in Barnes 1994) blurs the distinction that Aristotle is trying to make. According to Proclus (*apud* Philoponus *in APo.* 181.19) the study of rainbows belongs to catoptrics. Philoponus, *in APo.* 182.3–4 adds that the κατοπτρικός is not the ὀπτικὸς ἁπλῶς but κατὰ τὸ τῆς ἴριδος μάθημα. This is certainly correct.

for the subordination is that optics uses essentially geometrical principles to prove facts about visual rays, and iridology uses optical or catoptrical principles. To this extent the geometrical treatment of rainbows and haloes goes beyond the merely physical and makes this part of the treatise special. At the same time, geometrical methods are legitimate and appropriate, because the physical entity has the geometrical property *in its own right*. In the case of the rainbow, its semicircular form belongs to it in its own right, because its shape is one of the differentiae that distinguish the halo and the rainbow from other reflections, and a differentia must belong to a thing in its own right. The iridologist must, therefore, study the halo's and the rainbow's geometrical properties.

The notion of "belonging in its own right" is important in understanding how the subordinate science influences and limits the application of the superordinate science and in understanding why the geometrical proofs here take the form they do. It has sometimes been claimed that the higher, mathematical science provides the entirety of the cause, which is then merely applied to the physical situation.[36] The first geometrical proof in the *Meteorologica*, which attempts to show that the halo forms a circle, however, makes clear how much causal content is provided by the subordinate science and by the physical facts of the halo. There are after all many ways of constructing a circle consistent with the rules of geometry, but there is only one that is appropriate to this physical situation, where the visual ray is bent in a particular way. The physical substrate, that is, dictates how the geometrical cause is to be deployed, and this is just what makes the halo proof so problematic. As geometry, the proof is unimpeachable, but it appears to reverse the order of physical cause and effect, making the circle the cause of the reflection rather than the other way around. Geometry can only provide a cause if its order of explanation is parallel to that of the physical. I suggest, however, that this reversal of order is only apparent, and the result is a proof that, though hardly brilliant, nevertheless shows how important the physical situation is for the geometry that accounts for it.

The proof that the halo forms a circle around the sun requires that the lengths of rays emanating from the eye and striking the cloud be all of the same length and that likewise the rays reflecting from the cloud and striking the sun be all of the same length. These assumptions are sufficient to prove the circularity of the halo. But the proof seems to be little more than a *petitio*, since the assumption of the equality of length of the rays seems to rest on nothing other than the circular shape of the halo itself,

[36] McKirahan 1978: 201 and Lennox 1986: 48. See also Wilson 2000: 35.

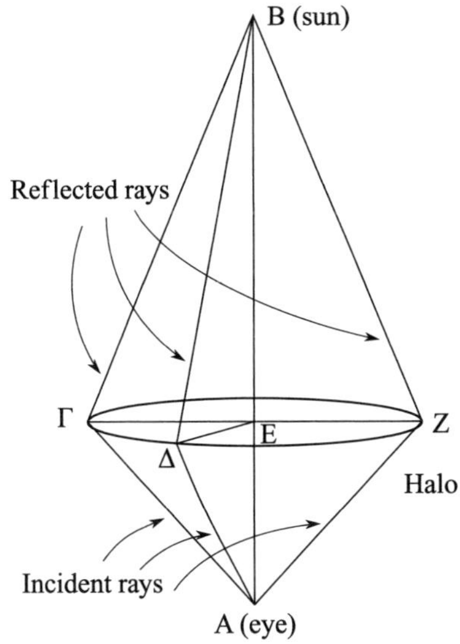

Figure 12.1 The halo

which is just what needs to be proved.[37] What we really want to know is why the incident rays and likewise the reflected rays should be of the same length in the first place (see Figure 12.1).

A close examination of the introductory statement shows that Aristotle is more careful in providing a physical basis for his assumptions. First, the reflection occurs from a mist gathered (συνισταμένης ἀχλύος 3.3.373a1) around the sun and, as the participle suggests, it is gathered in a determinate location. Aristotle is, of course, aware that mist is gathered far from the sun and much closer to the observer, and, given the great distance to the sun, the cloud can be thought of as localized at a uniform distance between sun and observer. Even if part of the cloud is nearer to the sun or nearer to the observer, the difference will be negligible in comparison with the distance between the observer and the sun. The further claim that "the ray is reflected equally on all sides (πάντοθεν 373a2)" is unclear, but almost certainly it refers to the assumption that the rays are all reflected to an equal degree: what Merker has called the principle of "homogeneity of

[37] Lee 1952: 249: "Here Aristotle in effect assumes what he is setting out to prove." This is cited approvingly by Strohm 1984: 207. Johnson 2009: 347–8 has recently defended the thesis that Aristotle's proof is strictly within the geometrical moment of the investigation, which abstracts from physical interpretations, and that Aristotle's middle term is no more than a definition of a geometrical circle.

reflection."[38] This principle is common to every kind of reflection treated in these chapters – it is essential to produce the arc of the rainbow – and is sufficient together with the assumption that the cloud is at a uniform distance between the sun and the observer to guarantee the geometrical result consistent with the physical cause as given. If all the rays are bent to an equal degree and at an equal distance between sun and eye, the incident rays must all be equal in length, as must be the reflected rays.[39] Aristotle's geometrical proof then proceeds straightforwardly from this point (373a6–19). On the basis of these physical assumptions there is no need simply to grant the principle that the rays are equal.

Nevertheless, even after the charge of *petitio* has been addressed, commentators have found the explanation unsatisfying, because they expect an explanation for the principle of homogeneity, and especially why visual rays are bent at three different and discrete angles (one for the halo, another for the first rainbow, and yet another for the second rainbow). Aristotle provides no hint of such an explanation, probably because he did not have one handy. For reasons considered below, however, it is unlikely that the peculiar angle is caused by the *ratio* between the lengths of the incident and reflected rays, as is suggested by the locus proposition in the first rainbow proof. Neither the shape of the cloud nor the ratio between incident and reflected rays is mentioned in the halo proof, and neither angle of reflection nor ratio of the rays has any role to play outside the geometrical proofs. Though the principle of homogeneity receives no deep physical explanation, the physical assumptions of the halo are precisely those that are modeled in the geometrical proof. The result is an unremarkable explanation, one in which geometry does very little explanatory work by modern standards.[40] However, it shows what it intends to show, that rays bent at the same point and at the same angle and arranged so that their respective ends touch one another can only be made to form a circle or circle segment. Aristotle clearly intended physics to drive the construction of the geometrical proof.

Before venturing any general observations, we should first turn to the rainbow proofs in chapter 3.5, where the most urgent question concerns their authenticity. Paul Tannery, in work that was repeated independently by Thomas Heath a couple of decades later, showed that at the core of the passage is a proof remarkably similar to one attributed to Apollonius

38 Merker 2002: 229.

39 As Ideler 1834-6: II.280 points out, the ray does not have to be bent in the middle, contrary to what the traditional diagrams suggest.

40 For a similarly banal proof, see *Pr.* 15.10.

of Perga by Eutocius, a locus proposition, that is, the determination of a set of points that satisfy certain conditions.[41] The proof in our text is rather sophisticated, certainly vastly more sophisticated than any other piece of Aristotelian mathematics, and though there has been a hope among scholars that this passage might provide a rare window into pre-Euclidean geometry, it has naturally been viewed with some skepticism.

Tannery along with Vitrac, who recently rediscovered Tannery's role in the interpretation of the passage, have offered several philological reasons for considering the locus proposition spurious.[42] These reasons are rather technical and are better described in the appendix below (pp. 258–70). It is sufficient for us to note here that philology never supplied the fundamental reason for suspecting the passage; it was the obscurity and redundancy of the proof and a mathematical sophistication untypical of Aristotle. With this in mind, Vitrac and his collaborator, Merker, identified natural junctures where an interpolation was likely, but like all the major modern scholars, they focused on the mathematical aspects of the proof. What has been largely lacking in the scholarship is a convincing *physical* justification for suspecting an interpolation, though one is close at hand.

But before we consider that, it will be useful to provide a brief overview of the argument. It is not my intention to discuss the locus proof in any detail, since that has been ably done elsewhere, especially by Vitrac, and would unnecessarily detain us. For the sake of clarity, I shall proceed as if Vitrac has correctly identified the interpolation as such, and this will allow us to concentrate our attention on evidence that confirms his identification. We will find a proof that proves what Aristotle sets out to prove, that the rainbow forms no more than a semicircle and does so in a way very similar to the halo proof.

I provide a translation of the proof from which the interpolations, as marked by Vitrac, have been eliminated (3.5.375b30–376b22; see Figure 12.2 below):

> Let [the sun] be placed first at the rising point, H, and let the line KM be reflected to H, and let the plane be produced from the triangle HKM into the circle A (ἐν ᾧ ἡ A).[43] Thus the section of the sphere will be a great circle. Let

[41] Tannery 1886; Heath 1925: II.197–200 and 1949: 181–90. [42] Vitrac 2002.

[43] The phrase, ἐν ᾧ ἡ A, is troublesome for several reasons (see Merker 2002: 210n84). It is entirely omitted in manuscripts H and N, and there is otherwise significant variation in the reading. ἡ A, conceived gender appropriately, could only refer to a sphere, which simply makes no sense. Moreover, the whole sentence is difficult to construe grammatically. Instead we might expect καὶ τὸ ἐπίπεδον ἐκβεβλήσθω τὸ ἐν ᾧ ἡ A ἀπὸ τοῦ τριγώνου ἐν ᾧ τὸ HKM ("Let the plane A be produced from the triangle HKM"). This requires only the transposition of the τὸ, for which, however, there is no authority.

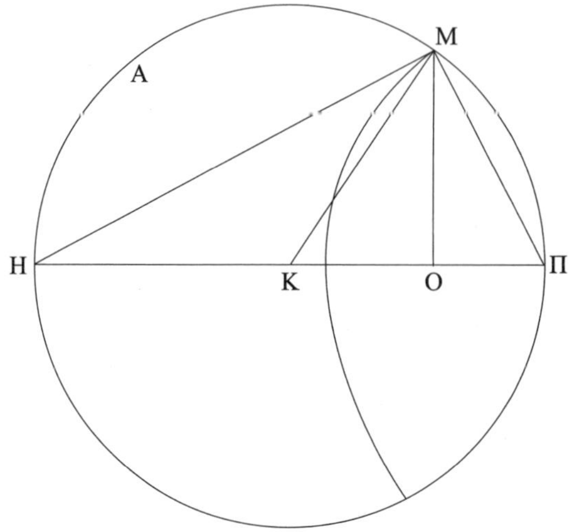

Figure 12.2 Reconstructed diagram (after Merker)

this be A. For it will make no difference whichever of the planes KMH on HK be produced [omit 376a1–b7 αἱ οὖν . . . πρὸς MK]. (1) So if, using Π as a pole and a distance MΠ, a circle is drawn, it will touch on all the angles, which the lines from the circle MA, when reflected, make [omit 376b10–12 εἰ δὲ μὴ . . . ἀδύνατον]. (2) So if you rotate the semicircle A around the diameter HKΠ, the lines reflected from HK to M will be similar in all planes, (3) and they will have the same angle as KMH. (4) And the angle that HΠ and MΠ make on HΠ will always be the same. (5) So the triangles on HΠ and KΠ are equal to HMΠ and KMΠ. (6) And their perpendiculars will fall on the same point of HΠ and they will be equal. Let them fall on O. (7) O then is the center of the circle, and the semicircle around MN is cut off by the horizon.

This passage lays out in rough fashion most of the stages of a classical Euclidean proof. At the beginning of 3.5 Aristotle states the πρότασις or enunciation, namely, that among its other properties (τῶν ἄλλων τῶν συμβαινόντων) the rainbow cannot form a circle nor anything larger than a semicircle (375b16–19). A sort of exposition (ἔκθεσις) follows (375b19–25), in which Aristotle situates a hemisphere upon a horizon circle (the horizon circle is almost certainly circle A, as I argue in more detail in the appendix) with a center K (where the observer is), and the rising point of the sun H presumably on the horizon. There is a curved line M, no doubt on the hemisphere, upon which lines from center K (observer) fall and are reflected (ἀνακλασθῶσιν 375b23) back to H (the sun). The rays from K to

M will form a cone with HK as its axis and M will be the circumference of the cone's base. A defective specification (διορισμός) follows, which interprets the diagram in physical terms and makes clear what is to be done (375b25–9). Finally, the proof (ἀπόδειξις) (375b30–376a1; 376b7–10, 12–22) shows how the point M when joined to the pole Π and rotated on the axis, generates the semicircle of the rainbow.

Although the passage in its present form is little more than a list of truths, those truths seem to suggest an orderly proof. We must suppose that there are a number of triangles in different planes arranged in a manner analogous to the halo proof (see Figure 12.3 below). All of these triangles share a base on the axis line HKΠ. I shall designate all these triangles by the same points and indicate where there is a plurality of them. It is not clear why step 1 is introduced first, nor what its significance is. It may be intended to posit some premises, that is, let a line be drawn from the pole Π to M on the horizon and let the pole be rotated on its axis HKΠ, then the line ΠM will describe a cone with a circle M as its base. All lines from Π to the base generated by the rotation of ΠM will therefore be of equal length and the angle KΠM will be fixed because of the rotation. On this interpretation, we can directly conclude that all triangles HΠM are equal

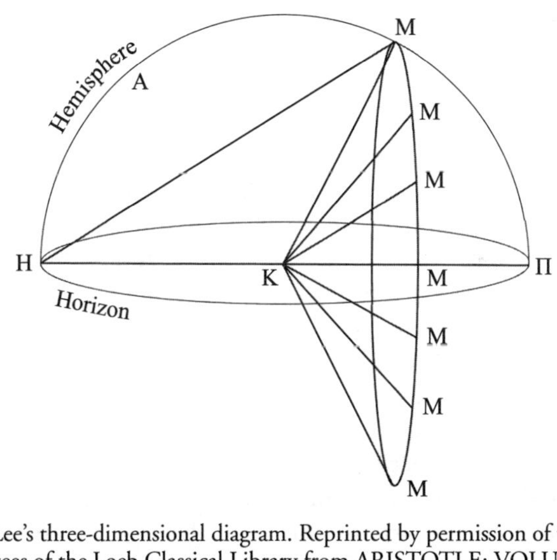

Figure 12.3 Lee's three-dimensional diagram. Reprinted by permission of the publishers and the Trustees of the Loeb Classical Library from ARISTOTLE: VOLUME VII, Loeb Classical Library Volume 397, with an English translation by H. D. P. Lee, pp. 269–70, Cambridge, Mass.: Harvard University Press, Copyright © 1952 by the President and Fellows of Harvard College. Loeb Classical Library ® is a registered trademark of the President and Fellows of Harvard College.

and proceed to prove the equality of all lines MO. This makes steps 2, 3, and especially 4 otiose.

Alternatively, step 1 could be understood as a κατασκευή, an additional construction that is intended to produce a figure similar to the halo figure and to facilitate the dropping of the perpendicular from M to O, which in turn will be proved equal in every such triangle and therefore be the radius of the circle. If this is what Aristotle has in mind, the proof (ἀπόδειξις) will be contained in steps 2–7. Now, step 2 is ambiguous: "the lines reflected from HK to M" properly makes no sense, since the ray is reflected not *to M* (the rainbow), but *to H* (the sun; and the text is not in dispute). But if the phrase is loosely interpreted as the rays between HK and M, it could refer to (i) line KM (in which case, it would at least be traveling to M) or (ii) HM (not likely, since that ray travels to H), or (iii) the rays between both H and M, and M, namely HM and KM. If (i) is correct, then together with (3), which states the equality of all angles KMH (the principle of homogeneity), and the fact that both HK and KM are radii of the circle A, we could conclude the equality of all triangles KHM. It is to be noted, however, that the text does not use the equality of the radii HK and KM. This is important, because the fact that KΠ is also a radius is sufficient to establish the equality of all triangles HΠM, and therefore all triangles KΠM. And yet Aristotle makes no mention of the equality of the radii. If (iii) is correct – (ii) being unlikely – then by Euclid, *Elements* 1.4 (equality through "side–angle–side") all triangles KHM are equal. This would avoid the use of the equality of radii and make the implausible physical implications of the circle A unnecessary (the sun is not on the circumference of the horizon circle). Next according to (4) – perhaps on the basis of (1) interpreted as a κατασκευή – all angles KΠM are equal. Adding to this the equality of all angles KHM (since all triangles KHM are equal) and the base HΠ, because it is the common axis, we can conclude the equality of all triangles HΠM. This together with the equality of all triangles KHM allows us to conclude the equality of all triangles KΠM, which finally would allow us to establish the equality of all lines MO the radius of the circle of the rainbow.

Vitrac's excision, therefore, allows us to reconstruct a sensible proof, which is remarkably similar to the proof for the halo. As with the halo, all the rays of incidence KM are of equal length, and likewise the reflected rays HM. Again this physical assumption is justified by the uniform distance of the vapor from the sun together with the principle of homogeneity (3), according to which the angles of reflection KMH are also all the same. Indeed, it is this similarity with the haloes that further corroborates our interpretation of step 2 (the rotation of the visual ray around the axis) as the core of the proof. Both kinds of reflection depend on an axis joining the observer and the sun with the phenomena situated in a circle around the axis. But the differences are just as important. The introduction of the plane of the horizon is an innovation of the rainbow proof, added obviously because it is required to cut off the semicircle of the rainbow, and in the second part of the proof to account for the rainbow's appearing as less than a semicircle.[44] Similarly new is the hemisphere placed atop the horizon, treated as if it were an independent hypothesis in the πρότασις, but perhaps merely the result of the rotation of the circle of the horizon around axis HΠ (again for reasons we shall consider below). The use of the horizon as the "stage" for the phenomena gives rise to the absurdity, as Olympiodorus rightly points out, that the sun H is placed on the edge of the horizon, when properly it should be placed much further away.[45] However, the falsehood thus introduced is no different from the traditional diagram of the halo, which places the halo midway between the observer and the sun and does not seriously hamper our understanding of the physical causes.[46]

It is clear that on this construction of the text Aristotle has attained all that he needs and that there is no reason to introduce anything beyond this simple apparatus. Most importantly, there is no need to introduce the ratio between the length of the ray of incidence KM and the length of the ray of reflection MN, without which the entire locus proposition would be unthinkable. In fact, Aristotle never mentioned the issue of the ratio or the fixity of the rainbow's position in the sky in his enunciation (πρότασις), and, though it could conceivably fall under "other properties" (3.5.375b17–18), this designation hardly does justice to its preponderance in the chapter.[47]

44 The second geometrical proof, that the rainbow appears as less than a semicircle when the sun is above the horizon, is dependent on the first proof and is uncontroversial. Aristotle merely tips the axis joining the sun and the viewer so that the appearance of the rainbow opposite the sun is cut off by the horizon. It is worth noting that the proof suggests that the chord representing the ray from the cloud to the sun is the same length whether the sun appears exactly at the horizon or whether it appears somewhat above the horizon. The result of this supposition is that the arc of the rainbow will have the same radius no matter how high or low it appears. This is in contradiction with what was claimed at 3.2.371b27–9: "at sunset and sunrise the circle is smallest and the segment largest; when the sun is higher the circle is larger, the segment smaller."

45 248.11–12. He goes on to say that K is not supposed to represent the center of the horizon.

46 The halo is drawn equidistant between the eye and the sun in all manuscripts examined (Parisinus 1853 [161r], Vindobonensis 100 [124r], Laurentianus 87.7 [294v], Paris. Coisl. gr. 166 [411v], Paris. suppl. gr. 314 [244r], Salamanca 2747 [Matrit. Reg. 41] [141v], Vaticanus 1027 [253r], Vaticanus 1387 [70v]).

47 Heath 1949: 188: "It seems probable, therefore, that Aristotle's object was to give mathematical form to a vague sense-impression of the phenomenon of the rainbow. Thus, without venturing to enunciate any law of refraction, he assumed as a probable hypothesis that the length of the incident ray would be in some constant ratio to the length of the inflected ray."

Moreover, the ratio between lengths of rays was not utilized in the proof concerning the halo, though it would have been very useful there. For if the visual ray had to conform to a given ratio between incident ray AZ and reflected ray ZB, and the distance between the eye and the sun is given, the point at which the ray is bent would also be given; and if the angle of reflection is given, then inevitably the shape of the halo would be a circle. As it is, the principle of ratio is not used elsewhere than in the locus part of the rainbow proof.

And even if Aristotle had been concerned with the ratio, he need not have indulged in such a complicated proof. He could have obtained all he needed from the very simple *Elements* 1.7, which, because it shows that there can be only one vertex of a triangle of given sides, could be used to prove that the rainbow can appear at only one elevation M in the sky. And if that is not sufficiently sophisticated, then *Elements* 3.7, which considers triangles drawn on a diameter to the circumference of the circle, is closer in spirit and can easily be adapted to prove that the ratio between the lengths of the rays in the circle A will vary, if M is moved away from the position of the rainbow arc. There is simply no need to introduce the locus.

But it is the physical implications of the locus that are most problematic, for all that they have been ignored. There is no need to enter here into the details of the locus proof. It is sufficient to note that it shows that in a triangle whose base KH remains constant, but whose unequal sides HM and KM grow or shrink while maintaining a fixed proportion to each other, the vertex M will trace a circle. Alternatively described, when two unequal circles with fixed centers grow toward one another, while maintaining their original proportion, they will first touch at a point, then, as they continue to grow, they will touch at two points. If we trace all the touching points (loci) as the circles grow, and connect them together, they will form yet another circle. When the intersecting circles are rotated around the axis joining their centers in *three* dimensions, the loci will form a sphere. In the *three*-dimensional context of a rainbow, we may imagine two unequal *spheres* not yet touching each other, the larger centered on the sun (suppose it is on the horizon), the smaller centered on the observer. Let these spheres grow at equal multiples, so that they maintain their original proportion the one to the other. Eventually they will come into contact at a point between the sun and the observer. Then, as they continue to grow through one another, their line of intersection will form a circle. As the spheres grow proportionally, the circle of their intersection will move back toward the observer and will grow until the circle of intersection passes through the zenith of the observer; then, continuing in the same direction, the circle of intersection will descend again, now generating the size and position of the actual rainbow opposite the sun, then decreasing again until the circle shrinks to a point on the horizon and ceases to exist.

These consequences are basic to the locus and should not have escaped the notice of anyone interested in both the geometry and the physics of the situation. Imagine for a moment that Aristotle intended the locus to be the sum of all the possible locations of the rainbow. Why else would he introduce the issue of ratio, except to identify it as a defining feature of the reflection? The locus will then describe the different positions in which a rainbow can appear relative to the observer and still maintain the ratio of lengths between the two parts of the ray. The fact that, according to the locus, the rainbow might appear toward the sun may be one that Aristotle welcomes, given that rainbow colors appear in the rods around the sun (3.4.374a13–20). But the problems with such an interpretation are numerous. The locus permits the rainbow to appear with any size, from tracing a great circle overhead to being a point on the horizon opposite the sun or between the sun and the observer. The indeterminacy of location is problematic enough by itself given where rainbows are actually observed, but it also seems to be contrary to the stated purpose of the locus proof, which is just to show that the rainbow can appear at only one angle relative to the observer (i.e., that the rainbow cannot appear, for example, at P). It is a premise of the argument that the rainbow can appear only on circle A, and the locus intersects that circle only at M. The rainbow can appear there and there only. Moreover, if the rainbow must appear on circle A, then not only must the rays have a fixed ratio to one another, they must in fact have determinate lengths. For the sun cannot move from H, and the eye cannot move from K, and those two points fix the diameter of the circle on which the rainbow appears. Effectively, by placing the eye and the sun where he does, the author undercuts the purpose of the locus, if it was to show the sum of all places where the rainbow can appear. Because the sun and the eye are fixed, the rainbow can appear in only one place, and the locus is useless.

Again, conceivably the ratio might be useful in accounting for the appearance of the rainbow's various colors. By adapting the principle of Hippocrates of Chios, that the reflected ray is especially weak and unable to travel far (1.6.343a13–15), Aristotle might have accounted for the order of the colors in the first rainbow, thereby avoiding the implausible "*en masse*" principle. Thus, if the rainbow can only appear within circle A, the outer band will be red, because the reflected ray is shorter relative to the incident ray than is the case with the inner blue band. This principle, however, would be no help in explaining the order of the second rainbow, which treats the outer blue as further away from the sun rather than closer. On this explanation the second rainbow is not, as it were, plastered on the hemispheric vault of the sky, but rather standing upright perpendicular to the horizon. Whereas in the physical chapters the distance the visual ray has to travel is greater the higher the band of the rainbow is in the sky, if the rainbow is plastered onto the hemisphere A, the distance of the reflected visual ray decreases the higher the rainbow is. These difficulties with the second rainbow prompt us to observe another difficulty, that contrary to the conclusion of the locus proof, more than one rainbow *does* appear on circle A. As we have seen in the physical chapters on reflections, Aristotle's care in planning his table of differentiae becomes apparent only after one sees the causal web of the entire genus. The physical implications of the locus proposition are so at odds with this scheme that even Aristotle would have balked at the task of reconciliation.

In sum, then, Vitrac's identification of the interpolation eliminates many problems that would arise in the physical nature of the reflections. The resulting proof provides a high degree of stylistic unity with the halo proof and continues the geometrical method that sets this section apart from the others in the treatise. Most importantly, it keeps the circular shape of the rainbow at the center of the proof, in accordance with the leitmotif of the treatise.

Appendix: the diagrams and the origin of the interpolation

At *Meteorologica* 3.5 Aristotle sets out a geometrical proof explaining why rainbows never appear as more than a semicircle. The transmitted text is difficult and has never been given a fully satisfactory interpretation. The scholarship of the twentieth century on this chapter was dominated by Thomas Heath, who in his edition of Euclid's *Elements* and again in his *Mathematics in Aristotle* drew attention, as he thought for the first time, to the similarity between this proof and a locus proof found in Eutocius' commentary on Apollonius.[48] The basic validity of Heath's observation has not been challenged, but he exaggerated the quality of the proof in our text and its similarity to that of Apollonius. Moreover, he never seriously doubted the authenticity of our passage and concluded that Aristotle had access to some high order, contemporary mathematical work. Wilbur Knorr, using

48 The details of the proof are to be found in Heath 1949: 181–9 and in his earlier 1925: 198–9 on Euclid, *Elements* 6.3, where he exaggerates the similarity ("exactly the same form") between Eutocius' construction (2.180–4 Heiberg) and Aristotle's.

stylistic features that had been first observed in the nineteenth century, noted that Aristotle makes use of an elaborate idiom in identifying points in diagrams (τὸ ἐφ' ᾧ τὸ Α), which is not found in Euclid (where instead we find τὸ Α) but is found in the earliest mathematical writing, evidenced, for example, in Hippocrates of Chios' work on lunules preserved in a citation of Simplicius.[49] Knorr argued that, since our *Meteorologica* passage makes use of this elaborate idiom, the locus proof was pre-Euclidean, though he thought it was too sophisticated to be Aristotelian.

Both Heath and Knorr, however, were unaware that the connection between our text and Apollonius had already been discovered by Paul Tannery, who had also observed the older idiom of designating points but had made more subtle use of it to distinguish passages of text that were old and pre-Euclidean from the newer non-Aristotelian passages.[50] Specifically he noted that the core of the passage (from πρὸς ἄλλῃ 376a7 to πρὸς ΜΚ] 376b7) makes use of the Euclidean convention and argued that it was not Aristotelian. Thus Tannery was able to excise the troublesome text without jettisoning the entire section on reflections.

The original contribution of Tannery was recently brought to light by Bernard Vitrac and Ann Merker, who also subjected the passage to a more rigorous analysis.[51] In particular Vitrac showed that the dating technique of Tannery and Knorr, concerning the convention of designating points, was seriously flawed. He argued that, in fact, the elaborate idiom for naming is evidenced in Euclid (citing the translations of Adelard of Bath and Gerard of Cremona, whose originals, he claimed, had not undergone the process of standardization suffered by the rest of the manuscript tradition) and in Archimedes. Vitrac argued that the elaborate idiom serves principally to "baptize" the points on their first introduction, after which the simple form of designation suffices. His study made use solely of passages of "style mathématique" and thereby passed over an important confirming example in the *Meteorologica* itself, where the points of the windrose (2.6) are baptized with the elaborate idiom and referred to thereafter in the simple style. The inconvenient result of Vitrac's research is that we now lack the stylistic scalpel adapted to remove the interpolated passage from its matrix. But fortunately the excision need no longer follow the naming convention, which, since our text mixes the elaborate and the simple convention, had caused some embarrassment for Tannery's interpretation.

49 Knorr 1986: 108. 50 Tannery 1886: 38n1 and n2.

51 Vitrac 2002: 242 mentions that Tannery's work had had "little echo" even in the French scholarship up to that time.

Tannery had also briefly noted that our passage uses the idiosyncratic formula ὅπερ ἡ Α πρὸς Β, ἡ Γ πρὸς Δ rather than the Euclidean ὡς . . . οὕτως to express proportional relations, though he failed to draw a clear significance from the fact.[52] Vitrac, however, argued that the use of ὅπερ in four-letter proportions is confined to eight uses in our *Meteorologica* passage and that this was sufficient to set this passage apart in the Aristotelian corpus and make it a good candidate for interpolation. But the strength of his argument is weakened by his failure to discuss *Nicomachean Ethics* 5.5.1133a32–b5, which, though strictly consistent with his contention, shows that the usage was not alien to the Aristotelian idiom.[53] In view of this passage, it is hard to see why it would be difficult for Aristotle to say ὅπερ Α πρὸς Β, Γ πρὸς Δ, and its omission elsewhere in the corpus could well be a matter of chance.

Vitrac concurred with Tannery that the core passage was not Aristotelian, and, though his study invalidated Tannery's major criterion for establishing a post-Aristotelian date, he nevertheless laid down precise parameters for any likely interpolation and viewed as an addition any passage that mentioned a ratio or proportion between the lengths of the incident and reflected rays.[54] He concluded that the interpolation arose from the successive addition of marginal notes but was unable to provide any motivation for the interpolation.

I agree with Vitrac that the core passage is an interpolation, and I propose a simple explanation for it that arises out of a combination of geometrical and physical considerations. Aristotle's text describes two diagrams, the first (which I call the "locus" diagram) showing that the rainbow never forms more than a semicircle, the second (which I call the "tilting" diagram) showing that, when the sun is elevated in the sky, the rainbow forms less than a semicircle. The second diagram accompanies a geometrical argument that is relatively simple, is consistent with Aristotle's stated purpose, and has not been widely suspected of interpolation (376b28–377a28). The two diagrams are clearly related, and examination of their similarities and differences reveals that they represent the horizon and the sky from two different perspectives, the aerial view and the side view. The diagrams that come down to us in the manuscript tradition both of Aristotle and

52 Tannery 1886: 43.

53 ὥστε ὅπερ γεωργὸς πρὸς σκυτότομον, τὸ ἔργον τὸ τοῦ σκυτοτόμου πρὸς τὸ τοῦ γεωργοῦ . . . γεωργὸς α, τροφὴ γ, σκυτότομος β, τὸ ἔργον αὐτοῦ τὸ ἰσασμένον δ.

54 Tannery starts the interpolation at πρὸς ἄλλῃ δέ γε (376a7) and ends it at τὸ ΜΗ πρὸς ΜΚ (376b7); Vitrac 2002 begins his interpolation at αἱ οὖν ἀπὸ τῶν (376a1) and ends it at τὸ ΜΗ πρὸς ΜΚ (376b7), and he adds to the interpolation the sentence εἰ δὲ μὴ (376b10) . . . ἀδύνατον (376b12). See Merker 2002 for other refinements.

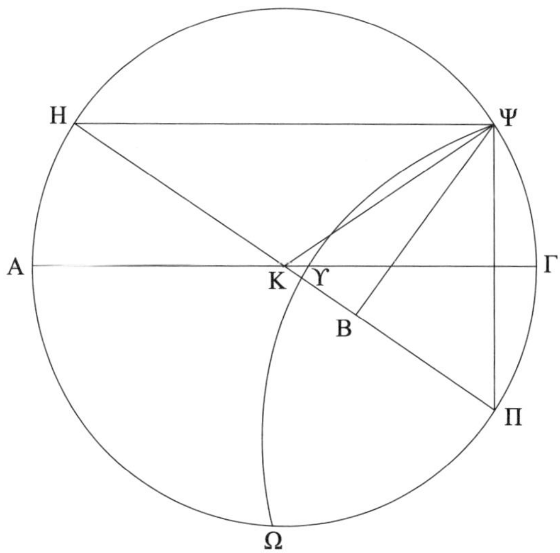

Figure 12.4 Tilting diagram

of the ancient commentators show how this change in perspective was misunderstood and how it could have given rise to the interpolation. My argument is thus an extension of those of Tannery, Vitrac, and Merker and seeks merely to give a plausible account for why within a geometrical proposition, which to all appearances should have been very simple, a proof of considerable and unnecessary sophistication was introduced.

It is not worthwhile here to rehearse the entire controversy of the locus. I shall begin by signaling my basic agreement with Vitrac and by assuming as a hermeneutic starting point the principle that the second, "tilting" proof (376b28–377a28; see Figure 12.4 above) and its accompanying diagram are genuine.[55] We may note first of all that three of the points in that

[55] The "tilting" diagram is not as well attested as the "locus" diagram. It does not have an embedded place in the Aristotelian manuscript E (Parisinus 1853) but rather was added later in the margin (164r). J has an embedded tilting diagram (though it is very schematic and could easily have been added by a provident scribe; Alexander [Parisinus 1880] does not have it; it is, however, as well embedded as the locus diagram in all the other MSS both of Aristotle and of the commentators that I have examined [besides those mentioned, Laurentianus 78.7 for Aristotle; Coislinianus 166 for Olympiodorus; Vaticanus 1387 for Olympiodorus has no rainbow diagrams at all]). Though it is reasonably easy to construct the diagram from the text, it probably existed originally, since point Γ and horizon AKΓ need to be interpreted. My thanks to the Aristoteles Archiv at the Freie Universität, to Dieter Harlfinger, Ingo Steinel, and Pantelis Golitsis for their generous help and guidance.

diagram are designated by initial letters, K (κέντρον), H (ἥλιος), and Π (πόλος).[56]

It is clear that Π is the pole of a sphere that rotates around axis HKΠ and thus is appropriately called a pole. We may note too that all three points occur in the locus diagram as well, K serving the same function as center, and H serving unambiguously for the sun but taking up a different position as dictated by the proofs. Π, however, serves ambiguously both as the pole of the sphere and, strangely given its name, as the center of the locus circle.[57] These three shared points suggest that the locus diagram and the tilting diagram are closely related and make reasonable the working hypothesis that they should be treated together.

Of course, there is one further common letter, A, which in the tilting diagram unambiguously designates a point on a diameter opposite Γ through the center K. In the locus diagram, A is not used as a point but probably designates the circle of the horizon (375b19–20).[58] In fact, it is far from clear how A is to be understood in this beginning section (375b16–376a1). However that may be, in the tilting diagram Aristotle makes A, which in the locus diagram seems to designate the whole circle of the horizon, into a single point on that same circle taking the place that has been vacated by the ascendant H.

If we understand that A designates the plane of the horizon in the locus diagram (see Figure 12.5 below), then we can begin to understand the difficult passage at 375b30–376a1:

> ἔστω [τὸ ἄστρον] γὰρ ἐπ' ἀνατολῆς πρῶτον, οὗ τὸ H, καὶ ἀνακεκλάσθω ἡ KM ἐπὶ τὸ H, καὶ τὸ ἐπίπεδον ἐκβεβλήσθω ἐν ᾧ ἡ A, τὸ ἀπὸ τοῦ τριγώνου ἐν ᾧ τὸ HKM. κύκλος οὖν ἡ τομὴ ἔσται τῆς σφαίρας ὁ μέγιστος. ἔστω ὁ ἐφ' ᾧ A· διοίσει γὰρ οὐδὲν ἂν ὁποιονοῦν τῶν ἐπὶ τῆς HK κατὰ τὸ τρίγωνον τὸ KMH ἐβληθῇ τὸ ἐπίπεδον.

> Let [the sun] be placed first at the rising point, H, and let the line KM be reflected to H, and let the plane be produced from the triangle HKM into

[56] Louis 2002: II.20n5 also suggests M stands for μέλαν, "nuage noir," though one had rather expected I (ἶρις, cf. Alexander in Parisinus gr. 1880 75r).

[57] As pole of the sphere implicitly at 376b8, 31; as center of the locus circle throughout the passage 376a10–b7 especially 376a18. It is remarkable that once in each environment (376a18; 376b31) he calls Π the πόλος τοῦ κύκλου. In the second proof it is called the pole of the circle probably by assimilation with the first instance.

[58] 375b19–20: ἡμισφαιρίου γὰρ ὄντος ἐπὶ τοῦ ὁρίζοντος κύκλου τοῦ ἐφ' ᾧ τὸ A. τοῦ ἐφ' ᾧ τὸ A is more readily understood modifying κύκλου than ἡμισφαιρίου. Lee understands it as a semicircle section of the hemisphere that rests on top of the horizon. Tannery 1886: 40n1 remarks: "we interpret this passage incorrectly if we take the circle A to represent the hemisphere and not just the horizon." He nevertheless believes that the plane KHM is inclined to the plane of the horizon (41). Merker 2002: 209n76 takes it, rightly I believe, as the horizon.

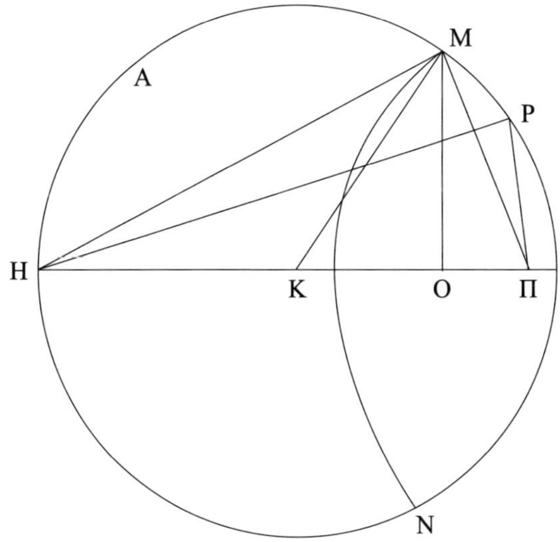

Figure 12.5 Lee's locus diagram = Lee's traditional diagram. Reprinted by permission of the publishers and the Trustees of the Loeb Classical Library from ARISTOTLE: VOLUME VII, Loeb Classical Library Volume 397, with an English translation by H. D. P. Lee, pp. 269–70, Cambridge, Mass.: Harvard University Press, Copyright © 1952 by the President and Fellows of Harvard College. Loeb Classical Library ® is a registered trademark of the President and Fellows of Harvard College.

> the circle A.[59] Thus the section of the sphere will be a great circle. Let this be A. For it will make no difference whichever of the planes KMH on HK be produced.

There is much that is obscure and difficult in this passage especially concerning how the plane is to be produced and the identity of the sphere. However, the last two sentences are reasonably clear. There is a great circle that is a section of a sphere. This section is called A and it will make as good a section for Aristotle's purposes as any. The easiest interpretation is to suppose that the great circle A is identical with the circle A laid out at the beginning as the circle of the horizon (375b19–20), and that Aristotle is placing the triangle of reflection HKM in that plane. Aristotle is aware that this is an odd way of proceeding and that is why he adds the sentence, διοίσει γὰρ οὐδὲν . . . , to acknowledge its oddity.[60]

[59] See n. 43 above.

[60] Alternatively, we may interpret A as the hemisphere, that the plane HKM is produced into it and that plane generates a great circle, and we might as well call that circle A since it is as serviceable as any other section of the sphere.

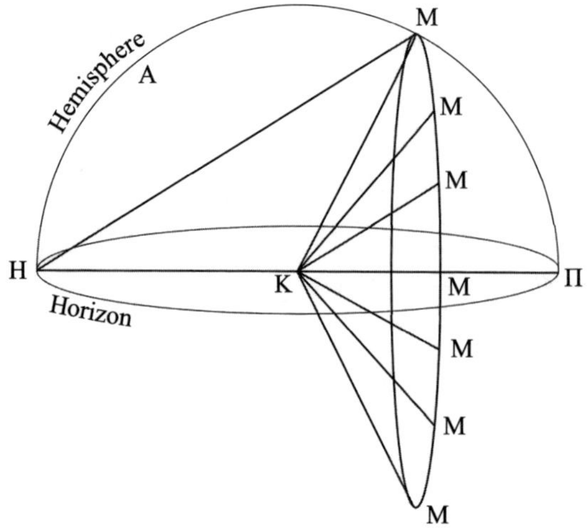

Figure 12.6 Lee's three-dimensional diagram. Reprinted by permission of the publishers and the Trustees of the Loeb Classical Library from ARISTOTLE: VOLUME VII, Loeb Classical Library Volume 397, with an English translation by H. D. P. Lee, pp. 269–70, Cambridge, Mass.: Harvard University Press, Copyright © 1952 by the President and Fellows of Harvard College. Loeb Classical Library ® is a registered trademark of the President and Fellows of Harvard College.

However A is to be understood, it has been the near universal interpretation, wherever an account of the diagram has been given, to treat the triangle of reflection HKM as placed at an angle, usually perpendicular, to the plane of the horizon.[61] None of these resulting diagrams is wildly wrong, but in almost every case the interpreter has failed to question in what plane the resulting figure is being viewed. It is clear from Lee's diagrams, for example, especially in the transition from his own three-dimensional diagram (p. 269; see Figure 12.6 above) to the locus diagram he reproduces (p. 270; see Figure 12.7 below), that the line HΠ of the traditional diagram represents the horizon as seen in side view, that is, as viewed from the edge of the plane of the horizon. Thus, M is above the earth and N is

[61] It is what Boyer 1959: 42 has called the "meteorological hemisphere." Heath 1949: 185: "GMR [i.e., HMΠ] is a semicircle in a plane which for clearness' sake we will suppose perpendicular to the horizon." Notwithstanding his diagram (186) representing the side view, it is an accurate representation of the aerial view as well. Tannery 1886: 39 produces a diagram similar in this respect. Lee simply reproduces Alexander's diagram and interprets A as the semicircle of reflection: Π is inside the circle. Strohm 1984: 211, bizarrely, appends the wrong diagram entirely. Fobes provides no diagrams. Knorr 1986 appends no relevant diagram; Vitrac 2002 provides no physically determinant diagrams. Merker 2002: 210 produces a diagram similar to Louis', but correctly without point N.

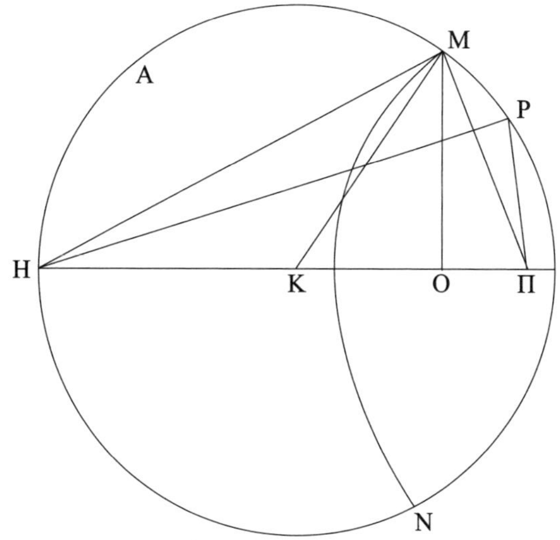

Figure 12.7 Lee's traditional diagram = Lee's locus diagram. Reprinted by permission of the publishers and the Trustees of the Loeb Classical Library from ARISTOTLE: VOLUME VII, Loeb Classical Library Volume 397, with an English translation by H. D. P. Lee, pp. 269–70, Cambridge, Mass.: Harvard University Press, Copyright © 1952 by the President and Fellows of Harvard College. Loeb Classical Library ® is a registered trademark of the President and Fellows of Harvard College.

below the earth; M is higher in the sky and P is lower in the sky. Accordingly, the line MN is supposed to represent the rainbow in a poor attempt at three-dimensional perspective, where MN as it appears in the diagram is the part of the rainbow *circle* that appears protruding from the page, while the undrawn part retreats into the book below the page.[62] The section of MN below HΠ is below the earth and hence invisible to the observer at K. Even Heath, who depicts the rainbow in correct aerial view as a straight, rather than a curved, line and names it MM' (to distinguish it from his locus circle MN), still says that his diagram is to be viewed for clarity's sake from the side view.[63]

It is clear from the manuscript tradition that Aristotle used only one figure, instead of Lee's two, to illustrate the geometry of his first proof that the rainbow is never more than a semicircle.[64] Let us suppose that

[62] For "foreshortening" in mathematical diagrams, see Netz 1999: 17–18; for the common practice of multiple perspectives in fourth-century art, see Richter 39–48.

[63] Heath 1949: 185–6.

[64] The best manuscripts of Aristotle, E (Parisinus gr. 1853) and J (Vindobonensis gr. 100), are in agreement in the placing of the "locus" diagram after the text of the locus proof (376b22–8 is

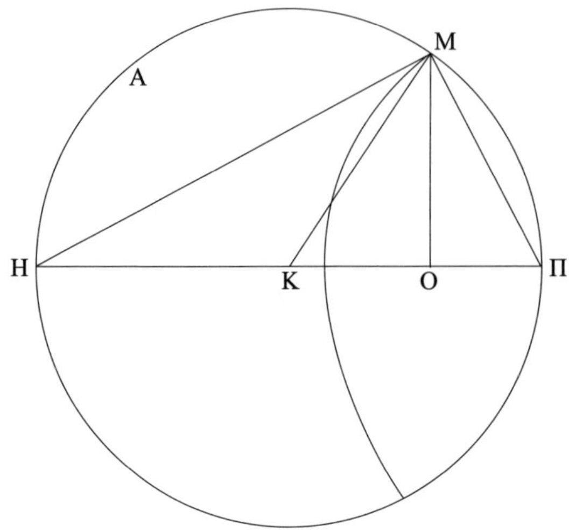

Figure 12.8 Reconstructed original diagram: aerial view of the horizon

in this diagram (see Figure 12.8 above) A originally designated only the plane of the horizon. From this perspective M is the point where the "leg" of the rainbow meets the horizon and KM is the visual ray of the observer looking along and parallel to the plane of the horizon toward its edge. The line MN in the traditional diagram is still in clumsy three-dimensional perspective, but now the visible semicircular rainbow is represented in its entirety as it actually appears above the horizon. No part of the area under the earth is represented here. We see what is meant to be proved: the rainbow appears at most as a semicircle above the horizon.

This perspective is different from what we see in the tilting proof, where the text clearly indicates the side view. There the line drawn to Ω in the tilting diagram need not and probably should not be so far extended; rather

interpolated between the text and the diagram in J, and is added around the diagram in E; it is missing from Alexander's commentary but is present in Olympiodorus). This suggests that the archetype of E and J placed the "locus" diagram after the first rainbow argument. But even this probably represents a displacement from its original, text-directed position at the beginning of the chapter (cf. διαγράμματος 375b18), where Alexander (Parisinus gr. 1880) places it. Tannery 1886: 41 hedges on the issue: "In fact, Aristotle constructs a new diagram like the preceding (we can use the same one), but where only the straight line HK maintains the same significance." The description of A as a semicircle at 376b13, is either, as Merker explains (2002: 216n102), a means of describing how the plane of the horizon can rotate around HKΠ, or a mistaken correction for τὸν κύκλον in light of 376a3. The "locus" diagram in all MSS (of Aristotle and commentators that I have examined) except J show the line MN as in Lee's traditional diagram.

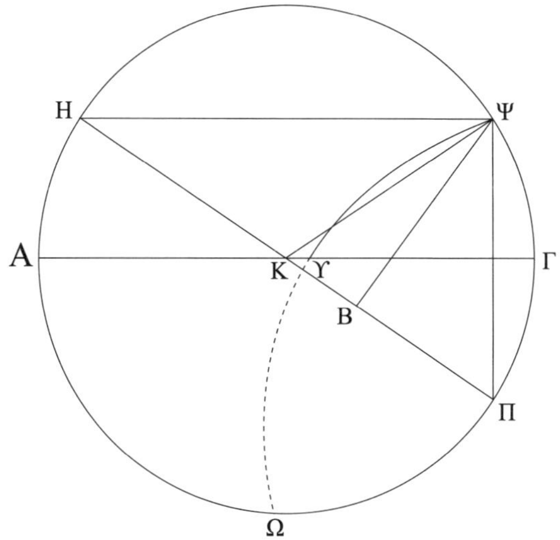

Figure 12.9 Reconstructed tilting diagram

the semicircle of the rainbow should terminate at line ΑΓ, as in Figure 12.9 above.[65]

I suggest that it was the curved line MN in the locus diagram and the understandable confusion about the reference of A that led to the introduction of the locus proposition. First, the line is not called MN until the locus proposition begins (first at 376a6–7), and, in fact, the letter M probably designated the whole rainbow, as is suggested by more than one line from K touching it (ἀπὸ τοῦ Κ ἐπὶ τὸ Μ ἐπιζευχθεῖσαι 375b23).[66] Once the locus proposition starts, the line is treated as MN (I shall refer to it as MN without prejudice as to its origin) and is the segment of the locus circle (376a7–9; 376b2; its significance at 376b21 at the end of the locus proof is obscure). I suggest that confusion arose because when a reader saw the original drawing, which was probably very similar to what we have, he

[65] I suggest that in the second diagram the line ΨΥ was extended to Ω inadvertently or by a reverse assimilation to the first diagram in the manner of MN.

[66] Manuscript diagrams are very consistent about the position of M, but much at odds about N. While the manuscripts of the commentators are reasonably consistent in placing point N where Lee does, MSS E and J are in disagreement with each other and with the commentator manuscripts. J is exceptional in having an MN line that could not even be a locus line at all, since its center seems to be O (radius OM) rather than Π. E places point N between M and Π on the circumference of circle A. The agreement on the basic shape of the diagram and the confusion as to the placement of N suggest that while the circle segment generally named "MN" was an original component of the diagram, N was not.

thought of A as representing the vault of the sky above the earth in a side view placed perpendicular to the horizon HKΠ, rather than, as I believe Aristotle intended, an aerial view of the plane of the horizon itself. From the former point of view the curve MN, since it extends below the horizon, makes a rather improbable rainbow. The fact that A is mentioned as a semicircle (376a2) in the locus proposition makes clear that our posited interpolator viewed the diagram in side view.

Moreover, if the section after Vitrac's proposed interpolation is genuine (376b7–10, 13–22), we find Aristotle telling us there to take the line ΠM and rotate it around the axis HΠ (376b7–10). It is clear that there is a line ΠM in the traditional diagram, and, since this line meets the curved line MN at M, Π might be mistaken for the center of a circle having MN as a segment; this mistake would be especially easy to make, since Π is (contrary to the modern commentators) situated on circle A in the manuscript diagrams.[67] If MN were drawn so as to approach K closely or even extend beyond and intersect HΠ between H and K, as Laurentianus 87.7 (see Figure 12.10 below: Laurentianus 87.7 [295v]) and Coislianus gr. 166 (423v) do, the mistake would be easier still.

I suggest that this misunderstanding occurred at some point early in the manuscript transmission and as a result MN came to look even less like a three-dimensional representation of the rainbow from above and more like the circle of radius ΠM in a side view.[68] Under such circumstances, it is not improbable that a reader with knowledge of geometry should be reminded of the locus proposition and make a note of it in the margin. It remained to write up the proof and insert it into the text. Point O would be an easy addition on analogy with point B of the second diagram. The position of Π should have been "corrected" from its original place at the pole of the sphere so as to become the center of the locus circle, but it was not.

Of course, this interpretation cannot solve all the problems of this vexed passage, and in particular it faces one major objection. Since the geometry of the rainbow and of the locus is based on the sphere, the diagrams

[67] All manuscripts examined agree on the placement of points H, M, and K. All show Π at the pole (i.e., on the circumference of the circle A). None, that is, show Π as the center of the circle of the locus. There is no authority for placing Π inside the circle as Vicomercatus 153D, Ideler 1834–6 II.307, and Lee do. Louis 2002: II.154 rightly provides the figure of Paris. gr. 1853 fol. 164r, which places Π on the circle A.

[68] If, as I contend, the curved line MN is original to the diagram and designates the rainbow, it must have been drawn in three-dimensional or "cavalier" perspective, no matter whether it was to be viewed from above or from the side. It is depicted as the circle segment of radius ΠM by Alexander (Parisinus gr. 1880 75r), Olympiodorus (Coislinianus gr. 166 421v), and Aristotle (Parisinus gr. 1853 [E] 164r and Laurentianus 87.7 295v [but very roughly drawn]; Vindobonensis 100 [J] does not have this same line).

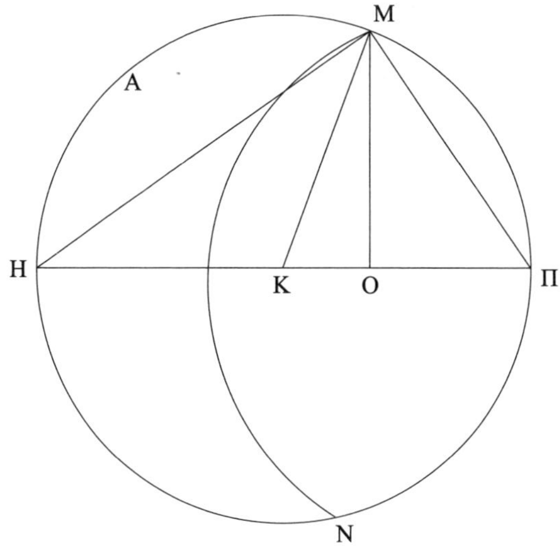

Figure 12.10 Laurentianus 87.7 (295v)

can be viewed equally well from the side and aerially. No confusion in interpreting the first diagram could arise, it might be said, because the views are geometrically equivalent. This is certainly true, but it has not stopped any of the commentators from imagining the locus diagram as a side view, and the reason stems from the way in which we imagine rainbows. The fact is that the diagrams are not merely geometrical; they are also schematic representations of physical situations. We think of rainbows as appearing above the earth, and though a side view may be clearer in this respect, as Heath suggests, it makes the line M (or MN) with its production of the rainbow below the horizon rather unnatural. Moreover, the natural physiology of our balance makes us better judges of altitude than of azimuth. For this reason and because the rainbow's junction with the ground is often obscured from our sight for various reasons, we more readily think of a rainbow as being higher or lower in the sky rather than as subtending a greater or lesser arc on the horizon. Mistaking Aristotle's aerial view for a side view, therefore, is completely natural.

In face of a possible objection that the change of perspective from the first to the second diagram is strange and pedagogically unsound, we may note that Aristotle engages in a thematically important change of geometrical perspective in his treatment of winds, in which a discussion of the zones of the earth is embedded (2.5). As we saw, Aristotle divides the *spherical* earth into zones according to tropical circles and the ever-visible circle of the stars. In the very next chapter (2.6) he shifts perspective and lays out the winds around a windrose, identical with the *plane* of the horizon with the same divisions of tropical circles and the ever-visible circle of the stars. The winds blow across the plane of the earth, which is divided in a way analogous to that of the sphere. It may not be going too far to suggest that this kind of change of perspective was part of Aristotle's geometrical style and completely in harmony with the grand bivalence of the treatise as a whole, balanced between the shifting perspectives of the unified cosmos and the multiplicity of terrestrial life.

CHAPTER 13

Metals and minerals (3.6)

Aristotle ends the treatise with a skeletal outline of the formation of metals and minerals and an unfulfilled promise of further elaboration. The application of his standard material principles, the exhalations, into this area is a testament to his passion for unity but was hardly required for a successful explanation. According to his predecessors, water and earth alone were sufficient to account for these subterranean stuffs, and Aristotle himself was content to treat the same subject in *Meteorologica* 4 without recourse to the exhalations.[1] Moreover, the fact that he wrote a περὶ μετάλλων indicates that the subject need not have been confined or even included in the *Meteorologica* in order to be effectively treated.[2] He might, then, have simply brought the treatise to a conclusion with the rods and mock suns.

There are other reasons, too, to believe that this coda was best omitted. Aristotle begins by claiming that he has described the effects of the exhalations above the earth and will now turn to what lies below (3.6.378a12–14). This simplistic and lopsided division between what is above and what is below ignores the careful opposition he had already drawn between effects wholly above the earth (kapnospheric and atmospheric effects) and those effects that involve movement across the earth–air interface (rivers, winds, earthquakes, and even the sea through subterranean passages) and that are consequently partly above and partly below the earth. Metals and minerals arise from exhalations that remain trapped inside the earth (ἐγκατακλειομένη τοῖς τῆς γῆς μέρεσιν 378a15–16). Earthquakes, though they have an atmospheric component lacking in minerals and metals, are also caused by exhalation trapped in the earth (κατακλείει 2.8.366a16). The distinction, then, cannot be so neatly drawn.

[1] *Mete.* 4.10. Plato too makes a distinction but treats the metals as a form of water whose parts are larger and regular, while water itself has parts small and irregular (*Ti.* 59). He also mentions dew and frost (59e5) but makes no attempt to elaborate the analogy with metals as Aristotle does.

[2] Olympiodorus makes mention of a monograph by that title (6.4–6).

But at the same time the lopsidedness of the division highlights the isolation of the subterranean phenomena. Aristotle had artfully rounded out the treatise with the reflections, treating them out of spatial order for methodological reasons. Their placement there provides a climax and unity to the treatise up to that point and makes the metals and minerals, which take no part in the circular leitmotif so characteristic of the rest of the treatise, the odd-men out.[3] These phenomena are indeed cut off from the rest. No doubt, too, his motivations for applying the exhalations in a limited way below the earth extend beyond the opportunity to subsume metal and mineral formation under the principles of the exhalations to include analogies – with frost and spring formation – that typify the etiological structure of the treatise as a whole.

Part of my method from the beginning has been to use Aristotle's structural comparisons and analogies to understand the nature of the various *meteora*. On this last occasion we see again how careful attention to other phenomena, here in the upper and middle band of the sublunary sphere, can bring to light significant facts concerning the subterranean materials that have been hitherto overlooked.

For clarity's sake it will be easiest to invert the order of the text and start with the metals. Aristotle explicitly likens the formation of metals to the deposition of dew and frost on the ground:

> the products of the vaporous exhalation are those that are got by mining and are fusible or ductile, such as iron, gold, and copper. The vaporous exhalation produces all of these when it is shut up [in the earth] and particularly inside rock when it is compressed and congealed into a unity by the dryness, like dew or frost when it is separated. Except here these [metals] are created before separation (πρὶν ἀποκριθῆναι). (378a26–32)

Recent commentators have focused their attention on the comparison with dew and frost that this passage makes explicit but have neglected the implicit comparison with spring formation.[4] They have claimed that the metals are distinguished from the upper waters because of the addition of the dry exhalation. While there is clear evidence for the presence of dry exhalation in *some* metals, as we shall see, I think that metals differ from

[3] Plato, *Sph.* 265c places them among orderly works: "Take animals and everything mortal including plants and everything on the earth that grows from seeds and roots and also all lifeless bodies made up inside the earth, whether fusible or not. Are we going to say that anything besides the craftsmanship of a god makes them come to be after previously not being? Or shall we rely on the saying and widespread belief that [they come from a spontaneous and unintelligent cause]?" (N. White trans. in Cooper and Hutchinson).

[4] Eichholz 1949; Strohm 1984: 214–15.

water primarily because metals condense before their wet exhalation has transformed into water and that Aristotle wants us to understand a scale of condensation descending from the clouds to the springs in the mountains and finally here to metal formation, the least distinct and least formed.

In the chapters on precipitation (1.10–11), dew and frost were presented as lower, shorter term, and smaller scale, analogues of rain and snow. Aristotle's emphasis there had been on the rain cycle and on the *common* effects of water and air, and for that reason it was essential for vapor to be generated from the wet exhalation before falling as rain and snow or being deposited as dew and frost. Aristotle's insistence there that frost congeal before it is condensed into water seemed a caviling point in that context (ὅταν ἡ ἀτμὶς παγῇ πρὶν εἰς ὕδωρ συγκριθῆναι πάλιν 1.10.347a16–17), but it becomes more intelligible in light of the new contrast with metals.[5] Dew and frost, and especially frost, are drawn into a new orbit of significance, as Aristotle emphasizes a new set of properties. Their role as junior partners in the rain cycle fades into the background here, since their new *comparanda*, metals, do not recycle into the exhalations from which they were formed.[6] The most obvious similarity is in the manner of their deposition: the dew and frost do not fall (in this sense they are also similar to the formation of springs) but form directly on the grass and leaves and grow there. Likewise the metals, giving the lie to Danaë's golden rain, form by direct congealment within rock. But the manner of their congealment is importantly different. Both frost and metals congeal, but metals are produced *before* the separation of the vapor has taken place (πρὶν ἀποκριθῆναι 378a32). But before the separation from what? According to modern commentators, it is before separation from the dry exhalation. The passage continues:

> So there is a sense in which these things are water and a sense in which they are not. For their material had the capacity for becoming water but has it no longer. Nor are they an affection of preformed water, as savors are. For neither does copper nor gold come about in this way – rather each of them is formed from the exhalation having congealed before water is generated. (378a32–b2)

In the formation of dew, it was vapor (ἀτμίς) that first separated from the air before condensing into water. Likewise, proto-frost had already attained the state of vapor (ἀτμίς), a pre-condensed form of water distinct from the ambient air, before it froze. By contrast, the pre-metallic exhalation

[5] Strohm 1984: 214.

[6] Metals and minerals are, like the rest of the phenomena of the treatise, localized, but they are not obviously homeostatic.

under the earth never becomes vapor in a separated form, and the metal formation occurs at a pre-vaporous stage, when the exhalation is only potentially water.[7] After it has been compressed and congealed into metal, it cannot become water again.[8] Metals are in this sense one step less formed than frost, which in turn is one step less formed than rain or snow.

The analogy with frost is not casual. Aristotle is clearly using it, as we have seen so often, to advance the architectonic unity of the treatise, and this becomes more evident when we compare how he accounts for metal formation in *Meteorologica* 4. While he acknowledges the role of exhalations (4.8.384b32) in a nod to our passage, in chapter 4.10 he directs his full attention to the "water" theory.[9] Here he mentions various homeomerous stuffs, including the metals (bronze, gold, silver, tin, iron, stone, as well as flesh, bone and other organic tissues 388a14–17), "whose material cause is the dry and moist, and therefore, *water and earth*" (388a22). Later in the same chapter (389a7–8) he offers a list of metals (gold, silver, copper, tin, lead, glass) that are specifically composed of *water*. There is no sign in this context of the "potential water" and no sign of the exhalations, which had played such a central and integrated role in the first three books. It is quite conceivable that Aristotle intended alternative explanations for metal formation and decided to separate them into different treatises according to their different principles. But by invoking the exhalations and drawing the analogy with frost, Aristotle introduces efficient, in addition to material, connections with other *meteora*. Metals are formed not just by heating and cooling as in *Meteorologica* 4 (388a24–5), but by a special form of condensation. More importantly, Aristotle makes clear that there are different stages of condensation and separation of the wet exhalation and these have distinct effects on what results.

Clarification of these distinct effects comes from the less well-recognized contrast Aristotle draws with the formation of subterranean springs in the mountains:

> just as in the space above the earth small droplets coalesce, and in turn these coalesce with others, and finally the rainwater descends in abundance, so also

[7] Eichholz (as Strohm 1984: 214 complains) ignores this point and attributes the inability of metals to turn back into water to the presence of the dry exhalation in them. But Strohm comes to a similar conclusion, that what distinguishes metals from frost is the presence of the earthy component.

[8] Strohm 1984: 214 is certainly correct to insist against Eichholz 1949: 142 that pressure rather than cold is the distinctive formative cause of the metals.

[9] Cf. 4.8.384b30–4: "the homeomerous bodies, therefore, both in plants and animals, are composed of water and earth; and likewise the metals, such as gold, silver, and the like are composed of water and earth as well as of the exhalation of each shut up in the earth."

> in the earth sources of rivers trickle together from small amounts of water at first, then, as it were, gush out from the earth together. (1.13.349b30–5)

As we noted in Chapter 9, Aristotle is here explicitly comparing subterranean condensation with rain that condenses in a cloud, and like a cloud, the mountain is spongy (σπόγγος 1.13.350a7). Metals likewise congeal under the earth, but not in the open textured and spongy earth of the mountains. Aristotle's description suggests that he has in mind a dry and hard environment: metals are compressed into one (εἰς ἓν συνθλιβομένη) because of the dryness and are thus solidified (πηγνυμένη 3.6.378a30).[10] By contrast, springs are formed in the absence of pressure by the free coalescence of tiny droplets. We may conclude that the matrix of the metal is denser and not spongy like the mountains that give rise to springs, and that Aristotle intended this density to contribute fundamentally to the difference in the process of formation.[11] The ultimate cause for this is that there is less space in the rocks and this greater density of the surrounding rock causes some of the dry exhalation to remain with the wet exhalation when it congeals (378b3–4).

In the standard interpretation the primary reason why the wet exhalation produces metals rather than water or ice is that the wet exhalation does not separate from the dry exhalation before congealing. Now, there is no question that Aristotle intends a role for the dry exhalation in metals: the dry exhalation accounts for the metals being oxidized when they are brought into contact with fire (with the exception of gold):[12]

> Each of the metals is formed from the exhalation having congealed before water is generated. For this reason they are susceptible to burning and contain earth. For they contain dry exhalation. Only gold is not susceptible to burning. (378b2–4)

But this does not imply that metals get their distinctive nature because they condense with dry exhalation in them. Gold is obviously a metal but is not subject to oxidation and therefore has no dry exhalation in it, as the other metals have. Moreover, it is not until the end of the description

[10] Strohm 1984: 214. Cf. *Ti.* 59b4 "gold solidifies when it has been filtered through the rock." A. E. Taylor 1928: 416 comments: "The suggestion is that the narrowness of the passage through the unyielding rock (quartz?), hinted at by the expression that it is 'strained' or 'filtered' through, forces the particles of the molten metal close together and so effects solidification in the fashion just described."

[11] *Mete.* 4.9.386b2–7: "All things can be squeezed that can contract into their own empty spaces or their own pores [for sometimes the spaces into which they contract are not empty], like a wet sponge (for its pores are full). But things whose pores are full of stuff that is softer than themselves contract into themselves."

[12] For Plato rust is the result of some earth present in the water that has separated (*Ti.* 59c3–5).

(378b3–4) that we learn that there is dry exhalation in the metals, where the fact is introduced as a confirmatory sign that the wet exhalation has not undergone a full transformation into water. It is clear, then, that the presence of the dry exhalation cannot be the reason why metals are not water. It is sufficient, as the analogy with spring formation shows, that the wet exhalation is not separated from the surrounding *earth* before it congeals and that the pressure of the earth forces it into another nature.

Nevertheless, most metals contain some dry exhalation. It is clear that one of the differences between minerals and metals is that with metals the dry exhalation remains in the congealment, until it is burned off later to form an oxide. With minerals the dry exhalation is burned in their very formation.[13] It is worth noting that at *Meteorologica* 4.10 iron differs from the other water-based metals (gold, silver, bronze, tin, lead, glass), in that earth is a major component just as it is in horn, nail, bone, and such like (389a12–14). In spite of the difference in fundamental theory, it is probable that in our passage the ease with which iron rusts is an indication of the large amount of dry exhalation it contains.[14]

Mineral formation is more difficult to understand from Aristotle's brief account. He tells us that minerals are made through the burning of the dry exhalation:

> It is the dry exhalation that by burning (ἐκπυροῦσα) creates all the minerals, for example, the infusible (ἄτηκτα) kinds of stones as well as realgar, ochre, ruddle, sulphur, and other such things.[15] Most minerals are colored dust, but some, like cinnabar, are stones generated from such dust. (378a21–6)

[13] Alexander 178.5–15 makes the connection that, because the metal is formed before separation, the metals have some dry exhalation in them still, which comes out when they are burned; the metals burn when they are smelted and undergo a weight reduction as a result, all metals except gold. According to Olympiodorus, they rust more or less according to the amount of dry exhalation in them (266.26–30).

[14] Eichholz 1949: 143.

[15] Lee certainly mistranslates as "all kinds of stones that are infusible (ἄτηκτα) – realgar, ochre, ruddle, sulphur and all other substances of this kind." Sulphur melts at 115°C, realgar at 320°C, cinnabar at 580°C, all well below the melting points of gold (1064°C) or copper (1085°C); ferric oxide (red ochre or ruddle) melts at 1566°C; yellow ochre is simply a hydrated form of ferric oxide. Iron (described as χυτά at 378a28) melts at 1538°C, not significantly lower than ochre. At *Mete.* 4.10.388b33 things that are melted by fire are in general composed largely of water. On this view, sulphur, being a mineral easily melted, would have to contain significant amounts of water. Given that Aristotle mentions that most metals contain dry exhalation, we might have expected him to mention that minerals contain wet, but he does not. Olympiodorus, whose lemma oddly reads τηκτά rather than ἄτηκτα) notes the contrast between the infusible stones and the specifically named minerals. (269.4–12; Alexander does not commit [177.16–18]). Plato (*Sph.* 265c) seems to treat the fusible–infusible distinction as the most basic to things that grow compacted (συνίσταται) below the earth. Plato's division in the *Timaeus* gives grounds for distinguishing between minerals that are fusible and those that are not, such as pottery (60d).

It is not clear what "heat" (ἐκπυροῦσα) implies here. The parallel with the formation of metals out of the wet exhalation would lead us to suppose that the dry exhalation is being treated as the material cause, but Eichholz has convincingly argued that πυρόω is always used transitively in Aristotle, and therefore the dry exhalation is not the material from which the minerals are formed, but rather the agent actively burning something else, which, Eichholz argues, is the earth.[16] Too great precision on this matter, however, may be more than Aristotle intended. According to his general theory of change, the form of the efficient cause is transferred to the patient. Thus the burning of the dry exhalation will produce in the stone that surrounds it qualities similar to the dry exhalation itself. This sort of process is exemplified in a previous context where Aristotle described how the inhabitants of Umbria burn reeds and rushes, and then they combine the ashes with water to produce salty water (2.3.359a35–b4). Because of the basic similarity between fire and salt, the combustion of the reeds has produced in the ashes a salty residue. In a similar manner the ignition of the dry exhalation under the ground renders the surrounding stone similar to dry exhalation, though in solid form.[17] What is important is that minerals are the result of the process of combustion that is now complete and no dry exhalation remains any longer in the mineral; in metals, by contrast, the dry exhalation remains latent and stands ready to burn when exposed to flame. This subtle distinction in the process of formation, as Eichholz points out, serves to reinforce a tendency that we have seen at work elsewhere in the *Meteorologica* and especially in the kapnosphere, where the dry exhalation is regularly treated as more formal or efficient and less material than the wet exhalation.

16 Eichholz 1949: 144. Solmsen's vague doubts (1960: 401n38) about the dry exhalation being merely the efficient cause are not persuasive. Theophrastus, *de Lapidibus* 8 says of ochre and ruddle, realgar and orpiment (and chrysocolla and cyanus) that their formation is due to conflux and filtering. But "some of them at least have obviously undergone exposure to fire and some process of complete combustion, as for example realgar, orpiment and the like; but it may be stated generally that all are products of the dry and smoky exhalation" (Eichholz trans.). This is the only mention of ἀναθυμίασις in the treatise.

17 When Aristotle says that most minerals are colored dusts or stones, it is clear that the dusts are the primary form and the stones are compressed variants of these.

A methodological postscript

We could sooner capture the rainbow and measure the sea than we could delimit the particular genius of a thinker as luminous and as vast as Aristotle. But if we were to seek the one source and principle that guides and directs the many achievements of the *Meteorologica*, we should find it in Aristotle's synthesis of internal multiplicity and external unity. From this ambiguous origin Aristotle established meteorology as an integral and semi-autonomous science within his series of physical sciences. Though the *Posterior Analytics* offers Aristotle's official theory of the structure and unity of science, the *Meteorologica* fills out this doctrine in a number of surprising ways.

We are told in the *Posterior Analytics* that each science has its own genus, the subject matter it studies. We are also told that a science displays the causes of its subject matter, and that these causes supply the definitions of that subject and its parts. The existence of the genus and the definitions of its parts are the principles, characterized by necessary and binding relationships, from which the properties and attributes of the subject matter are demonstrated. If the subject matter and its causes are specific and unique to the science, as they should be, the science will be autonomous and independent of others. If it shares principles with another subject matter, it may be absorbed into the other and together they will form a single super-science.

We have now seen how the *Meteorologica* is neither wholly autonomous and independent, nor absorbed into the larger whole which is general physics. The *Meteorologica* is dependent on the general physical doctrines of the *Physics*, *de Caelo*, and *de Generatione et Corruptione*, but it modifies and adapts its specific principles in such a way that they are both causally related to more general principles and irreducible to them. As such, the *Meteorologica* provides a model, hitherto unappreciated, of an integral genus or subject matter. We do not have to choose, therefore, between autonomy and dependence. Though the *Posterior Analytics* presses hard in its defense of autonomy, the *Meteorologica* shows the more practical and satisfying middle path. Any discussion of Aristotle's theory of the unity of science must take account of this structure.

The *Analytics* presents the doctrine that a science proves its conclusions through definitions and causes, and in the *Meteorologica* we learn the significance of this doctrine. Of course, at a basic level, this is a familiar lesson, but what is remarkable about the *Meteorologica* is the way in which Aristotle deploys the semi-autonomous set of causes to establish the autonomy and internal multiplicity of meteorology. His three main causes – the material principles, the dry and the wet exhalations; the efficient cause, the heat of the sun and its effects; and place, divided into zones between the heavens and the earth – are all adapted from general physics to their specific role in meteorology; together they define the genus and domain; and through the repeated interactions of their differences they give rise to the multiplicity of phenomena, the greatest groups and their species, which constitute the subject matter of meteorology.

We learned from the *Analytics* about the *per se*, or necessary, relationships that bind terms of demonstrative premises together, and we learned why they are a condition for demonstrative knowledge. But we could not have anticipated from the *Analytics* the profound effect these relationships can have on the unity of the subject matter. In fact, it is perfectly consistent with the doctrine of the *Analytics* that, depending on one's choice of principles, the subject matter and its demonstrations be loose and episodic. But, as we have seen repeatedly, all meteorological phenomena are defined by the same three generic causes, which are in turn divided by differentiae ranked in hierarchical order. And because of the hierarchical order among the causal differentiae, where one material, one place, one efficient cause is superior to another, Aristotle is able to produce a set of hierarchies among the phenomena that unifies them as a group and prefigures the *scala naturae* among the living beings of the biological works. When we add to this the principle, according to which one cause is the cause of a different kind of cause, the result is a profound unity arising out of an internally organized multiplicity, of which the *Analytics* provides only the slightest hints.

We learn from the *Analytics* the importance of division in establishing definitions. In the *Meteorologica* we see how thoroughly the plan of division and differentiae structures its characteristic hierarchies and correspondences, so much so that Aristotle's plan of differentiae, which I have dubbed the *Historia Meteorōn*, must clearly have been a guiding force. As Aristotle examined the phenomena he wanted to explain, he divided according to differentiae that promoted the unity of the discipline. We saw how each of the three principal causal factors were divided according to their internal differentiae and subdivided again, how high and low were used in the vertically oriented kapno-atmosphere, how solstitial points differentiated the radial rivers and the winds, how the differences in the visual ray, not a purely meteorological entity, contributed to the unusual status of reflective phenomena. There is no reason to suppose that a *Historia Meteorōn* existed complete before the *Meteorologica* was drafted, but the divisionary introductions to precipitations and to reflections makes clear that Aristotle was making regular use of this mode of organization. The differentiae in the *Meteorologica* are relentlessly disciplinary: Aristotle divided in order to unify.

All these methodological features, background issues in the *Analytics*, are emphasized in the *Meteorologica* and the biological works. Together they make up what might be called the "sublunary method," since they are common both to *meteora* and to living things. In the *Meteorologica* these elements – the divisions and the causes – are bound tightly together and contained within a single treatise. In the biology, by contrast, they are dispersed among several. The isolation of the facts concerning animals (in the *Historia Animalium*) from the several relevant causes in the biological works (the final in the *Parts of Animals*, the material and efficient in the *Generation of Animals*) is a symptom of the increasing disunity in the world of terrestrial life.

But in addition to creating an internal multiplicity, Aristotle also intended his meteorology to form an external unity reaching back to general physics and the unified cosmos as a whole and forward to the manifold world of terrestrial life. The *Meteorologica* relates to general physics by adapting its principles and by being a part of its cosmic whole; it relates to biology by analogy and structural similarity. More remarkably, meteorological phenomena, poised between the teleologically directed natures of simple bodies on the one hand and the souls of terrestrial living things on the other, have as such no final cause. Their nature is adequately determined by the efficiency of the heavens and the place-directed natures of the exhalations and elements and needs no additional drives or impulses. The impressive order and structure of meteorological phenomena is generated entirely out of the mechanical interaction of its principles, for all that it prefigures the order and structure of the teleologically directed living world. When viewed under the unifying force of the efficient, heavenly cause, the sublunary world of meteorology can be understood as a single integrated being, a kind of body, quasi-organic in its articulations, a single animal, as it were, ruled by the heavens as by a soul. On this conception the sublunary realm of meteorology completes the view of the world as a single complete whole begun in the *Physics* and realized in the *de Caelo*. The sublunary world is analogous to the animal body and similar to it in structure. But Aristotle's technique of addition and abstraction allows him to deny of it what we might consider to be its most important role in the cosmic animal, that it should serve the heavens as an organ or instrument. The heavens have no need of such a body, so under this conception a final cause is inappropriate.

But inasmuch as the meteorological phenomena form "greatest groups," similar to the great animal genera such as birds and mammals, which in turn break down into species and separate individuals, the sublunary world appears to be populated by a great plurality of independent individuals. Thus the *meteora* are drawn into analogy with the realm of biology. But the relationship goes beyond the analogies. Aristotle's doctrine of spontaneous generation provides the missing link between meteorology and biology (and this link in turn explains this otherwise bizarre Aristotelian doctrine). The sun plays a common unifying role in the formation both of meteorological phenomena and of spontaneous animals, and it is only with sexual reproduction that the unified generative force of the sun is dispersed into the individual seeds of animals. In this way the *Meteorologica* constitutes the vital step on the *scala naturae* that reaches from the divine mind down to the plants.

The relationship between celestial and terrestrial souls, which is scarcely alluded to in the *de Anima*, is a vexing and tantalizing fact. The obscurity of that relationship has created a chasm between the two realms and little progress has been made in bridging it. It is my hope that this study has revealed some of the substantive ways in which Aristotle bridged that chasm, and that this study may inspire a more concerted effort to bring the macrocosm together with the microcosm in a thoroughly Aristotelian way.

More generally, I hope to have made clear that if one wants to understand the great project of Aristotelian physics, it is impossible to ignore the *Meteorologica*. Aristotle may have been "wrong in nearly all his conclusions," but the treatise remains the focus and pivot point between general physics and biology, between the living cosmos and terrestrial life, between macrocosm and the microcosms, and between celestial perfection and a more disorderly nature.

Bibliography

TEXTS AND TRANSLATIONS

Bekker, I. (1831) *Opera Aristotelis*. Academia Regia Borusica. Berlin.

Ross, W. D. (1964) *Aristotelis Analytica Priora et Posteriora*. L. Minio-Paluello auxit. Oxford.

(1956) *Aristotelis Physica*. Oxford.

Allan, D. J. (1936) *Aristotelis de Caelo*. Oxford.

Joachim, H. H. (1922) *Aristotle: On Coming-to-be and Passing-away*. Oxford.

Rashed, M. (2005) *Aristote: de la génération et la corruption*. Paris.

Fobes, F. H. (1919) *Aristotelis Meteorologicorum Libri Quattuor*. Cambridge, Mass.

Tricot, J. (1941) *Aristote. Les Météorologiques*. Paris.

Lee, H. D. P. (1952) *Aristotle: Meteorologica*. Cambridge, Mass.

Louis, P. (2002) *Aristote. Météorologiques*. 2 vols. Paris.

Ross, W. D. (1965) *Aristotelis de Anima*. Oxford.

(1955) *Aristotle: Parva Naturalia. A Revised Text with Introduction and Commentary*. Oxford.

Hett, W. S. (1957) *Aristotle: On the Soul. Parva Naturalia. On Breath*. Cambridge, Mass.

Peck, A. L. (1965) *Aristotle: Historia Animalium. Books i–iii*. Cambridge, Mass.

(1970) *Aristotle: Historia Animalium. Books iv–vi*. Cambridge, Mass.

Balme, D. M. (1991) *Aristotle: History of Animals. Books vii–x*. Cambridge, Mass.

(2002) *Aristotle: Historia Animalium*, vol 1: *Books I–X: Text*, prepared by A. Gotthelf. Cambridge.

Peck, A. L. and E. S. Forster (1961) *Aristotle: Parts of Animals; Movement of Animals; Progression of Animals*. Cambridge, Mass.

Peck, A. L. (1942) *Aristotle: Generation of Animals*. Cambridge, Mass.

Mayhew, R. (2011) *Aristotle: Problems Books 1–21*. Cambridge, Mass.

Mayhew, R. and D. Mirhady (2011) *Aristotle: Problems Books 22–38. Rhetoric to Alexander*. Cambridge, Mass.

Jaeger, W. (1957) *Aristotelis Metaphysica*. Oxford.

Bywater, L. (1894) *Aristotelis Ethica Nicomachea*. Oxford.

Walzer, R. and J. Mingay (1991) *Aristotelis Ethica Eudemia*. Oxford.

Ross, W. D. (1957) *Aristotelis Politica*. Oxford.

Rose, V. (1886) *Aristotelis Fragmenta*. Stuttgart.

Barnes, J. (1984) *The Complete Works of Aristotle*. The Revised Oxford Translation. Princeton.

Webster, E. W. (trans.) (1931) *Meteorologica*, in *The Works of Aristotle*, vol. III, ed. W. D. Ross. Oxford.

Alexander of Aphrodisias (1899) *In Aristotelis Meteorologicorum Libros Commentaria*, ed. M. Hayduck. Berlin.

Olympiodorus (1900) *In Aristotelis Meteora Commentaria*, ed. W. Stüve. Berlin.

Philoponus, J. (1901) *In Aristotelis Meteorologicorum Librum Primum Commentarium*, ed. M. Hayduck. Berlin.

(1909) *In Aristotelis Analytica Posteriora Commentaria*, ed. M. Wallies. Berlin.

Simplicius (1894) *In de Caelo Commentaria*, ed. I. L. Heiberg. Berlin.

(2002) *On Aristotle: On the Heavens 1.1–4*, trans. R. J. Hankinson. London.

Diels, H. (1929) *Doxographi Graeci*. Berlin.

Diels, H. and W. Kranz (1952) *Die Fragmente der Vorsokratiker*, 6th edn. Hildesheim.

Heiberg, I. L. (1893) *Apollonius Pergaeus*, 2 vols. Leipzig.

Hicks, R. D. (1925) *Diogenes Laertius: Lives of the Philosophers*, 2 vols. Cambridge, Mass.

Jones, W. H. S. (1923) *Hippocrates*, vols. I and II. Cambridge, Mass.

Wachsmuth, C. (1897) *Johannes Lydus: Ostentis*. Leipzig.

Burnet, J. (1900–7) *Platonis Opera*, 5 vols. Oxford.

Cooper, J. and D. Hutchinson (1997) *Plato: Complete Works*. Indianapolis.

van Raalte, M. (1993) *Theophrastus: Metaphysics*. Leiden.

Coutant, V. and V. Eichenlaub (1975) *Theophrastus: de Ventis*. Notre Dame.

Eichholz, D. (1965) *Theophrastus: de Lapidibus*. Oxford.

Rackham, H., W. H. S. Jones, and D. Eichholz (1938–63) *Pliny: Natural History*, 10 vols. Cambridge, Mass.

Corcoran, T. (1971) *Seneca: Naturales Quaestiones*, 2 vols. Cambridge, Mass.

SECONDARY LITERATURE

Abbreviations of journal titles follow those of *L'Année philologique*

Aubenque, P. (1961) "Sur la notion aristotélicienne d'aporie," in Mansion (1961a): 3–19.

Badawi, 'A. (1971) *Commentaires sur Aristote perdus en grec et autres épîtres*. Beirut.

Bailey, C. (1947) *Titi Lucreti Cari: De Rerum Natura*. Oxford.

Balme, D. M. (1962) "Development of biology in Aristotle and Theophrastus: theory of spontaneous generation," *Phronesis* 7: 91–104.

Barnes, J. (1981) "Aristotle and the method of ethics," *RIPh* 34: 490–511.
(1994) *Aristotle: Posterior Analytics*, 2nd edn. Oxford.
Barrett, A. A. (1977) "Aristotle and averted vision," *Journal of the Royal Astronomical Society of Canada* 71: 327.
Barthélemy Saint-Hilaire, J. (1863) *Météorologie d'Aristote*. Paris.
Beare, J. (1906) *Greek Theories of Elementary Cognition from Alcmaeon to Aristotle*. Oxford.
Betegh, G. (2007) "On the physical aspect of Heraclitus' psychology," *Phronesis* 52: 3–32.
Bodnár, I. (1997) "Movers and elemental motions in Aristotle," *OSAPh* 15: 81–117.
(2004) "The mechanical principles of animal motion," in *Aristote et le mouvement des animaux: Dix études sur le De motu animalium*, ed. A. Laks and M. Rashed. Villeneuve d'Ascq: 137–47.
(2005) "Teleology across natures," *Rhizai* 2: 9–29.
Böker, R. (1958a) "Winde," in *RE* 2nd Series 8A2: 2211–65.
(1958b) "Windrosen," in *RE* 2nd Series 8A2: 2325–81.
Bolchert, P. (1908) *Aristoteles Erdkunde von Asien und Libyen*. Berlin.
Bolton, R. (1991) "Aristotle's method of natural science: *Physics* I," in *Aristotle's Physics: A Collection of Essays*, ed. L. Judson. Oxford: 1–29.
(2009) "Two standards for inquiry in Aristotle's *De Caelo*," in Bowen and Wildberg (2009): 51–82.
Bone, N. (1993) *Meteors*. Cambridge, Mass.
(1996) *The Aurora: Sun–Earth Interactions*. Chichester and New York.
Bonitz, H. (1961) *Index Aristotelicus*. Berlin.
Bourgey, L. (1955) *Observation et expérience chez Aristote*. Paris.
Bowen, A. and C. Wildberg (eds.) (2009) *New Perspectives on Aristotle's De Caelo*. Leiden.
Boyer, C. (1946) "Aristotelian references to the law of reflection," *Isis* 36: 92–5.
(1956) "Refraction and the rainbow in antiquity," *Isis* 47: 383–6.
(1959) *The Rainbow from Myth to Mathematics*. New York.
Brahe, T. (1876) *Meteorologiske Dagbok. Holdt paa Uraniborg for Aarene 1582–1597*. Copenhagen.
Brekke, A. and A. Egeland (1983) *The Northern Light: From Mythology to Space Research*. Berlin.
Brisson, L. (2004) *How Philosophers Saved Myths: Allegorical Interpretation and Classical Mythology*, trans. C. Tihanyi. Chicago.
Bunbury, E. H. (1883) *A History of Ancient Geography*, 2 vols. 2nd edn. New York.
Burkert, W. (1972) *Lore and Science in Ancient Pythagoreanism*, trans. E. L. Minar, Jr. Cambridge, Mass.
Burnyeat, M. (2004) "Introduction: Aristotle on the foundations of sublunary physics," in *Aristotle's On Generation and Corruption I. Symposium Aristotelicum*, ed. Frans de Haas and J. Mansfeld. Oxford: 7–24.
Camaterus, Joannes (1908) *Introductio in Astronomiam*. Leipzig.
Cantù, P. (2010) "Aristotle's prohibition rule on kind-crossing and the definition of mathematics as a science of quantities," *Synthese* 174: 225–35.

Capelle, W. (1908) "Erdbeben im Altertum," *Neue Jahrbücher für das klassische Altertum*. 21: 603–33.

(1912a) "Μετέωρος – μετεωρολογία," *Philologus* 71: 414–48.

(1912b) "Das Proömium der Meteorologie," *Hermes* 47: 514–35.

(1913) "Zur Geschichte der meteorologischen Litteratur," *Hermes* 48: 321–58.

(1919) "Anaxagoras," *Neue Jahrebücher für das klassische Altertum* 44: 81–102; 169–98.

(1924) "Erdbebenforschung," in *RE Supplementband* 4: 344–74.

(1935) "Meteorologie," in *RE Supplementband* 6: 315–58.

Carapiperis, L. N. (1951) "On the periodicity of the Etesians in Athens," *Weather* 6: 378–9.

(1956) "Some appearances of the aurora borealis in Greece," *Pure and Applied Geophysics* 35: 139–42.

(1962a) *The Etesian Winds. The Opinions of Ancient Greeks on the Etesian Winds*. Ethnikon Asteroskopeion Athēnōn. Hypomnemata. Seria II. Meteorologia 9. Athens.

(1962b) *The Etesian Winds. On the Frequency of the Etesian Winds*. Ethnikon Asteroskopeion Athēnōn. Hypomnemata. Seria II. Meteorologia 10. Athens.

Case, T. (1910) "Aristotle," reprinted in *Aristotle's Philosophical Development: Problems and Prospects*, ed. W. Wians. Lanham, Md.: 1–40.

Chantraine, P. (1968) *Dictionnaire étymologique de la langue grecque*, 2 vols. Paris.

Charles, D. (2000) *Aristotle on Meaning and Essence*. Oxford.

Cherniss, H. (1935) *Aristotle's Criticism of Presocratic Philosophy*. Baltimore.

Chitwood, A. (2004) *Death by Philosophy*. Ann Arbor.

Code, A. (1997) "The priority of final causes over efficient causes in Aristotle's *PA*," in Kullmann and Föllinger (1997): 127–43.

Cohen, S. M. and P. Burke (1990) "New evidence for the dating of Aristotle *Meteorologica* 1–3," *CPh* 85: 126–9.

Cole, S. (2010) "'I know the number of the sand and the measure of the sea': geography and difference in the early Greek world," in Raaflaub and Talbert (2010): 197–214.

Coles, A. (1997) "Animal and childhood cognition in Aristotle's biology and the *scala naturae*," in Kullmann and Föllinger (1997): 287–323.

Colvin, M. (2005) "Heraclitus and material flux in Stoic psychology," *OSAPh* 28: 257–72.

Cooper, J. (1982) "Aristotle on natural teleology," in Schofield and Nussbaum (1982): 197–222.

Coutant, V. and V. Eichenlaub (1975) *Theophrastus: de Ventis*. Notre Dame.

Curd, P. (2007) *Anaxagoras of Clazomenae*. Toronto.

Daiber, H. (1975) *Ein Kompendium der aristotelischen Meteorologie in der Fassung des Hunain ibn Ishaq*. Amsterdam.

Denniston, J. D. (1950) *The Greek Particles*, 2nd edn. Oxford.

Descartes, R. (1637) *Les Météores. Oeuvres de Descartes*, vol. VI, ed. C. Adam and P. Tannery. Paris.

Deslauriers, M. (1991) "Plato and Aristotle on division and definition," *AncPhil* 10: 203–19.
Detel, W. (1997) "Why all animals have a stomach. Demonstration and axiomatization in Aristotle's *Parts of Animals*," in Kullmann and Föllinger (1997): 63–84.
Dilke, O. (1985) *Greek and Roman Maps*. Ithaca.
Dorandi, T. (2007) *Nell' officina dei classici. Come lavoravano gli autori antichi*. Rome.
Drossaart Lulofs, H. J. (1930) *Aristoteles over de Zee. Specimen van antieke Hydrographie*. Utrecht.
(1955) "The Syriac translation of Theophrastus' *Meteorology*," in *Autour d'Aristote. Festschrift A. Mansion*. Louvain: 433–49.
Ducos, J. (1998) "La météorologie en français au Moyen Age (xiiie et xive siècles)." Paris.
Düring, I. (1944) *Aristotle's Chemical Treatise: Meteorologica Book IV*. Göteborg.
(1957) *Aristotle in the Ancient Biographical Tradition*. Göteborg.
(1966) *Aristoteles. Darstellung und Interpretation seines Denkens*. Heidelberg.
Eichholz, D. E. (1949) "Aristotle's theory of the formation of metals and minerals," *CQ* 43: 141–6.
Elders, L. (1965) *Aristotle's Cosmology. A Commentary on the de Caelo*. Assen.
Falcon, A. (1997) "Aristotle's theory of division," in *Aristotle and After*, ed. R. Sorabji. *BICS*, suppl. vol. 68: 127–46.
(2005) *Aristotle and the Science of Nature: Unity with Uniformity*. Cambridge.
(2011) "General and special physics: Aristotle and the Aristotelian tradition on the project of an integrated study of motion." Unpublished paper.
(2012) *Aristotelianism in the First Century BCE: Xenarchus of Seleucia*. Cambridge.
Fobes, F. H. (1913) "A preliminary study of certain MSS of Aristotle's *Meteorology*," *CR* 27: 249–52.
(1915a) "Textual problems in Aristotle's *Meteorologica*," *CPh* 10: 188–214.
(1915b) "Mediaeval versions of Aristotle's meteorology," *CPh* 10: 297–314.
Fontaine, R. (1995a) *Otot ha-Shamayim: Samuel Ibn Tibbon's Hebrew Version of Aristotle's Meteorologica*. Leiden.
(1995b) "Why is the sea salty? The discussion of salinity in Hebrew texts of the thirteenth century," *Arabic Studies and Philosophy* 5: 195–218.
(2001) "The reception of Aristotle's *Meteorology* in Hebrew scientific writings of the thirteenth century," *Aleph: Historical Studies in Science and Judaism* 1: 101–39.
Fränkel, H. (1973) *Early Greek Poetry and Philosophy*, trans. M. Hadas and J. Willis. New York and London.
Freeland, C. (1990) "Scientific explanation and empirical data in Aristotle's *Meteorology*," *OSAPh* 8: 67–102.
Freudenthal, G. (1995) *Aristotle's Theory of Material Substance: Heat and Pneuma, Form and Soul*. Oxford.

(2009) "The astrologization of the Aristotelian cosmos: celestial influences on the sublunary world in Aristotle, Alexander of Aphrodisias, and Averroes," in Bowen and Wildberg (2009): 239–81.

Frisk, H. (1960) *Griechisches Etymologisches Wörterbuch*, 2 vols. Heidelberg.

Furley, D. (1976) "Aristotle and the atomists on motion in the void," in *Motion and Time, Space and Matter*, ed. P. K. Machamer and R. J. Turnbull. Columbus: 83–100; reprinted in Furley (1989): 77–90.

(1983) "The mechanics of *Meteorologica* IV: a prolegomena to biology," in *Zweifelhaftes in Corpus Aristotelicum*, Akten des 9. Symposium Aristotelicum, ed. P. Moraux and J. Wiesner. Berlin and New York: 73–93; reprinted in Furley (1989): 132–48.

(1986) "The rainfall example in *Physics* II.8," in *Aristotle on Nature and Living Things*, ed. A. Gotthelf. Pittsburgh: 177–82; reprinted in Furley (1989): 115–20.

(1989) *Cosmic Problems: Essays on Greek and Roman Philosophy of Nature*. Cambridge.

Gilbert, O. (1907) *Die meteorologischen Theorien des griechischen Altertums*. Leipzig.

Gildersleeve, B. (1900) *Syntax of Classical Greek*. New York.

Gill, M. L. (1997) "Material necessity and *Meteorology* IV 12," in Kullmann and Föllinger (1997): 145–61.

Gotthelf, A. (1987a) "First principles in Aristotle's *Parts of Animals*," in Gotthelf and Lennox (1987): 167–98.

(1987b) "Aristotle's conception of final causality," in Gotthelf and Lennox (1987): 204–42.

(1997) "The elephant's nose: further reflections on the axiomatic structure of biological explanation in Aristotle," in Kullmann and Föllinger (1997): 85–96.

Gotthelf, A. and J. Lennox (eds.) (1987) *Philosophical Issues in Aristotle's Biology*. Cambridge.

Graham, D. (1997) "Heraclitus' criticism of Ionian philosophy," *OSAPh* 15: 1–50.

Graux, C. and A. Martin (1900) "Figures tirées d'un manuscript des *Météorologiques* d'Aristote," *RPh* 24: 5–18.

Guthrie, W. K. C. (1962–81) *A History of Greek Philosophy*, 5 vols. Cambridge.

Hankinson, R. J. (2003) "Xenarchus, Alexander and Simplicius on simple motions, bodies and magnitudes," *BICS* 46: 19–42.

(2005) "Aristotle on kind-crossing," in *Philosophy and the Sciences in Antiquity*, ed. R. W. Sharples. Aldershot: 23–54.

(2009) "Natural, unnatural and preternatural motions: contrariety and the argument for the elements in *De Caelo* 1.2–4," in Bowen and Wildberg (2009): 83–118.

Harley, J. B., D. Woodward, and G. Aujac (1987) "The foundation of theoretical cartography in archaic and classical Greece," in *The History of Cartography*, vol. 1, ed. J. B. Harley and D. Woodward. Chicago: 130–47.

Harry, B. (1971) "A defence of Aristotle, *Meteorologica* 3, 375a6ff.," *CQ* N.S. 21: 397–401.

Heath, T. (1925) *The Thirteen Books of Euclid's Elements*, 3 vols. Cambridge.
(1949) *Mathematics in Aristotle*. Oxford.
Hecquet-Devienne, M. (2000) "Les mains du Parisinus graecus 1853," *S&C* 24: 103–71.
Heidel, W. A. (1937) *The Frame of the Ancient Greek Maps*. American Geographical Society Research Series 20. New York.
Higgins, M. D. and R. Higgins (1996) *A Geological Companion to Greece and the Aegean*. Ithaca.
Horden, P. and N. Purcell (2000) *The Corrupting Sea: A Study of Mediterranean History*. Oxford.
Huffman, C. (1993) *Philolaus of Croton: Pythagorean and Presocratic*. Cambridge.
Hussey, E. (1991) "Aristotle's mathematical physics: a reconstruction," in Judson (1991a): 213–42.
Ideler, J. L. (1832) *Meteorologica veterum Graecorum et Romanorum*. Berlin.
(1834–6) *Aristotelis Meteorologicorum Libri Quattuor*, 2 vols. Leipzig.
Inwood, B. (2001) *The Poem of Empedocles*, 2nd edn. Toronto.
Irwin, T. (1988) *Aristotle's First Principles*. Oxford.
Janko, R. (2003) "God, science, and Socrates," *BICS* 46: 1–18.
Johansen, T. K. (1999) "Myth and logos in Aristotle," in *From Myth to Reason? Studies in the Development of Greek Thought*, ed. R. Buxton. Oxford: 279–91.
(2004) *Plato's Natural Philosophy: A Study in the Timaeus-Critias*. Cambridge.
Johnson, M. R. (2005) *Aristotle on Teleology*. Oxford.
(2009) "The Aristotelian explanation of the halo," *Apeiron* 42: 325–57.
Jones, A. (1994) "Peripatetic and Euclidean theories of the visual ray," *Physis* 31: 47–76.
Judson, L. (ed.) (1991a) *Aristotle's Physics: A Collection of Essays*. Oxford.
(1991b) "Chance and 'always or for the most part' in Aristotle," in Judson (1991a): 73–99.
(2005) "Aristotelian Teleology," *OSAPh* 29: 341–66.
Kahn, C. (1979) *The Art and Thought of Heraclitus*. Cambridge.
(1984) *Anaximander and the Origins of Greek Cosmology*, 2nd edn. Philadelphia.
Kidd, I. G. (1992) "Theophrastus' meteorology, Aristotle and Posidonius," in *Theophrastus: His Psychological, Doxographical, and Scientific Writings*, ed. W. W. Fortenbaugh and D. Gutas. Rutgers University Studies in Classical Humanities 5. New Brunswick, N.J.: 294–306.
Kingsley, P. (1995) *Ancient Philosophy, Mystery, and Magic: Empedocles and Pythagorean Tradition*. Oxford.
Kirk, G. S. (1962) *Heraclitus: The Cosmic Fragments*, 2nd edn. Cambridge.
Knorr, W. R. (1986) *The Ancient Tradition of Geometric Problems*. Boston.
(1992) "When circles don't look like circles: an optical theorem in Euclid and Pappus," *AHES* 44: 287–329.
Kullmann, W. (1974) *Wissenschaft und Methode: Interpretationen zur aristotelischen Theorie der Naturwissenschaft*. Berlin.
(1985) "Different concepts of the final cause in Aristotle," in *Aristotle on Nature and Living Things*, ed. A. Gotthelf. Pittsburgh and Bristol: 169–75.

(1997) "Die Voraussetzungen für das Studium der Biologie nach Aristoteles," in Kullmann and Föllinger (1997): 43–62.

Kullmann, W. and S. Föllinger (eds.) (1997) *Aristotelische Biologie: Intentionen, Methoden, Ergebnisse*. Stuttgart.

Lang, H. S. (1998) *The Order of Nature in Aristotle's Physics: Place and the Elements*. Cambridge.

Lee, H. D. P. (1952) *Aristotle: Meteorologica*. Cambridge, Mass.

Lehoux, D. (2007) *Astronomy, Weather, and Calendars in the Ancient World*. Cambridge.

Lennox, J. (1982) "Teleology, chance, and Aristotle's theory of spontaneous generation," *JHPh* 20: 219–38; reprinted in Lennox (2001): 229–49.

(1984) "Aristotle on chance," *AGPh* 66/1: 52–60; reprinted in Lennox (2001): 250–8.

(1986) "Aristotle, Galileo, and 'Mixed Sciences,'" in *Reinterpreting Galileo*, ed. W. A. Wallace. Studies in Philosophy and the History of Science 15. Washington: 29–51.

(1987a) "Divide and explain: the *Posterior Analytics* in practice," in Gotthelf and Lennox (1987): 90–119.

(1987b) "Kinds, Forms of kinds, and the more and the less in Aristotle's biology," in Gotthelf and Lennox (1987): 399–59.

(2001) *Aristotle's Philosophy of Biology: Studies in the Origins of Life Science*. Cambridge.

(2005) "The place of zoology in Aristotle's natural philosophy," in *Philosophy and the Sciences in Antiquity*, ed. R. W. Sharples. Aldershot: 58–70.

(2009) "*De Caelo* 2.2 and its debt to the *de Incessu Animalium*," in Bowen and Wildberg (2009): 187–214.

(2010) "Aristotle's Natural Science: the Many and the One," in *From Inquiry to Demonstrative Knowledge*, ed. J. H. Lesher. Kelowna, BC: 1–23.

Lettinck, P. (1999) *Aristotle's Meteorology and its Reception in the Arab World*. Leiden.

Leunissen, M. (2007) "The structure of teleological explanations in Aristotle: theory and practice," *OSAPh* 33: 145–78.

(2010) *Explanation and Teleology in Aristotle's Science of Nature*. Cambridge.

Lewis, E. (trans.) (1996) *Alexander of Aphrodisias. On Aristotle's Meteorology 4*. Ithaca.

Littmann, M. (1998) *The Heavens on Fire*. Cambridge.

Lloyd, G. E. R. (1961) "The development of Aristotle's theory of the classification of animals," *Phronesis* 6: 59–81.

(1966) *Polarity and Analogy: Two Types of Argumentation in Early Greek Thought*. Cambridge.

Louis, P. (2002) *Aristote. Météorologiques*, 2 vols. Paris.

Lovejoy, A. O. (1936) *The Great Chain of Being*. Cambridge, Mass.

Madigan, A. (1999) *Aristotle: Metaphysics Book B and Book K 1–2*. Oxford.

Mansion, S. (ed.) (1961a) *Aristote et les problèmes de méthode. Symposium Aristotelicum 2*, Louvain.

(1961b) "Le rôle de l'exposé et de la critique des philosophies antérieures chez Aristote," in Mansion (1961a): 35–56.

Marcovich, M. (1967) *Heraclitus*. Mérida.

Mariolopoulos, E. G. (1961) *An Outline of the Climate of Greece*. Dēmosieumata Ergastēriou Meteōrologias Panepistēmiou Athēnōn; Publications of the Meteorological Institute of the University of Athens. Athens.

Mariotte, E. (1718) *A Treatise of the Motion of Water and Other Fluids*, trans. T. J. Desagulier. London.

Martin, C. (2010) "The ends of weather: teleology in renaissance meteorology," *JHPh* 48: 259–82.

(2011) *Renaissance Meteorology: Pomponazzi to Descartes*. Baltimore.

Martini, E. (1897) "Lucubrationum Posidonianarum Specimen II," *RH* 52: 348–76.

McKirahan, R. (1978) "Aristotle's subordinate sciences," *BJHS* 11: 197–220.

(1992) *Principles and Proofs. Aristotle's Theory of Demonstrative Science*. Princeton.

Merker, A. (2002) "Aristote et l'arc-en-ciel: enjeux philosophiques et étude scientifique," *AHES* 56: 183–238.

Missiakoulis, S. (2008) "Aristotle and earthquake data: a historical note," *International Statistical Review* 76: 130–3.

Moraux, P. (1967) "Le Parisinus graecus 1853 (Ms. E) d'Aristote," *Scriptorium* 21: 17–41.

(1973) *Der Aristotelismus bei den Griechen: von Andronikos bis Alexander von Aphrodisias*, vol. 1. Berlin.

Mugler, C. (1964) *Dictionnaire historique de la terminologie optique des Grecs*. Paris.

Müller, K. (1965) *Geographi Minores Graeci*. Hildesheim.

Mulvany, C. M. (1926) "Notes on the legend of Aristotle," *CQ* 20: 155–67.

Myres, J. L. (1896) "An attempt to reconstruct the maps used by Herodotus," *GJ* 8: 605–29.

Negri, F. (1929) *Il Viaggio settentrionale*. Milan.

Netz, R. (1999) *The Shaping of Deduction in Greek Mathematics: A Study in Cognitive History*. Cambridge.

(2004) "Eudemos of Rhodes, Hippocrates of Chios and the earliest form of a mathematical text," *Centaurus* 46: 243–86.

Neumann, J. (1973) "The sea and land breezes in the classical Greek literature," *Bulletin of the American Meteorological Society* 54: 5–8.

(1985) "Climatic change as a topic in classical Greek and Roman literature," *Climatic Change* 7: 441–54.

Neumann, J. K. (1884) "Die Fahrt des Patrokles auf dem Kaspischen Meere und der alte Lauf des Oxos," *Hermes* 19: 165–85.

Ninck, M. (1921) *Die Bedeutung des Wassers im Kult und Leben der Alten: eine symbolgeschichtliche Untersuchung. Philologus*, Suppl. vol. 14, no. 2. Leipzig.

Nussbaum, M. C. (1978) *Aristotle's de Motu Animalium*. Princeton.

(1982) "Saving Aristotle's appearances," in Schofield and Nussbaum (1982): 267–93.

Oder, E. (1898) "Ein angebliches Bruchstück Demokrits über die Entdeckung unterirdischer Quellen," *Philologus Supplementband* 7: 229–384.

Oleson, J. P. (2006) "Herodotus, Aristotle, and sounding weights: the deep sea as a frontier in the classical world," *2nd International Conference on Ancient Greek Technology*. Athens: 32–43.

Owen, G. E. L. (1960) "Logic and metaphysics in some earlier works of Aristotle," in *Aristotle and Plato in the Mid-fourth Century*, ed. I. Düring and G. E. L. Owen, *Symposium Aristotelicum*. Göteborg: 163–90; reprinted in Owen (1986): 180–99.

(1961) "*Tithenai ta phainomena*," in Mansion (1961a): 83–103; reprinted in Owen (1986): 239–51.

(1970) "Aristotle: method, physics, and cosmology," in *Dictionary of Scientific Biography*, ed. C. C. Gillespie, vol. 1. New York: 250–8; reprinted in Owen (1986): 151–64.

(1986) *Logic, Science and Dialectic*. Ithaca.

Owens, J. (1978) *The Doctrine of Being in the Aristotelian "Metaphysics,"* 3rd edn. Toronto.

Panessa, G. (1981) "Oscillazioni e stabilità del clima nella Grecia antica: introduzione ad una ricostruzione paleoclimatoligica," *ASNP* 11: 123–58.

Perry, B. E. (1952) *Aesopica*. Urbana.

Poss, H. L. (1980) "The date of an astronomical observation in Aristotle's *Meteorologica*," *Bulletin of the American Astronomical Society* 12: 885.

Raaflaub, K. and R. Talbert (2010) *Geography and Ethnography: Perceptions of the World in Pre-modern Societies*. Chichester.

Rashed, M. (2005) *Aristote: de la génération et la corruption*. Paris.

Rehm, A. (1916) *Griechische Windrosen*. Sitzungsberichte der königlichen bayerischen Akademie der Wissenschaften, Philosophisch-Philologische und Historische Klasse. Abhandlung 3. Munich.

Richter, G. M. A. (undated) *Perspective in Greek and Roman Art*. London and New York.

Riginos, A. S. (1976) *Platonica: The Anecdotes Concerning the Life and Writings of Plato*. Leiden.

Romm, J. (2010) "Continents, climates, and cultures: Greek theories of global structure," in Raaflaub and Talbert (2010): 215–35.

Ross, W. D. (1949) *Aristotle*, 5th edn. London.

Rubino, E. (ed.) (2010) *Aristoteles Latinus (ALPE X, 1). Meteorologica. Liber quartus. Translatio Henrici Aristippi*. Corpus Philosophorum Medii Aevi. Turnhout.

Sambursky, S. (1962) *The Physical World of Late Antiquity*. Princeton.

Sayili, A. (1939) "The Aristotelian explanation of the rainbow," *Isis* 30: 65–83.

Scharle, M. (2008) "Elemental Teleology in Aristotle's *Physics* 2.8," *OSAPh* 34: 147–83.

Schofield, M. and M. Nussbaum (eds.) (1982) *Language and Logos*. Cambridge.

Schoonheim, P. L. (2000) *Aristotle's Meteorology in the Arabico-Latin Tradition*. Leiden.

(2003) "Aristote de Stagire, Météorologiques. Tradition syriaque et arabe," in *Dictionnaire des philosophes antiques, Supplément*, ed. R. Goulet. Paris: 324–8.

Sedley, D. (1991) "Is Aristotle's teleology anthropocentric?," *Phronesis* 36: 179–96.

Sharples, R. W. (1990) "The school of Alexander?," in *Aristotle Transformed: The Ancient Commentators and their Influence*, ed. R. Sorabji. Ithaca: 83–111.
Shcheglov, D. (2006) "Posidonius on the dry west and the wet east: fragment 223 EK reconsidered," *CQ* 56: 509–27.
Solmsen, F. (1950) "Chaos and apeiron," *SIFC* N.S. 24: 235–48.
(1960) *Aristotle's System of the Physical World*. Ithaca.
Sorabji, R. (1972) "Aristotle, mathematics, and colour," *CQ* N.S. 22: 293–308.
Sorof, G. (1886) *De Aristotelis Geographia Capita duo*. Halle.
Stothers, R. (2009) "Ancient meteorological optics," *CJ* 105: 27–42.
Strohm, H. (1936) "Untersuchungen zur Entwicklungsgeschichte der aristotelischen Meteorologie," *Philologus Supplementband* 28: 1–85.
(1937) "Zur Meteorologie des Theophrast," *Philologus* 92: 249–68; 401–28.
(1984) *Aristoteles Meteorologie, Über die Welt*, 3rd edn. Aristoteles Werke in Deutscher Übersetzung 12. Berlin.
Sudhaus, S. (1898) *Aetna*. Leipzig.
Takahashi, H. (2004) *Aristotelian Meteorology in Syriac: Barhebraeus, Butyrum Sapientiae, Books of Mineralogy and Meteorology*. Leiden.
Talbert, R. (ed.) (2000) *Barrington Atlas of the Greek and Roman World*. Princeton.
Talbert, R. and R. Unger (eds.) (2008) *Cartography in Antiquity and the Middle Ages: Fresh Perspectives, New Methods*. Boston.
Tannery, P. (1886) "Aristote, 'Météorologie,' livre III, ch. v," *Revue de philologie, de littérature et d'histoire anciennes* N.S. 10: 38–46.
Taub, L. (2003) *Ancient Meteorology*. London.
Taylor, A. E. (1928) *A Commentary on Plato's Timaeus*. Oxford.
Taylor, C. C. W. (1999) *The Atomists: Leucippus and Democritus: Fragments. A Text and Translation with a Commentary*. Toronto.
Theiler, W. (1965) *Zur Geschichte der teleologischen Naturbetrachung bis auf Aristoteles*, 2nd edn. Berlin.
(1982) *Posidonios: die Fragmente*. Berlin.
Thompson, D. W. (1918) "The Greek winds," *CR* 32: 49–56.
Thorp, J. (1982) "The luminousness of the quintessence," *Phoenix* 36: 104–23.
Thurot, C. (1869–70) "Observations critiques sur les *Météorologica* d'Aristote," *RA* 20: 415–20; 21: 87–93, 249–55, 339–46, 396–407.
Tuan, Yi-Fu. (1968) *The Hydrological Cycle and the Wisdom of God: A Theme in Geoteleology*. University of Toronto Department of Geography Research Publications. Toronto.
Usener, H. (1887) *Epicurea*. Leipzig.
Vallisnieri, A. (1715) *Lezione Accademice intorno all'origine delle fontane*. Venice.
van der Waerden, B. L. (1952) "Das grosse Jahr und die ewige Wiederverkehr," *Hermes* 80: 129–55.
Verlinsky, A. (2006) "The flood in Aristotle's Meteorologica (I. 14)," in *Antike Naturwissenschaft und ihre Rezeption*, vol. XVI, ed. J. Althoff, B. Herzhoff, and G. Wöhrle. Trier: 51–68.
Viano, C. (2006) *La matière des choses: le livre iv des Météorologiques d'Aristote et son interprétation par Olympiodore*. Paris.

Vicomercatus, F. (1565) *In quattuor libros Aristotelis Meteorologicorum Commentarii*. Paris and Venice.

Vita, V. (1983) "Two geometric loci in Aristotle's 'Meteorologica,'" *BSSM* 3: 3–18.

Vitrac, B. (2002) "Note textuelle sur un (problème de) lieu géométrique dans les *Météorologiques* d'Aristote (III.5 375b16–376b22)," *AHES* 56: 239–83.

Vuillemin-Diem, G. (ed.) (2008) *Meteorologica. Translatio Guillelmi de Morbeka, Praefatio – Editio textus*. Aristoteles Latinus (ALPE X, 2.1–2). Corpus philosophorum Medii Aevi. Turnhout.

West, M. L. (1966) *Hesiod: Theogony*. Oxford.

(1971) *Early Greek Philosophy and the Orient*. Oxford.

(1988) *Hesiod: Theogony, Works and Days*. Oxford.

Williams, G. D. (2006) "Greco-Roman seismology and Seneca on earthquakes in *Natural Questions* 6," *JRS* 96: 124–46.

Wilson, M. (2000) *Aristotle's Theory of the Unity of Science*. Toronto.

(2008) "Hippocrates of Chios' theory of comets," *JHA* 39: 141–60.

(2009) "A somewhat disorderly nature: unity in Aristotle's *Meteorologica* I–III," *Apeiron* 42: 63–88.

Woodbury, L. (1965) "The date and atheism of Diagoras of Melos," *Phoenix* 19: 178–211.

Index of principal passages discussed

Aristotle
de Anima
1.1.403a4–b19, 148
1.1.403a10–16, 149
1.2.405a25–7, 58
1.4.408a5–23, 170
2.4.415a26–b1, 112
2.4.415b2–3, 99
2.4.415b2–7, 100
de Caelo
1.3.269a9–17, 127
2.7.289a19–32, 48
2.12, 108
2.12.292a14–b25, 109
3.2.301a4–6, 4
3.3.302b4–5, 44
de Generatione et Corruptione
2.6.333a20–34, 43
2.10.336b30–2, 103
2.11, 97
de Motu Animalium
10.703a31–b2, 109
11.703b20–2, 109
11.703b20–3, 108
de Mundo
2.392b1–5, 130
2.395b10–15, 130
de Somno
3.456b20–1, 189
3.456b20–4, 225
Generation of Animals
1.18.725a33–b3, 190
2.1.732b15–31, 76
2.4.739a2–4, 190
3.11.762a19–23, 112
3.11.762b13–18, 111
Historia Animalium
1.1.486a14–487a1, 76
1.1.488a14–16, 82
1.1.488a20–1, 82
1.1.488a25–6, 82
1.1.488a31–2, 82
Metaphysics
1.7.988a18–24, 26
7.15, 143
7.15.1039b20–1040a7, 144
7.15.1040a27–b4, 144
12.08, 108
12.10.1075a19–22, 109
Meteorologica
1.1, 9
1.1.338a20, 35
1.1.338a26, 25
1.1.338a26–b21, 4
1.1.338b25, 238
1.2.339a11–20, 36
1.2.339a23, 104
1.2.339a23–4, 83
1.2.339a25, 36
1.3.339a33–6, 36
1.3.339a36–b2, 36, 43
1.3.339b2–340a18, 43
1.3.339b2–340b4, 42
1.3.339b6–9, 43, 60
1.3.339b9–10, 43
1.3.339b9–13, 186, 188
1.3.339b16–19, 43
1.3.339b19–21, 44
1.3.339b25–6, 44
1.3.339b30–1, 44
1.3.339b30–4, 44
1.3.340a1–2, 44
1.3.340a3–6, 44
1.3.340a13–17, 43
1.3.340a19–341a12, 43, 45
1.3.340a24–6, 46
1.3.340a32–b3, 45
1.3.340b4–341a36, 42
1.3.340b5, 43
1.3.340b6, 43
1.3.340b6–10, 47, 141
1.3.340b8, 46
1.3.340b14–19, 46
1.3.340b21–9, 40

1.3.340b23–9, 118
1.3.340b28–9, 47
1.3.340b32–341a2, 128
1.3.340b32–341a9, 46
1.3.340b32–6, 85, 88
1.3.340b36–341a1, 86
1.3.341a12–36, 43
1.3.341a17–18, 118
1.3.341a17–22, 48
1.3.341a30–1, 50, 61
1.4.341b3, 120
1.4.341b6–10, 79, 83, 84, 88
1.4.341b6–8, 52
1.4.341b9–10, 84
1.4.341b10, 40, 118
1.4.341b10–12, 88
1.4.341b13–14, 79
1.4.341b18–22, 79, 88
1.4.341b18–24, 119
1.4.341b19–29, 119
1.4.341b22, 125
1.4.341b24, 124
1.4.341b24–35, 121
1.4.341b25–7, 122
1.4.341b27–8, 122, 124
1.4.341b30, 121
1.4.341b32–5, 122
1.4.341b33, 234
1.4.341b35–342a13, 90
1.4.341b36–342a3, 41
1.4.341b36–342a16, 123
1.4.342a4, 52
1.4.342a13–14, 234
1.4.342a24–7, 203
1.4.342a30–3, 126
1.5.342a34, 129
1.5.342a34–5, 129
1.5.342a36–b1, 132
1.5.342b1–4, 134
1.5.342b3–4, 132
1.5.342b3–5, 128
1.5.342b4, 199
1.5.342b5–11, 132
1.5.342b6, 241
1.5.342b11–13, 30, 132
1.5.342b13–14, 129
1.5.342b14–18, 133
1.5.342b23, 129
1.6.342b27–9, 30, 135
1.6.342b29–35, 135
1.6.342b35–343a20, 30, 135
1.6.342b35–343a4, 54
1.6.343a4–8, 30
1.6.343a13–15, 257
1.6.343a22–5, 136
1.6.343a22–35, 30
1.6.343a23–5, 32
1.6.343a24–5, 138
1.6.343a25–6, 32
1.6.343a26–30, 32
1.6.343a30–5, 32
1.6.343a35–b7, 32
1.6.343b4–7, 32
1.6.343b8–14, 32
1.6.343b14–16, 138
1.6.343b14–25, 32
1.6.343b28–32, 32
1.6.343b30, 9
1.6.343b32–344a2, 136
1.7.344a5, 174
1.7.344a5–8, 137
1.7.344a6–7, 141
1.7.344a8–22, 119
1.7.344a14, 41
1.7.344a17–20, 140
1.7.344a18, 119
1.7.344a20, 41
1.7.344a22–3, 142
1.7.344a29, 119
1.7.344a35–b8, 140
1.7.344b1, 140
1.7.344b2, 140
1.7.344b8, 141
1.7.344b12–17, 32
1.7.344b17–18, 238
1.7.344b18–345a10, 120
1.7.344b26–8, 142
1.7.344b28–34, 142
1.7.345a1, 9
1.7.345a5 10, 54, 142
1.8.345a13–18, 135
1.8.345a18–25, 32
1.8.345a25–31, 134, 135
1.8.345a25–36, 32
1.8.345a36–b9, 33, 136
1.8.345b9–12, 135
1.8.345b9–25, 32, 34
1.8.345b10–12, 30
1.8.345b32–5, 119
1.8.345b32–6, 49
1.8.346a1–2, 140
1.8.346a8–9a, 140
1.8.346a10–11, 141
1.8.346a11–14, 140
1.8.346a11–16, 120
1.8.346a17–19, 145
1.8.346a19–23, 195
1.8.346a22, 120
1.8.346a22–3, 120, 140
1.8.346a28, 145

Aristotle (*cont.*)
1.8.346b6, 119
1.8.346b7–10, 151
1.8.346b8–10, 140
1.9.346b16–19, 79
1.9.346b17–18, 147
1.9.346b18, 148
1.9.346b20–1, 146
1.9.346b20–3, 85, 88
1.9.346b20–31, 79
1.9.346b23–4, 149
1.9.346b23–6, 50, 84, 146
1.9.346b24, 88
1.9.346b24–31, 84
1.9.346b26, 49
1.9.346b26–31, 147
1.9.346b31, 88
1.9.346b32–3, 149
1.9.346b36, 103
1.9.347a2–8, 85
1.9.347a3, 148
1.9.347a6–8, 147
1.10.347a14, 147
1.10.347a16–17, 273
1.10.347a29–34, 90
1.10.347a30–4, 49
1.10.347a34–5, 90
1.10.347a34–7, 150
1.10.347a35–b7, 154
1.11.347b19–21, 152
1.12.347b35–6, 154
1.12.347b37–348a2, 153
1.12.348a4–13, 153
1.12.348a14–20, 153
1.12.348a18–20, 30
1.12.348a20–30, 153
1.12.348a20–36, 32
1.12.348a27, 153
1.12.348b12–14, 153
1.12.348b12–15, 30
1.12.348b26–8, 154
1.12.349a4–7, 155
1.13.349a13–14, 157
1.13.349a16, 157
1.13.349a16–20, 86
1.13.349a20–6, 86
1.13.349a26, 157
1.13.349a26–7, 158
1.13.349a33–b2, 160
1.13.349b15–19, 32, 161
1.13.349b19–35, 161
1.13.349b19–350a15, 194
1.13.349b27–35, 79
1.13.349b27–350a13, 32, 162
1.13.349b30–5, 275
1.13.349b30–5, 86, 88, 162
1.13.350a2–4, 86
1.13.350a7, 275
1.13.350b10–12, 169
1.13.350b15–18, 169
1.13.350b30–351a18, 86
1.13.350b36–351a18, 162
1.13.351a7–16, 184
1.14.351a26, 175
1.14.351a36–b8, 172
1.14.351b8–28, 174
1.14.351b28, 174
1.14.352a9–14, 174
1.14.352a22–5, 32
1.14.352a25–8, 177
1.14.352a26–8, 34
1.14.352a28–33, 175
1.14.352a29–31, 178
1.14.352a31–3, 188
1.14.352b1, 178
1.14.352b22–31, 173
1.14.352b31–353a1, 173
1.14.353a1–7, 173
2.1.353a32, 183, 186
2.1.353a34–5, 180, 182
2.1.353b1–2, 180
2.1.353b5–6, 180
2.1.353b18–35, 34, 182
2.1.353b19–20, 183
2.1.353b19–35, 226
2.1.353b20–3, 183
2.1.353b23–4, 190
2.1.353b24–5, 184
2.1.353b30–1, 183
2.1.354a3–5, 184
2.1.354a5–11, 225
2.1.354a8, 185, 226
2.1.354a11–13, 186
2.1.354a11–26, 168
2.2.354b11–13, 32
2.2.354b14, 187
2.2.354b15–16, 189
2.2.354b29, 187
2.2.355a15–21, 33
2.2.355a21–32, 33, 186
2.2.355a25–6, 30
2.2.355a25–8, 4
2.2.355a32–4, 187
2.2.355a33–4, 190
2.2.355b11–15, 188
2.2.355b13, 187
2.2.355b19, 187
2.2.355b22, 187
2.2.355b29–30, 187
2.2.356a14–19, 33

2.2.356a19, 185
2.2.356a19–33, 33
2.2.356a35, 194
2.3.356b9–17, 180
2.3.356b17–19, 28
2.3.356b17–21, 188
2.3.356b21–5, 33
2.3.357a6–12, 33
2.3.357a15–21, 192
2.3.357a15–24, 33
2.3.357a24–8, 180
2.3.357a24–32, 192
2.3.357b18–21, 188
2.3.358a21–6, 61, 193
2.3.358a25–6, 5
2.3.358b12–16, 193
2.3.358b34–359b21, 192
2.3.359a24–b21, 192
2.3.359a35–b4, 277
2.3.359b1–7, 195
2.4.359b32–4, 197
2.4.360a6, 200
2.4.360a6–8, 200
2.4.360a8–13, 197
2.4.360a15–17, 49
2.4.360a21–7, 197
2.4.360a27–33, 85
2.4.360a29–31, 86
2.4.360b3–5, 198
2.4.360b5–7, 216
2.4.360b31–2, 198
2.4.360b34–6, 198, 203
2.4.361a4–22, 207
2.4.361a17–20, 199
2.4.361a22–5, 88
2.4.361a23, 203
2.4.361a24, 202, 203
2.4.361a24–5, 201
2.4.361a27–32, 203
2.4.361a28–30, 201
2.4.361a31–2, 222
2.4.361a35–6, 89
2.4.361b1–5, 86
2.4.361b1–8, 205
2.4.361b2–5, 204
2.5.361b14–20, 199
2.5.361b17–20, 206
2.5.361b27–30, 206
2.5.362a2–7, 206
2.5.362a5–7, 199
2.5.362a9–11, 199
2.5.362a11–13, 207
2.5.362a20–3, 206
2.5.362a26–8, 207
2.5.362a32–b5, 211
2.5.362a32–b9, 165
2.5.362a35, 213
2.5.362b12–30, 167
2.5.362b30–363a1, 209
2.5.362b35–363a4, 209
2.5.363a4–8, 215
2.5.363a13–18, 214
2.6.363a25–6, 166
2.6.363a26–8, 211
2.6.363a29, 211
2.6.363a29–30, 211
2.6.363a30–1, 215
2.6.364a4, 212
2.6.364a13–22, 216
2.6.364a24–7, 216
2.6.364a31–b3, 215
2.6.365a3–5, 233
2.7.365a19–35, 219
2.7.365a25–33, 33
2.7.365a33–5, 30
2.7.365b1–6, 33, 218
2.7.365b1–7, 23
2.7.365b6–12, 218
2.7.365b6–20, 33
2.7.365b12–14, 30
2.8, 86
2.8.365b28–366a5, 218, 226
2.8.366a5–8, 223
2.8.366a7–8, 222
2.8.366a16, 222, 271
2.8.366a17–23, 224
2.8.366a18–23, 223
2.8.366a23–5, 219
2.8.366a23–b2, 224
2.8.366a27, 219
2.8.366b2, 97
2.8.366b2–3, 220
2.8.366b2–4, 221
2.8.366b9–14, 226
2.8.366b9–15, 220
2.8.366b20–1, 223
2.8.366b26, 223
2.8.366b31–367a11, 233
2.8.367a9–11, 90, 233
2.8.367a9–12, 219
2.8.367b8–12, 222
2.8.367b14–19, 226
2.8.367b19–30, 223
2.8.367b20, 97
2.8.368a14–17, 232
2.8.368a26–34, 220
2.8.368a34–b12, 225
2.8.368a34–b13, 102
2.8.368b12–20, 224
2.8.368b19–22, 221

Aristotle (*cont.*)
2.9.369a12–24, 88
2.9.369a12–29, 79, 228
2.9.369a13–24, 84
2.9.369a21–4, 228
2.9.369a28–9, 228
2.9.369b3–7, 228
2.9.369b15–18, 121
2.9.369b21–2, 33, 49, 125
2.9.369b22–4, 33, 125
2.9.369b27–34, 230
2.9.370a1–5, 33, 231
2.9.370a10–21, 34
2.9.370a24–6, 230
2.9.370a25–7, 217
2.9.370a25–32, 227
2.9.370a27–32, 217
3.1.370b4–6, 234
3.1.370b5, 233
3.1.370b5–10, 80
3.1.370b6–7, 232
3.1.370b7–8, 232, 235
3.1.370b14–17, 59
3.1.370b17–31, 234
3.1.370b27–371a1, 234
3.1.370b29–31, 232
3.1.370b30–1, 232
3.1.371a1–3, 232
3.1.371a3, 233
3.1.371a3–8, 221
3.1.371a11, 235
3.1.371a15–17, 235
3.1.371a16, 232
3.1.371a17–b14, 228
3.1.371a18, 232
3.1.371a19, 232
3.1.371a31, 9
3.2.371b21–2, 247
3.2.371b22–3, 247
3.2.372a5, 152
3.2.372a21–9, 245
3.2.372a24–9, 97
3.2.372a29–32, 241
3.2.372b6–9, 243
3.2.372b8, 240, 241
3.2–6, 91
3.3.372b12–17, 30
3.3.372b15–17, 244
3.3.373a1, 249
3.3.373a6–19, 250
3.3.373a23–5, 244
3.3.373a24–5, 236
3.3.373a27–32, 245
3.4.373b2–10, 241
3.4.373b13–17, 243
3.4.373b34–374a10, 242
3.4.374a1–3, 243, 244
3.4.374a3–4, 133
3.4.374a4–8, 243
3.4.374a11–15, 244
3.4.374a13–20, 257
3.4.374a16–20, 243
3.4.374a20–9, 238
3.4.374a23, 240
3.4.374b24–8, 243
3.4.374b30–3, 242
3.4.374b33–375a28, 245
3.4.375a17–20, 245
3.4.375a30–b15, 242
3.4.375a32–b3, 245
3.4.375b3–9, 245
3.5.375b16–19, 252
3.5.375b16–376a1, 262
3.5.375b17–18, 255
3.5.375b19–20, 262, 263
3.5.375b19–25, 252
3.5.375b25–9, 253
3.5.375b30–376a1, 253, 262
3.5.375b30–376b22, 251
3.5.376b28–377a28, 260, 261
3.6.377a31–b4, 243
3.6.377a34–b2, 240
3.6.377b4–13, 243
3.6.377b22–3, 244
3.6.377b30–2, 244
3.6.377b32–4, 244
3.6.378a6–8, 244
3.6.378a8–11, 245
3.6.378a12–14, 271
3.6.378a15–16, 271
3.6.378a21–6, 276
3.6.378a25–b3, 79
3.6.378a26–32, 272
3.6.378a28–31, 91
3.6.378a30, 275
3.6.378a32, 273
3.6.378a32–b2, 273
3.6.378b2–4, 275
3.6.378b3–4, 275
3.6.378b5–6, 9
4, 9
4.8.384b32, 274
4.9.387a24–5, 199
4.9.387a30–1, 199
4.9.387a32–b1, 199
4.9.387b6–9, 199
4.9.387b8–9, 199

4.10.388a14–17, 274
4.10.388a24–5, 274
4.10.389a7–8, 274
4.12, 98
Nicomachean Ethics
5.5.1133a32–b5, 260
Parts of Animals
1.2.642b5–9, 75
1.3.643a3–5, 150
1.3.643b17–26, 76
1.3.643b9–13, 76
1.3.643b9–18, 75
1.3.644a6–8, 75
1.4.644a19–21, 76
1.4.644a21–23, 76
1.4.644b7–15, 76
3.4.666b13–16, 96
Physics
1.2.193b25–30, 247
2.2.194a33–b9, 99
2.8.198b16–199a5, 94
2.8.199a3–5, 97
7.2.243a11–244b2, 64
8.1.252a12, 94
Politics
1.5.1254a32–3, 170
Posterior Analytics
1.13.79a10–13, 247
2.1–2, 77
2.11.94b33–4, 96
2.12, 97
2.13.97b26–8, 143
Sense and Sensibilia
2.438a25–32, 240
5.443a21–2, 148
5.443a23, 61
fr. 250 Rose, 216

Heraclitus
DK 22A1.9, 60
DK 22A1.9–11, 54
DK 22B31, 56
Hesiod
Theogony 740, 182
Works and Days 547–53, 53
Homer
Odyssey 11.13–15, 179

Olympiodorus
In Mete.
120.6–10, 173
123.2–7, 173

Philoponus
In Mete.
59.12–15, 124
59.32–3, 124
Plato
Phaedo
90c4–6, 180
97b–98, 22
112b3, 185
Phaedrus
229d3–4, 180
Sophist
220b1, 81
Timaeus
36b–d, 107
49b7–c7, 64
58b6–8, 64
59e, 24

Seneca
Quaestiones Naturales
6.10, 218

General index

Achelous River, 178
Aigospotamoi, meteorite of, 143
air pressure, 224
Ammon, 173
analogy, 59
  between genera of *meteora* and animals, 105
  biological, 175, 223
  digestive, 189, 225
  in precipitations, 152–3, 155
  of exhalations with blood and ichor, 105
  of kapnosphere and atmosphere, 92, 150–2
  of the bladder and the sea, 192–3
  of winds and rivers, 86, 160, 196, 201, 205, 213, 214: abuse of, 160
Anaxagoras, 20, 21, 22
  on aether, 44
  on comets and the Milky Way, 134–9
  on earthquakes, 30, 33, 219
  on hail, 32, 154
  on lightning, 121
  on rainbows, 239
  on shooting stars, 120–6
  on stormy phenomena, 229
  on the Milky Way, 32, 33
  on the sea, 191
Anaximander, 150
  on aer, 53
  on feeding the sun, 53
Anaximenes, 19, 148
  on earthquakes, 30, 33, 218
  on rainbows, 239
antiperistasis, 154, 224
antipodes, 207
Apollonius of Perga, 251
Archelaus
  on earthquakes, 220
architectonic structure, 146–7, 156, 158, 163, 190, 192, 195, 209, 214, 216, 222, 226, 234, 235, 237, 271–2
aurora borealis, 5, 91, 128–34
  ancient confusion of terminology, 130
  color, 132
  identity of, 128–30
  reflection, 132
  three causes of, 132–3
  torches, 132–3

belonging in its own right, 248
biological method, 105

Caecias, 67
cameo roles, 92
Capelle, W., 21
catalyst, 197–200
cause
  and definition, 279
  causes of, 90, 92
  efficient, 79
  long and short chains of, 51, 148, 158, 172
  material, 78, 83–4
  sidereal vs. solar, 146
  unity and plurality of, 111
caverns, 161
civilization
  rise and fall of, 147
classification
  dichotomous division, 75–6, 81
  division, 279
  elegance of, 139
  greatest groups, 76, 83, 281
  "more and less," 80, 153
  multiple differentiae system, 76, 78, 90
  slipshod division, 81
Cleidemus
  on lightning, 34
climate change, 169–78
clouds, 69
  dispersing of, 67
  withered, 68
coast and inland, 174
cold, 68, 70
collective memory, 176, 177

colliquescences, 190
comets, 66, 71, 134–43
as planets, 21
in zodiac, 67
long haired and bearded, 142
commensurate universals, 29
common place of air and water, 147–9, 188
condensation and separation, 65
in Presocratics, 64
contraction, 65
cycles
redundant, 194

date of composition, 9–10
death, Aristotle's, 179
definition
causal, 78
Democritus, 20, 22
on comets and the Milky Way, 134–9
on earthquakes, 33, 218
on the Milky Way, 32, 33
on the salinity of the sea, 23, 33
on winds, 23, 208
on winds and rivers, 159
Deucalion's flood, 177, 188
differentiae
biological, 82
in series, 82
per se, 82
Diogenes of Apollonia, 20
on the sea, 191
on the turnings of the sun, 208
doxography, 157
on earthquakes, 219
on reflections, 236
on stormy phenomena, 227
dualizers, 107, 113–14, 117, 126, 141, 164, 234
dunamis, 39, 43

earth
disk and sphere, 216, 270
flat, 165, 169
minuscule size of, 43, 136, 177
relative and absolute points on, 211
source of generative heat, 200
spherical, 208
tambourine shaped, 213
zones, 209
earthquakes, 66, 68, 69
confinement of, 225
local, 224
sea and, 225
seasons for, 221
self-impulsive winds of, 222
violence of, 218
winds associated with, 68, 69, 218, 221–3: ebb and flow of, 225
eclipse, 143
efficient cause
diurnal and annual, 88, 150–1
locomotion, 88
sidereal, 88
sun as, 36
Eichholz, D., 277
ejection, 227
elements
as principles too general, 51
changes of, 68
final cause of, 94
first body, 46, 47
mutual transformation of, 36, 37, 42, 43, 44–7, 51, 52, 94
emigration, 177
Empedocles, 22
on condensation and separation, 63
on stormy phenomena, 229
on the salinity of the sea, 33
empiricism, 6, 26, 31, 37
comets, 135–6, 137
hail, 153
Milky Way, 141
waters, 181
endoxa, 26–8
entropist theory, 157, 170, 173, 175, 176, 177, 192
Ephorus, 166
Epicurus, 20
ever-visible circle, 211
exhalations
anathumiasis: ambiguity of the term, 52
as mixtures, 197
as principles less applied, 156, 171
cooperating with one another, 197, 200
dry: and salinity, 193; burning, 277; different manifestations of, 86, 118–20; fuel (*hypekkauma*), 71, 118; ignition, 71; non-recycling, 62, 85, 148; oxidation, 275; rising, 70; well mixed, 71, 125
mutual transformation of, 198
specific to meteorology, 26, 83
specification view, 39–41
sustasis view, 41–2
vs. elements, 49
wet: feeding the heavens, 61, 146, 181, 186; in Presocratic philosophy, 52

final cause, 191, 281
accidental, 99
automaton, 112
beneficiary and aim, 99
better than that of which it is the cause, 102

final cause (*cont.*)
  distinct from that of which it is the cause, 99
  identity of formal and, 101
  meteora do not strive, 103
  meteora not final causes, 102
  steering, 104
firesphere, 117
flames (*phlox*), 121
fossils, 175
Freeland, C., 28, 30
freezing, 66, 67
friction
  and the Milky Way, 119
  vs. ignition, 118–19

geography, 169
Gerard of Cremona, 2
goats (*aiges*), 121
great winter, 172, 174, 178, 188

hail
  doxography, 153
  paradoxical nature of, 153, 154
halo, 248–50
harmonia, 170
heat, 68, 69
  boiling, 71
  rising, 67
  scattered downward, 50
Heath, T., 259
heavens, 31
  as soul, 108
  guiding the sublunary realm, 47
Heraclitus
  on exhalations, 60
  on feeding the stars, 56
  on homeostasis, 25
  on mutual transformation of matter, 57
  on soul as exhalation, 58
  on three-element view, 59
Hesiod
  on *aer*, 53
hierarchy, 163, 279
  degree of separation, 87
  not used as explanations in *Meteorologica*, 83, 87, 96
  of circles, 105, 110, 234, 237
  of directness, 89–90
  of posture, 87, 105, 165
  of simplicity and complexity, 89
  of waters, 181
Hippocrates of Chios, 20, 134, 257
  on comets, 30, 32
  on comets and the Milky Way, 139
  on *hypoleipsis*, 142
  on moisture, 54
  on southern vapor, 208
  on the Milky Way, 32, 33
Historia Meteorōn, 279
homeostasis, 156, 177
hurricane, 69

imitation, 104

Jaeger, W., 37

Kahn, C., 21
kapnosphere
  imitating the heavens, 117
  order of treatment of subjects in, 138
  rotary motion of, 127–8
katabasis, 179
Knorr, W., 246, 259

lagoons, 172
Lake Maeotis, 169, 173
land–sea exchange
  role of rivers in, 172, 174
Lee, H. D. P., 1
lightning, 69
locus proposition, 251
  authenticity of, 238, 246–7

Mansion, S., 26
material
  quantity of, 151
mathematics
  arguments from, 136
  Aristotle's attitude toward, 21
  physics and, 247
Mediterranean, 168, 185, 211
Merker, A., 249, 251, 259
metabasis, 175
metals, 68, 86
  compression, 275
  condensation, 273
  congealing, 67
  dew and frost, 272
  pre-vapor, 274
meteors, 126
  apparent motion of, 126, 141
  as activities, 126
  differentiae of, 122, 142
  directions of, 122–3
  modern correspondences, 121–2
Metrodorus, 19
  on rainbows, 239
  on the sea, 191
Milky Way, 5, 33, 66
  and aurora borealis, 141

and friction, 140
and halo, 140
and reflection, 140
and upper comets, 140
causes of, 140–1
color, 141
minerals, 68, 86, 276
mock suns, 66
Moerbeke, 2
motion
unnatural, 72
mountains
as sponges, 161
Chremetes, Parnassus, Pyrene, 168

necessity
hypothetical, 98
material, 97
Nile Delta, 173

Ocean, river of, 147
order
caused by elements, 94

parallax, 143
Parmenides, 4, 22
per se relations, 279
peri metallōn, 271
Philolaus, 20
Philoponus, 123–4, 143
physical works
unity in the, 114
place
as cause, 79–80, 84–6
Plato, 20, 21
on condensation and separation, 63
on rivers, 33
points, method of identifying, 259
Pontus, 69
prediction, 3
preternatural motion, 126–7
principles
application of higher principles, 36
proper, 39, 50
proportions, method of identifying, 260
Pythagoreans, 136
on comets, 135
on the Milky Way, 32

rain
violent, 66, 69
rain cycle, 2, 60, 79, 85, 147
and reflections, 152
and storms, 149–50
differentiae of, 149
rainbows
around lamps, 238
colors, 257
double, 258
interpolation of passage, 260
physical considerations, 269
rays, 88, 146, 154, 230; *see also* visual ray
reflections, 88
differentiae, 239, 242, 258: distance of the reflected ray, 244; goal of the visual ray, 245; physics of vapor, 244; principle of contrast, 245; reasons for, 246; reflecting surface, 243
halo and rainbow proof compared, 254
outside the subject matter, 237
physical implications of the locus proposition, 256–7
place in the *Meteorologica*, 236
principle of homogeneity of, 250
reproduction
sexual, 110
spontaneous generation, 111–13, 281
reservoir theory, 157, 159, 161, 178, 230
residues, 190
rivers, 69, 85, 164
Ister and Nile, 167
multiple causes of, 163
redundancy of the vapor theory, 163

scala naturae, 113, 138, 234, 279
sea, 68
compared with heavens, 170
confinement of, 181–91
flowing and standing, 183
fresh vs. saline, 189
importance of in the cosmos, 179–80
isolation from the rain cycle, 187
isolation of, 186
natural place of, 187
salinity of, 5, 149, 186, 191–5
salt as an addition to, 192
sources of, 182–5
taste and weight, 193
Seneca, 23
shape, 152
shooting stars, 65, 67, 120–6
ejected, 68
silt, 172
smoky fumes, 199
society, 176
Socrates, 21, 22
Solmsen, F., 24, 38
solstitial points, 166
soul, 93
as part and as form, 100, 108

celestial and terrestrial, 281
functions of, 110
parts of heavenly, 108
unity of the, 108
southern hemisphere, 207
springs, 273
saline, 192
Stoic cataclysm, 178
stormy phenomena
correspondences of earthquakes and kapnosphere, 232
kapnosphere and, 233
primary and secondary differentiae, 231
primary differentiae: quantity of dry exhalation, 232; texture, 232
priority of unignited over ignited among, 228, 229
seasons and, 233
species of, 229
winds associated with, 227
straits, 189, 219, 223–7
Strohm, H., 1, 37
subject genus, 278
sublunary method, 280
sublunary world
as a single animal, 106, 107
dependent on heavens, 103
distinct from heavens, 138
instrumental parts of, 110
subordinate science, 247
subterranean passages, 219, 220
sun, 68
as a unified cause, 112
as efficient cause (friction), 79
heat of, 88, 111

Tannery, P., 250
teleology, 6, 114
always and for the most part, 96–7
anthropocentric, 96
no examples in *Meteorologica*, 96
of seeds, 98
that for the sake of which, 104
Theophrastus, 55
on winds, 202
thunder, 67, 228
thunderbolts, 69, 79
time
great lengths of, 174
speed of change, 151
torches (*daloi*), 121
Tycho Brahe, 131

unity
causes with in the genus, 91
external, 280
of cause, 5
of the *Meteorologica*, 31, 81
unmoved mover, 109

vapor
rising of the, 72
visual ray, 239–42, 256
Vitrac, B., 251, 259
volcanoes, 90, 219, 233

water
dispersed nature of terrestrial, 43, 67
West, M. L., 182
whirlwinds, 67, 69
windrose, 166, 211, 212, 227
winds, 66, 69, 85
causes of motion of, 86, 128, 201–5
celestial cause, 201
definition, air in motion, 160, 197, 202
ductility of, 71, 204, 222, 228
Etesian, 196, 204, 206
horizontal motion of, 88, 201, 202
local, 221
opposition of, 215
rain and, 199
role of wet exhalation in, 198
seasonal cause of motion, 203
southern Etesians, 214
times and seasons of, 206
withered, 68
withering, 206

Xenarchus, 126
Xenophanes, 148, 150
on feeding the sun, 54
on St. Elmo's fire, 54
on the sea, 191

zodiac, 136, 138